Arctic Modernities

Arctic Modernities:

The Environmental, the Exotic and the Everyday

Edited by

Heidi Hansson and Anka Ryall

Cambridge Scholars Publishing

Arctic Modernities: The Environmental, the Exotic and the Everyday

Edited by Heidi Hansson and Anka Ryall

This book first published 2017. The present binding first published 2021.

Cambridge Scholars Publishing

Lady Stephenson Library, Newcastle upon Tyne, NE6 2PA, UK

British Library Cataloguing in Publication Data
A catalogue record for this book is available from the British Library

ISBN (10): 1-5275-7103-3
ISBN (13): 978-1-5275-7103-7

TABLE OF CONTENTS

ACKNOWLEDGEMENTS

It is a pleasure to acknowledge the numerous people and institutions that have made this volume possible. It began as a three-year research project financed by the Polar Research Programme of the Research Council of Norway, with additional financing from the Faculty of Humanities, Social Sciences and Education at UiT – The Arctic University of Norway. Umeå University, Sweden, also provided support in the form of travel grants.

The project, led by Anka Ryall, was located at and generously supported by the Centre for Women's and Gender Research at UiT, where the project leader, a post-doctoral research fellow and two PhD candidates have had their offices. Many thanks are due to the director of the Centre, Ann Therese Lotherington, as well as to Siri Gerrard, Torunn Berger, Ingrid Hovda Lien and Lise Nordbrønd.

Thanks also to the project's invaluable steering group, Johan Schimanski and Henning Howlid Wærp, to all the project members for crucial input at seminars and workshops, to the anonymous referees for constructive feedback, to Liz Sourbut for expert language editing, and to Per Hansson and Tom Kristiansen for moral and technical support.

Finally, sincere thanks to the institutions that gave us permission to use photographs, and to Marja Helander for the cover image.

Umeå and Tromsø, June 2017
Heidi Hansson and Anka Ryall

CONTRIBUTORS

CATHRINE BJERKNES is a PhD Candidate in Comparative literature at UiT – The Arctic University of Norway. Her thesis project concerns gender and intertextuality in modern Arctic crime fiction. She is the editorial assistant of the literary journal *Norsk Litteraturvitenskapelig Tidsskrift*.

JAN BORM is Professor of British Literature at the University of Versailles Saint-Quentin-en-Yvelines (UVSQ) and co-director of the international Masters 2 programme "Arctic Studies" affiliated with the excellence cluster "University of Paris-Saclay" that UVSQ is a founding member of. He has published widely on travel literature in English, French and German, and has co-edited nine collections of articles, with a tenth titled *Cold: Adaptation, Production, Representations, Effects* forthcoming.

FREDRIK CHR. BRØGGER is Professor Emeritus of American Literature and Civilization at UiT – The Arctic University of Norway. Most of Brøgger's work has been concerned with American Studies. His research before 2000 centred mainly on American literature and the field of American Culture Studies, and particularly on the literary modernism of the 1920s. In subsequent years, however, nature writing, ecocriticism and Arctic Studies have served as main fields of interest.

ADRIANA CRACIUN is the Emma MacLachlan Metcalf Chair of Humanities and Professor of English at Boston University. Her most recent book, *Writing Arctic Disaster: Authorship and Exploration* (2016), was shortlisted for the 2016 Kendrick Book Prize by the Society for Literature, Science & the Arts. With Simon Schaffer she edited *The Material Cultures of Enlightenment Arts and Sciences* (2016). She has published widely on Arctic history and culture, in academic journals, as well as in newspapers and magazines regarding contemporary Arctic issues.

SUSI K. FRANK is Professor of Eastern Slavic Literatures and Culture at Humboldt University, Berlin. Her recent publications include *Imperiale Aneignung. Diskursive Strategien der Kolonisation Sibiriens durch die russische Kultur* (2010), *Evidenz und Zeugenschaft*, edited with Schamma Schahadat (2013), as well as articles on the Soviet Arctic.

HEIDI HANSSON is Professor of English Literature at Umeå University, Sweden. In the last few years, her research has concerned the representation of the North in travel writing and fiction. She was the leader of the interdisciplinary research programme *Foreign North: Outside Perspectives on the Nordic North* and has contributed to several research projects on aspects of the North and the Arctic in literature and culture. She is a member of the board of Arcum, the Arctic Research Centre at Umeå University.

ELIN HAUGDAL is Associate Professor of Art History at UiT – The Arctic University of Norway. Her research areas include contemporary architecture in Sami and Northern areas, architecture and regionalism, northern landscapes, and discourses of the Arctic. Her publications include edited books and articles related to architecture, landscape, photography, and contemporary art. She has been a partner in projects concerning indigenous studies and the cultural Arctic, and is currently leading the research group *Worlding Northern Art*.

RENÉE HULAN is Professor of English Literature at Saint Mary's University in Halifax, Nova Scotia, where she teaches Canadian literature. She is the author of *Canadian Historical Writing: Reading the Remains* (2014) and *Northern Experience and the Myths of Canadian Culture* (2002). Between 2005 and 2008, she served as co-editor of the *Journal of Canadian Studies/Revue d'études canadiennes*. She has also edited *Native North America: Critical and Cultural Perspectives* (1999) and, with Renate Eigenbrod, *Aboriginal Oral Traditions: Theory, Practice, Ethics* (2008).

SIGFRID KJELDAAS is a PhD candidate at the Department of Language and Culture, UiT – The Arctic University of Norway. Her research project concerns the depiction of the Arctic in North-American nature writing. Her fields of interest are ecocriticism, animal studies, nature writing, and arctic exploration narratives.

AUDUN J. MØRCH is Associate Professor of Russian Literature at the University of Oslo. His publications include *The Novelistic Approach to the Utopian Question: Platonov's Chevengur in the Light of Dostoevskij's Anti-Utopian Legacy* and articles on the works of Platonov, Dostoevskii,

Gogol, Pelevin, Rykheu and others. His current interests are mythological aspects of Post-Soviet Russian literature and literature of the Russian north including literature by indigenous writers.

ANKA RYALL is Professor Emerita of Gender Research in the Humanities at UiT – The Arctic University of Norway, where she has been co-leader and leader of two international interdisciplinary research projects, *Arctic Discourses* and *Arctic Modernities*. She is the author/editor of several books on travel writing and gender. Her most recent book is a feminist study of Virginia Woolf's writings (2011). She is working on a book about women in Norwegian polar history.

ROSWITHA SKARE is Professor of Media and Documentation Studies at UiT – The Arctic University of Norway. Her research interests include documentation theory, the impact of paratexts on literature and film and the performance practices of silent films. She is the author of *Nanook of the North From 1922 to Today: The Famous Arctic Documentary and Its Afterlife* (2016).

ULRIKE SPRING is Professor of History at the Western Norway University of Applied Sciences. Her latest publications include *Expeditions as Experiments: Practising Observation and Documentation* (co-edited with Marianne Klemun, 2016) and *Passagiere des Eises. Polarhelden und arktische Diskurse 1874* (co-written with Johan Schimanski, 2015). She is currently leading the research project *TRAUM–Transforming Author Museums*.

KIRSTEN THISTED is Associate Professor at Copenhagen University, Institute of Cross-Cultural and Regional Studies, Minority Studies Section. Her research areas include minority-majority relations, cultural and linguistic encounters, cultural translation and post-colonial relations. She has published widely on Greenlandic oral traditions, modern Greenlandic literature and film, Arctic explorers and Scandinavia seen in a post-colonial perspective. She is currently a partner in the "Centre of Excellence for Resources, Extractive Industries and Sustainable Arctic Communities (REXSAC)" and leads the research project *Denmark and the New North Atlantic*.

HENNING HOWLID WÆRP is Professor of Scandinavian Literature at UiT – The Arctic University of Norway. He has published books on Norwegian novels, nature poetry and the prose poem. In recent years his research has mainly concerned the Arctic, and his latest book is *Arktisk litteratur* (2017), on Nordic Arctic literature. He is a literary critic at the Norwegian newspaper *Aftenposten*, and in the period 2006–2010 served as co-editor of *Edda: Scandinavian Journal of Literary Research*.

Introduction: Environmental, Exotic, Everyday Arctic

Heidi Hansson and Anka Ryall

The Finnish artist Marja Helander's short video work *Trambo* (2014) opens with a cliché of the Arctic sublime: an almost naked white landscape, a snow-covered expanse devoid of life apart from some stunted trees in the horizon where snowfield and sky meet. This is the Arctic as the ultimate other of modernity – cold, empty, desolate and unpeopled. For a few seconds the viewer of *Trambo* sees only this landscape, its immobility and stillness broken by almost imperceptibly falling snow and the sound of the wind blowing. Then a scraping sound, accompanied by heavy breathing, precedes the entrance from the right of a woman in a bright red Sami costume adorned with silver brooches, laboriously hauling a large trampoline, an entirely incongruous and manifestly man-made object, across the snowfield. When she has brought it into the centre of the frame, she stops and climbs onto it. After a few seconds of standing still, she starts jumping, and the camera zooms in to show her head and upper body – bands and fringes swinging, jewellery clanking – as it appears over the top of the trampoline. Then it zooms back, and after bouncing straight up and down for a few more seconds her pose freezes. Next we see her on the ground outside the trampoline, before she hauls it towards the left and out of the picture, leaving the white desolation without any visible trace of human activity.

Helander's work is inspired by the contradictions of being caught between the natural and the urban and between a Sami identity and modern Finnish culture. She has called *Trambo* a self-portrait in which "I am dragging a big trampoline, a burden I hope will bring a bit of joy to the monotonous journey and life. The trampoline is a reference to the modern age, but can also be seen as a prison or a wall that is hard to see through" ("Grosses Treffen"). In its playful way, both the video itself and the still we have used on the cover of this book bring together the major concerns of Arctic modernity: the paradoxical amalgamation of exoticized indigenous tradition and a mundane everyday, as well as the fallacy of an undisturbed Arctic landscape at a time when the anthropogenic impact on it can no longer be denied. Significantly, *Trambo* also turns the conventional gendering of the Arctic on its head. Rather than being a testing ground for conquering male polar explorers, Helander's Arctic is both feminized and deliberately unheroic.

Trambo represents a distinctively personal and ironic take on the Arctic. It references and then undermines the common stereotype of the Arctic as one of "the vast 'premodern' spaces, uninhabited by urbanized peoples" (Smith 2001, 31). But if the Arctic is not outside modernity, then what exactly is it? Carina Keskitalo, who prefers putting "the Arctic" in quotation marks, has described the modern Arctic as the result of an international political project of region-building (Keskitalo 2007). As she points out, the Arctic Council, established in 1996, has been instrumental in redefining and enlarging the Arctic. Not only the frozen polar area of traditional Arctic discourse, it is now viewed as a "region" that includes the northern territories of the eight member states, Canada, Denmark (Greenland), Russia, the United States (Alaska), Iceland, Norway, Sweden and Finland, although by natural-science definitions much of this territory is considered subarctic. Although its boundaries have been drawn in various ways, Keskitalo refers to the Arctic Council's Arctic Monitoring and Assessment Programme (AMAP), which has delineated the Arctic on the basis of "simple, latitudinal definitions" (Keskitalo 2007, 193). According to such definitions, the Arctic would encompass up to 15 per cent of the world's land area from the North Pole down to between 60° northern latitude in North America, Greenland, Iceland and eastern Russia and the Arctic Circle (at 66° 33'N) in Norway, Sweden, Finland and northwest Russia – a vast area with a population of about four million, of whom some 13 per cent are indigenous peoples.

As is obvious in writings about and images of the Arctic, however, the exact demarcations of the area are less important than the cultural notions surrounding it as representing a generalized North. The question, then, is not where the Arctic actually begins, but where we imagine that it begins. Moreover, while Keskitalo writes that the politically motivated extension of the Arctic has taken place mainly since the 1970s (Keskitalo 2007, 188), earlier writings show that the term "Arctic" has for centuries been applied very loosely to any perceived northern periphery (Ryall, Schimanski and Wærp 2010, xii). Indeed, the distinction between "the Arctic" and "the North" were and are fuzzy. Like the cultural ideas surrounding such Arctic facts as cold, snow and ice, their meanings are always "context-dependent and far from stable" (Hansson and Norberg,

2009b, 13). Whereas ice was something to be conquered during the so-called heroic age, it has transformed into something to be preserved in an era of ecological awareness, even into an agent in itself. Even locations shift. Hence, the North Cape was perceived by late nineteenth-century tourists as *both* the boundary between the Arctic and Europe or Norway and as Arctic sublimity itself, as Ulrike Spring shows in her chapter in this volume. Likewise, Lapland represents the Arctic for some, like the British writer Olive Murray Chapman, discussed by Jan Borm, while others simply view it as an area of northern Fennoscandia. Thus, the Arctic is understood as both the exotic other of modernity and the mundane home-place of its inhabitants.

Currently, the area that is generally defined as the Arctic region is undergoing rapid changes in both its physical state and in the way it is perceived, creating imbalances and blind spots in our understanding of the processes involved. In many respects, the Arctic is under threat from modernity, represented by the depletion of natural resources and industrial pollution. Powerful political and commercial interests are closing in on the region in ways that are often both contradictory and simplifying. As Adriana Craciun argues in this volume, the Arctic is again in danger of appearing as an empty waste where opportunism reigns and the indigenous experience is muted. Yet while the melting of the Arctic ice is taken as a warning of ecological crisis, it is also viewed as a harbinger of economic development in terms of transportation, tourism and resource exploitation (cf. Grant 2010). Research on the Arctic has addressed both these aspects. The focus of natural scientists has been on the understanding of climate change, the management of biodiversity, and the mapping and extraction of resources, while social-science research has investigated the political and social implications of climate change, industrial development and technology for the future of the Arctic, as well as the problems in indigenous societies. Much of this research has a short-term perspective, and much is characterized by a simplistic way of thinking that Mike Hulme names "climate reductionism" (Hulme 2011, 264).

The chapters in this volume offer a different and more historically oriented perspective. By turning the attention to the cultural and aesthetic field, they address the very basis of our Arctic decisions, that is, the interaction between environment and people that has contributed to new visions of the Arctic. Even the Arctic sublime has proved susceptible to the changing relationship between humans and nonhuman nature, as Sigfrid Kjeldaas shows in her analysis of the ice and icebergs in Barry Lopez's *Arctic Dreams*. The Arctic represented in writings, visual images and films is of course less tangible than melting glaciers or the changing social conditions of Arctic peoples, and therefore to a large extent neglected in scholarship and policy-making. Yet, informed by the historical and cultural knowledge necessary to produce what Clifford Geertz calls "thick description" (Geertz 1983), analyses of such material have the potential for a "de-simplifying" critique exposing the underlying attitudes and grey zones ignored in much present-day decision-making about the Arctic.

For all the contributors to this volume, the Arctic is not only a geographical area, but also a discursive formation. As defined by Michel Foucault, discourse is a network of enunciative functions or statements that combine into a meaning-bearing sign system that regulate how a topic can be talked about (Foucault 1972). For Edward Said, authoritative texts create both the knowledge and the reality they seem to describe (Said 1978). Following these theories, language cannot express any pre-existing knowledge but constitutes what counts as knowledge or truth. By regulating the terms of the conversation, discourse may function as a tool for oppression, but since discursive formations constantly incorporate new enunciations they are also incomplete and inherently contradictory. While discourse exerts control, it also produces opposition, which creates the need to ask whose interests are served by the continued circulation of a particular discursive formation (Grace 2001; Hulan 2002). In the chapters that follow, three major Arctic discourses are interrogated: the Arctic understood as threatened environment, the Arctic perceived as the exotic opposite of modernity and the Arctic described as the everyday, lived reality of its inhabitants.

During the age of polar exploration, the Arctic was typically represented in atemporal terms, as a battling ground for the superhuman struggles of a few exceptional men against the forces of nature – in which the values and masculine virtues of the hero figures were usually seen to symbolize those of the nation. While remnants of this masculinist rhetoric survive in many contemporary narratives of technology-supported polar adventures, it has been deconstructed in several studies of nineteenth- and early twentieth-century polar exploration (e.g. Bloom 1993; Spufford 1996; Moss 2006). We have also seen the publication of studies of the Arctic in national cultures, usually connected to specific geographical locations, Canada in particular (Atwood 1995; Grace 2001; Hulan 2002). The militarization of the Arctic during and after the Cold War is another topic that has been thoroughly documented in recent years (e.g. Tamnes 1991; Grant 2010; Sale and Potapov 2010; Williams 2010). Several collections of articles encompassing Arctic discourses from the whole circumpolar area are also evidence of an expanding field (Bravo and Sörlin 2002; Hansson and Norberg 2009a; Ryall, Schimanski and Wærp

2010; Schimanski, Theodorsen and Wærp 2011; Kjeldaas and Ryall 2015). While some of this work deals with recent or contemporary Arctic discourses, the focus has rarely been on the Arctic as a site of discourses of modernity. Likewise, with a few exceptions (e.g. Bloom 1993; Thisted 2002; Ryall 2007; Hansson 2009; Ryall 2009; Aarekol 2016; Høvik 2016; Thisted 2016; Thisted 2017), there has been little extended emphasis on gender and indigeneity in studies of Arctic discourses.

In terms of Arctic exploration, 1930 – the year the famous Norwegian explorer Fridtjof Nansen died – marks a symbolic end of the masculinist heroic age. But for over half a century, at least since the 1870s, the Arctic had increasingly been placed under the sign of the modern, and even the famous polar explorers had made themselves increasingly dependent on modern technologies of transportation, as when Roald Amundsen, Lincoln Ellsworth and Umberto Nobile concluded the quest for the North Pole in 1926 by flying over it in an airship (Fleming 2001). During the past century the Arctic has gradually become a space of scientific investigation, tourism and industrial developments rather than geographical discovery and "firsts". As Renée Hulan points out in this volume, Goose Bay in Labrador was the world's largest airport in 1923, greatly adding to the accessibility of Arctic Canada and symbolizing what Zygmunt Bauman has called "liquid modernity", in which remoteness is no longer an obstacle to human mastery. Susi K. Frank, in her chapter, notes a parallel development in Soviet literature and film of the 1930s. There a central idea is that polar and transpolar flights integrate the Arctic into Soviet space by transforming it from an alien, threatening space to a friendly, inhabited one (see also Frank 2010). In this way technologies of transportation go hand in hand with technologies of representation, as Roswitha Skare shows in her discussion of the film *The Romance of the Far Fur Country* (1920), where the trade activities of the Hudson Bay Company are portrayed and romanticized. As Bauman discusses, modern technical developments like aviation and film collapse time and space and create a sense of availability that paves the way for national expansion, exploitation of the region's natural resources and a budding tourist industry, to name but a few (Bauman 2000).

In fact, if tourism is a defining feature of modernity, as Dean MacCannell has claimed (MacCannell 1989), it may be argued that the Arctic became part of modernity through its discovery as a tourist destination from the late nineteenth century onwards. Ulrike Spring's study of early mass tourism at the North Cape shows that these processes of modernization are not only related to the development of infrastructure, but also, in complex ways, to changes in the practices and conceptions of gender. The impact of such factors as air travel, industry, tourism, urgent environmental concerns and changing gender norms on discourses of Arctic modernity has until recently remained largely unexamined. It is a central theme, however, in several chapters in this volume.

If the exact delimitation of the Arctic is contested, the same could with even more justification be said about the terms "modern", "modernization", "modernity" and "modernism". It has often been taken for granted that modernity, in the words of Anthony Giddens, is "distinctively a Western project" with roots in European history and radiating to other parts of the world (e.g. Giddens 1990; Giddens 1991; Punter 2007). Models of power relations based on a centre/periphery binary still sometimes form an unexamined framework of studies of modernity/modernism. However, many theorists now argue in favour of a polycentric view in which every location represents distinct forms of modernity (e.g. Felski 1995; Eisenstadt and Schluchter 1998; Wittrock 1998; Booth and Rigby 2000; Eisenstadt 2000; Felski 2000; Friedman 2006). In these respects, preservation discourses, which are in many ways part of a modern, environmental project, occasionally reject modern, social developments in peripheral regions, as discussed by Kirsten Thisted in relation to Greenlandic modernity.

Theorists of both culture and literature therefore call for a new geography of modernity in which modernity/modernism is seen as transnational breaks with tradition (Appadurai 1996; Doyle and Winkiel 2005; Friedman 2006; Ramazani 2009). Such spatialization of modernities, emphasising "the temporal rupture of before/after wherever and whenever such ruptures might occur in time and space" (Friedman 2006), has necessarily had a crucial effect on historical periodization. Although this new work on plural modernities has usually focused on southern postcolonial discourses, its findings encourage investigations of discourses of Arctic modernity outside the framework of a conventional centre/periphery perspective. A case in point is Audun J. Mørch's reading of the Chukchi writer Yuri Rytkheu, whose work spans the transition from primordiality via modernity to postmodernity but in ways that make it necessary to fundamentally rethink those concepts.

The move from a temporal to a spatialized view of modernity has sparked a reconsideration of the meaning of the term itself. In her influential provisional definition of modernity Susan Stanford Friedman rejects "nominal" (noun-based) definitions of modernity that connect it to a particular historical moment in Western history and argues instead in favour of a strategic "relational" (adjective-based) definition (Friedman 2001). In this view the term "modernity", like its correlates "modern", "modernization" and "modernism", acquires its meaning through negation, as a rebellion – often figured as a "new", emancipated and secularized, woman – against a presumed past (Felski 1995; Friedman 2001; Friedman 2006). While modernity is conceived as a global, shifting and unstable category, it can only be understood in terms of a geographically specific comparative "modern"/"traditional" binary (Subrahmanyam 1998). Across the circumpolar Arctic, too, there are many "alternative modernities" (Gaonkar 1999) that each requires a different contextualization.

Some theorists, most notably Bruno Latour, have argued against definitions of modernity that rely on the assumption that there is a rupture between modernity and tradition (Latour 1993). Modernity in Latour's definition depends on the work of "purification", that is, the creation of two distinct ontological zones, that of humans/culture on the one hand and nonhumans/nature on the other. However, he also argues that the practices of purification are always concomitant with practices of "translation" and "hybridization". The proliferation of hybrids between nonhuman and human creates mixed categories, while modernity is dependent on keeping the two sets apart. Global warming is a good example of this; another is the militarization of the Arctic, where the pristine-looking ice cap conceals an arsenal of far-reaching nuclear submarines. Latour's attempt to find a position beyond the either/or of modernity via an emphasis on the non-separability of the common productions of societies and natures holds great promise for the study of Arctic modernities. Susi K. Frank's deconstruction of Soviet modernity, as well as Fredrik Chr. Brøgger's study of Helge Ingstad's anti-modernist project of entering a pristine and pre-modern wilderness while at the same time re-enacting pioneering and conquering modernity, illustrate the non-separability of these processes.

The image of the circumpolar Arctic as a cold, empty, dangerous region imbued with notions of heroic masculinity and frontier rhetoric still persists. Although the Arctic has been incorporated into modernity in an array of locally specific ways and is home to some four million people, many of whom live in modern cities and urban areas, it remains a place apart. Hence, the two chapters by Anka Ryall and Elin Haugdal that deal with emergent modernity in the Norwegian Svalbard archipelago, show how the harsh High-Arctic environment endows even everyday family life and mundane chores with a sense of drama and adventure. Like the other chapters in this volume, they examine to what extent the processes of modernization have changed the discursive signification of the modern Arctic.

In terms of "a new geography of modernity" the chapters investigate some *specificities* of Arctic modernity. In general, however, the modern

Arctic seems to have a double discursive signification. On the one hand it signifies something accessible, everyday and mundane; on the other it is a never-never land of romance and adventure. While some of the exploration narratives of Arctic explorers like Fridtjof Nansen, Knud Rasmussen and Helge Ingstad may be viewed as quests for a pastoral sanctuary from modernity, as Henning Howlid Wærp's chapter shows, much recent writing on the Arctic has made it synonymous with modernity itself. Earlier versions of this view are found in representations of the Nordic North as exemplary of progressive gender politics in women's travel writings from around 1900 (Ryall 2009; Hansson 2011). As a utopian space the Arctic points to a different future (Smith 2012). At the same time it often seems to have become an expression of the continued survival – within modernity – of the past as nostalgia, longing, dream and myth. In the words of Sherrill Grace, "unlike the permafrost and ice, myths are less amenable to climate change" (Grace 2010). Like other marginal places, the Arctic often appears as an exotic counter-modernity "at once threateningly strange and reassuringly familiar" (Huggan 2001, 22), embodying a mythic or romantic dimension existing beyond the modern everyday world while remaining thoroughly imbricated in modernity.

One hypothesis linking many of the chapters in this volume is that the Arctic may be seen as a stark embodiment of the paradoxes of modernity. Unsurprisingly, this is most obvious in Arctic popular fiction, exemplified in Cathrine Bjerknes' chapter on Melanie McGrath's crime novel *White Heat* (2011). In that sense, they show that the discursive significations of the modern Arctic throw light on general processes of modernization. In different ways they illuminate both how discourses of the Arctic have been inflected by various kinds of modernization and how earlier Arctic images – whether they are affirmed, contested or repudiated – have shaped, influenced and informed a wide range of modern discourses of the Arctic. The geographical span of the material is equally wide. The perspectives on the Arctic in the selected material are both external/touristic, as exemplified by travelogues, expedition narratives and sociological reports, and internal/indigenous and based on intimate local knowledges. Although six of the book's fourteen chapters deal specifically with Canadian Arctic discourses, reflecting the central role Canada has played in articulating the Arctic as a region (Keskitalo 2009), they are thrown into relief by the very different discourses of Northern Russia, Greenland and Svalbard. As Kirsten Thisted emphasizes in her chapter, Greenlandic experiences cannot be understood when viewed through a Canadian lens.

The chapters of this book have been written within a three-year international research project, *Arctic Modernities*, located at UiT – The Arctic University of Norway (UiT). Completed in July 2016, the project was financed by the Polar Research Programme of the Research Council of Norway, with additional funding from the Faculty of Humanities, Social Sciences and Education at UiT. Over twenty researchers from seven countries and many disciplines were involved. Several articles by project members were published in a special *Arctic Modernities* issue of the open-access journal *Nordlit* (Kjeldaas and Ryall 2015), while another four were published in a special issue of *Acta Borealia: A Nordic Journal of Circumpolar Societies* guest-edited by Anka Ryall (issue 2, 2016). Further contributions will be published in a volume on Arctic children's literature. Because the different studies vary widely in scope and approach, as a whole they demonstrate not only that Arctic modernity is a multifarious phenomenon constituted by different transnational variants with strong local roots and geographical specificity, but also that research on the Arctic from a humanities perspective is a vital, burgeoning and necessary field.

References

Aarekol, Lena. 2016. "Arctic Trophy Hunters, Tourism and Masculinities, 1827–1914." *Acta Borealia* 33 (2): 123–139.

Appadurai, Arjan. 1990. "Disjuncture and Difference in the Global Cultural Economy." *Theory, Culture & Society* 7: 295–310.

—. 1996. *Modernity at Large: Cultural Dimensions of Globalization*. Minneapolis, MN: Minnesota University Press.

Arlov, Thor B. 2003. *Svalbards historie* [The history of Svalbard]. Trondheim: Tapir Akademiske Forlag.

Atwood, Margaret. 1995. *Strange Things: The Malevolent North in Canadian Literature*. Oxford: Clarendon Press.

Bauman, Zygmunt. 2000. *Liquid Modernity*. Cambridge: Polity Press.

Bloom, Lisa. 1993. *Gender on Ice: American Ideologies of Polar Expeditions*. Minneapolis, MN: University of Minnesota Press.

Booth, Howard J., and Nigel Rigby, eds. 2000. *Modernism and Empire*. Manchester: Manchester University Press.

Bravo, Michael, and Sverker Sörlin, eds. 2002. *Narrating the Arctic: A Cultural History of Nordic Scientific Practices*. Canton, MA: Science History Publications.

Doyle, Laura, and Laura Winkiel. 2005. "Introduction: The Global Horizons of Modernism." In *Geomodernisms: Race, Modernism,*

Modernity, edited by Laura Doyle and Laura Winkiel, 1–14. Bloomington, IN: Indiana University Press.
Eisenstadt, Shmuel N., and Wolfgang Schluchter. 1998. "Introduction: Paths to Early Modernities: A Comparative View." *Daedalus* 127 (Summer): 1–18.
Eisenstadt, S.N. 2000. "Multiple Modernities." *Daedalus* 129 (Winter): 1–29.
Felski, Rita. 1995. *The Gender of Modernity*. Cambridge, MA: Harvard University Press.
—. 2000. *Doing Time: Feminist Theory and Postmodern Culture*. New York: New York University Press.
Fleming, Fergus. 2001. *Ninety Degrees North: The Quest for the North Pole*. New York: Grove Press.
Foucault, Michel. 1972. *The Archeology of Knowledge*. (*Archéologie du savoir*.) Translated by A.M. Sheridan Smith. London: Routledge.
Frank, Susi K. 2010. "*City of the Sun on Ice*: The Soviet (Counter-) Discourse of the Arctic in the 1930s." In *Arctic Discourses*, edited by Anka Ryall, Johan Schimanski and Henning Howlid Wærp, 106–131. Newcastle: Cambridge Scholars Publishing.
Friedman, Susan Stanford. 2001. "Definitional Excursions: The Meanings of Modern/Modernity/Modernism." *Modernism/Modernity* 8 (Sep.): 493–513.
—. 2006. "Periodizing Modernism: Postcolonial Modernities and the Space/Time Borders of Modernist Studies." *Modernism/Modernity* 13 (Sep.): 425–443.
Gaonkar, Dilip Parameshwar. 1999. "On Alternative Modernities." *Public Culture* 11: 1–18.
Geertz, Clifford. 1983. *Local Knowledge: Further Essays in Interpretive Anthropology*. New York: Basic Books.
Giddens, Anthony. 1990. *The Consequences of Modernity*. Stanford, CA: Stanford University Press.
—. 1991. *Modernity and Self-Identity: Self and Society in the Late Modern Age*. Cambridge: Polity Press.
Grace, Sherrill. 2001. *Canada and the Idea of the North*. Montréal: McGill-Queen's University Press.
—. 2010. "From the 'Hand of Franklin' to *Frobisher*: Opera in the Canadian North." In *Arctic Discourses*, edited by Anka Ryall, Johan Schimanski and Henning Howlid Wærp, 283–300. Newcastle: Cambridge Scholars Publishing.
Grant, Shelagh D. 2010. *Polar Imperative: A History of Arctic Sovereignty in North America*. Vancouver: Douglas & McIntyre.

"Grosses Treffen." https://www.grossestreffen.org/marja-helander Accessed 26 Jan. 2017.

Hansson, Heidi. 2009. "Feminine Poles: Josephine Diebitsch-Peary's and Jennie Darlington's Polar Narratives." In *Cold Matters: Cultural Perceptions of Snow, Ice and Cold*, edited by Heidi Hansson and Cathrine Norberg, 105–123. Umeå: Umeå University/Royal Skyttean Society.

—. 2011. "Between Nostalgia and Modernity: Competing Discourses in Travel Writing about the Nordic North." In *Iceland and Images of the North*, edited by Sumarlidi Isleifsson, 255–282. Québec: Presses de l'Université du Québec.

Hansson, Heidi, and Cathrine Norberg, eds. 2009a. *Cold Matters: Cultural Perceptions of Snow, Ice and Cold.* Umeå: Umeå University/Royal Skyttean Society.

Hansson, Heidi, and Cathrine Norberg. 2009b. "Revisioning the Value of Cold." In *Cold Matters: Cultural Perceptions of Snow, Ice and Cold*, edited by Heidi Hansson and Cathrine Norberg, 7–22. Umeå: Umeå University/Royal Skyttean Society.

Huggan, Graham. 2001. *The Postcolonial Exotic: Marketing the Margins.* London and New York: Routledge.

Hulan, Renée. 2002. *Northern Experience and the Myths of Canadian Culture*. Montréal: McGill-Queen's University Press.

Hulme, Mike. 2011. "Reducing the Future to Climate." *Orisis* 26: 245–266.

Høvik, Ingebjørg. 2016. "Reproducing the Indigenous: John Møller's Studio Portraits of Greenlanders in Context." *Acta Borealia* 33 (2): 166–188.

Keldaas, Sigfrid, and Anka Ryall, eds. 2015. *Arctic Modernities*. *Nordlit* 35.

Keskitalo, Carina. 2007. "Development of the Arctic as an International Region." *Cooperation and Conflict: Journal of Nordic International Studies Association* 42 (2): 187–205.

Keskitalo, E. Carina H. 2009. "'The North' – Is There Such a Thing?" In *Cold Matters: Cultural Perceptions of Snow, Ice and Cold*, edited by Heidi Hansson and Cathrine Norberg, 23–39. Umeå: Umeå University/Royal Skyttean Society.

Latour, Bruno. 1993. *We Have Never Been Modern*. (*Nous n'avons jamais été modernes*, 1991.) Translated by Catherine Porter. Hemel Hempstead: Harvester Weatsheaf.

MacCannell, Dean. 1989. *The Tourist: A New Theory of the Leisure Class*. New York: Schocken Books.

Moss, Sarah. 2006. *The Frozen Ship: The Histories and Tales of Polar Exploration*. New York: BlueBridge.

Punter, David. 2007. *Modernity*. Basingstoke: Palgrave MacMillan.

Ramazani, Jahan. 2009. *A Transnational Poetics*. Chicago, IL: University of Chicago Press.

Ryall, Anka. 2007. "A Humbling Place: Tests of Masculinity in Early Nineteenth-Century Travel Narratives from Lapland." In *Northbound: Travels, Encounters and Constructions 1700–1830*, edited by Karen Klitgaard Povlsen, 261–279. Aarhus: Aarhus University Press.

—. 2009. "Europe's Northern Periphery and the Future of Women in the Travel Narratives of Ethel Tweedie." *Nora – Nordic Journal of Feminist and Gender Research* 17: 273–288.

Ryall, Anka, Johan Schimanski, and Henning Howlid Wærp, eds. 2010. *Arctic Discourses*. Newcastle: Cambridge Scholars Publishing.

Said, Edward W. 1978. *Orientalism*. New York: Pantheon Books.

Sale, Richard, and Eugene Potapov. 2010. *The Scramble for the Arctic: Ownership, Exploration and Conflict in the Far North*. London: Frances Lincoln.

Schimanski, Johan, Cathrine Theodorsen, and Henning Howlid Wærp, eds. 2011. *Reiser og ekspedisjoner i det litterære Arktis* [Travels and expeditions in the literary Arctic]. Trondheim: Tapir Akademiske Forlag.

Smith, Laurence. 2012. *The New North: The World in 2050*. London: Profile Books.

Smith, Sidonie. 2001. *Moving Lives: Twentieth-Century Women's Travel Writing*. Minneapolis, MN: University of Minnesota Press.

Spufford, Francis. 1996. *I May Be Some Time: Ice and the English Imagination*. London: Faber and Faber.

Subrahmanyam, Sanjay. 1998. "Hearing Voices: Vignettes of Early Modernity in South Asia, 1400–1750." *Daedalus* 127 (Summer): 75–104.

Tamnes, Rolf. 1991. *The United States and the Cold War in the High North*. Aldershot: Dartmouth Publishing.

Thisted, Kirsten. 2002. "The Power to Represent: Intertextuality and Discourse in *Miss Smilla's Sense of Snow*." In *Narrating the Arctic: A Cultural History of Nordic Scientific Practices*, edited by Michael Bravo and Sverker Sörlin, 311–342. Canton, MA: Science History Publications.

—. 2016. "The Specter of Danish Empire: *The Prophets of Eternal Fjord* and the Writing of Danish-Greenlandic History. In *New Dimensions of*

Diversity in Nordic Culture and Society, edited by Jenny Bjørklund and Ursula Lindqvist, 151–173. Newcastle: Cambridge Scholars Publishing.

—. 2017. "The Greenlandic Reconciliation Commission: Ethnonationalism, Arctic Resources, and Post-Colonial Identity." In *Arctic Environmental Modernities from the Age of Polar Exploration to the Era of the Antropocene*, edited by Lill-Ann Körber, Scott MacKenzie and Anna Stenport Westerstahl, 231–246. Basingstoke: Palgrave Macmillan.

Williams, Lisa. 2010. "Telling an Arctic Tale: Arctic Discourses in Canadian Foreign Policy." In *Arctic Discourses*, edited by Anka Ryall, Johan Schimanski and Henning Howlid Wærp, 240–258. Newcastle: Cambridge Scholars Publishing.

Wittrock, Björn. 1998. "Early Modernities: Varieties and Transitions." *Daedalus* 127 (Summer) 19–40.

Environmental Arctic

ICE AS A LITERARY MOTIF IN SOVIET ARCTIC MODERNITIES

SUSI K. FRANK

To think of nature and humankind as completely separate, "purified" of each other, to understand humans' correlation towards nature as appropriation and empowerment, as a "relationship of mastery and possession" (Serres 1995, 39), and to see nature as an object under humans' sovereign control, are undisputed aspects of modernism. The narrative of humans' emancipation from nature and of human mastery over nature is one of the main narratives of this era. "Mastery and possession: These are the master words launched by Descartes at the dawn of the scientific and technological age, when our Western reason went off to conquer the universe," writes Michel Serres. "Our fundamental relationship with objects comes down to war and property" (1995, 32).[1] Horkheimer and Adorno on the one hand and Bruno Latour on the other were two of the most important critical voices to criticize this project, each in his/their own way, as a fatal utopia. Whereas Horkheimer and Adorno in their *Dialectic of Enlightenment* warned that this attitude will come at a price: "World domination of Nature turns against the thinking subject itself" (Horkheimer and Adorno 2002, 20)[2], Latour elaborated on its illusionism, stating that the strict distinctions – e.g. between human and nature – that are at the core of modernism's thinking are thoroughly false: we have never been modern (Latour 1993, 9ff.).

Climate is one essential field where independence from nature had to be won. As the geographer Ellsworth Huntington phrased it in 1915, "If we'll be able to master climate, the whole world will become stronger and nobler" (1915/2001, 294). Or take "the anthropology of climate", which developed from geodeterminism (environmentalism) into a (utopian) understanding of climate as man-made. The Arctic – the geographical zone where environmental conditions are harshest and most hostile towards humans – has for centuries been one of the most significant image-reservoirs for a narrative "vocabulary" of this concept of a human-nature correlation. Images of men's heroic and victory-bound fight against nature, of their power to overcome all limits set by nature, could not be better visualized and/or narratively developed than in an Arctic setting. Another reason for the attractiveness of this setting was certainly the fact that, up to the 1930s, the Arctic was one of the last white spots on Earth that European colonial exploration had not yet managed to conquer and transform into a zone of culture.

This is true of the anthropological imaginary of modernism in general and of Russia and the Soviet Union in particular. In this respect, a certain continuity between pre-Soviet, so-called "cosmism" and Soviet and even Stalinist anthropology must be emphasized. Cosmists like Nikolai Federov (1828–1903) and Konstatin Ciolkovskii (1857–1935) had already at the very beginning of the twentieth century developed ideas of transcending the supposedly natural limitations of human life, such as mortality or human confinedness to the earth, and thereby paved the way for the Soviet world view. By the 1930s, nature – along with capitalism – was declared to be the ultimate enemy that humans had to defeat. The understanding of humans as subject and nature as object could not be clearer than at this time. One of the main arenas to symbolically enact and implement this human–nature relationship was the Soviet Arctic.

The best demonstration of the entanglement between these historical directions of thinking and the topic of the Arctic can be found in the reflections of Vladimir Vernadsky (1863–1945), himself a cosmist, founding father of bio-, geo-, and cosmochemistry and also precursor of environmentalism and the earth-system approach. In an article published not long before his death, Vernadsky developed his suggestion that, with humankind in the age of modernism, a new epoch of earth's and cosmic history had begun, the "noosphere": "Mankind taken as a whole is becoming a mighty geological force," he claims (2005, 19). For the first time in human history, Vernadsky emphasizes:

> man [...] knew and embraced the whole biosphere, completed the geographic map of the planet Earth, and colonized its whole surface. Mankind became a single totality in the life of the Earth. There is no spot on Earth where man cannot live if he so desires. Our [i.e. Soviet] people's sojourn on the floating ice of the North Pole in 1937–1938 has proved this clearly. (2005, 19)

Continuing with these reflections, Vernadsky enthusiastically predicts that "new species and races of animals and plants are being created by man. Fairytale dreams appear possible in the future: man is striving to emerge beyond the boundaries of his planet into cosmic space" (2005, 21). His conclusion is that "historians and political leaders are only beginning to approach a comprehension of the phenomena of nature from this point of view" (2005, 21).

In the present day, Paul Crutzen, the coiner of the term "Anthropocene", who refers to Vernadsky's "noosphere", can be seen as the ultimate vocalizer of this point of view. Even though Crutzen admits that, again and again, human interventions cause unexpected, unintended and unforeseeable environmental changes, he does not call human force itself into question. On the contrary, he claims that it should be consolidated. Mankind, according to Crutzen, "will remain a major environmental force for many millennia" (2002, 23). As he puts it, "A daunting task lies ahead for scientists and engineers to guide society towards environmentally sustainable management during the era of the Anthropocene. This [...] may well involve internationally accepted, large-scale geo-engineering projects, for instance to 'optimize' climate" (Crutzen 2002, 23). I will return to Vernadsky and Crutzen at the end of this chapter.

Modernism has brought forward at least three strategies for handling the Arctic climate: firstly, technological devices that have helped to handle climate conditions like permafrost soil; secondly, climate capsules aimed at guaranteeing independence from the outside climate; thirdly – and most utopian – strategies for transforming the Arctic into a tropical garden of Eden; for instance, by heating it and thereby totally transforming the region's climate. This is a vision that first found expression in Charles Fourier's "Théories des Quatre movement" [The theory of the four movements, 1808], can be traced in Soviet thinking from the 1930s through to the 1950s, and has found a gloomy reflection in Michel Houellebecq's 2005 novel *La possibilité d'une île* (translated as *The possibility of an island*).

In this context, the Soviet anthropology of modernism and the concrete attempts to realize the project of man's final victory over nature appear most radical. Countless efforts were made to create fertile soil out of dry steppes or to grow tropical fruit anywhere. Soviet strategies to settle in the Arctic North, to transform the Polar region into the frontline of the Soviet world, and in the long run into an ordinary settled and civilized region, can be seen as the most consistent manifestation of this project.

Elaborating on some significant examples from different periods of the twentieth and beginning of the twenty-first century, I first outline how the anthropology of modernism, by which I mean the conceptualization of humans in relation to nature, developed and changed. I consider the significance that the motif of ice had in this context. This is important because, as I see it, the way the motif of ice is developed in (Russian) literature indicates a shift from a classical (first) modernism to positions that revise the anthropocentric worldview by decentring and multifariously entangling humankind with its environment and by re-ascribing non-human nature an agency that had been stripped from it within modernism. Choosing examples from different epochs of Soviet and post-Soviet Russian literature, I then trace not only one single historical process of transcoding, but also divergences that were motivated by different social or cultural positions.

What is it about ice? Why is ice so important as a literary motif in this context? In my opinion, it is not due only to the fact that it is the most obvious symptom of an Arctic climate (of cold in general), but also to the symbolic potential of its material quality. Ice kills by its cold; or rather, cold kills by transforming living matter into ice, but at the same time ice preserves living matter. Therefore, ice could on the one hand be developed as a metaphor of death (at least since Dante's "visualization" of hell's ultimate circle as an icy polar world) and, on the other hand, it could be enacted as a means of overcoming death. Because ice represents hostile and inhospitable environmental conditions for humans, it could be enacted as a setting for man's heroic attempts to withstand nature, to prevail and to conquer the Arctic as Earth's most resistant place, which thus became a symbol of humanity's mastery over the Earth.

In this context, ice was enacted not only as a medium into which the whole history of conquest is inscribed, but also as an indicator of the ultimate limits of human conquest that had to be transgressed by every newcomer; that is, it became a medium where traces could be transcribed and thereby extinguished. As a medium of preservation, ice blurs the line between life and death. Therefore, it can become a means of the human will to be master of life and death. But, as a natural archive it can also become a symbol of nature's own agency, because it conserves beyond human intention and despite the human will to forget or to delete. Last but not least, as we in these days know well, ice is a medium that sequesters bad matter that otherwise – if it melted – would have fatal consequences.

During the course of the twentieth and early twenty-first centuries, fictional narration has made use of all the above connotations of ice. Generally, two main directions of semantic development can be observed. On the one hand, the motif is developed in order to substantiate the narrative of conquest. This involves its use as an inscription metaphor in which ice is an instrument that enables humans to reawaken the dead and thus to finally become master of life and death. On the other hand, works of fictional narration elaborate ice as matter that possesses a certain agency and cannot be subjected to human will. It is the condition of permafrost that, even as it conserves, hides, and sequesters, it can also –

for instance, in the case of its unintended melting – reveal. Melting permafrost may bring to light what humans have done (and do not want to be reminded of) or it may set something free: energetic power, powers of cognition or methane that pollutes the atmosphere.

I believe that different literary stagings of the motif of ice reflect quite clearly the historical process of remodelling the human–nature relationship during the course of Soviet modernism and in its aftermath. Let me elaborate on that point in four steps: firstly, Soviet modernism and socialist realism of the 1930s; secondly, multinational Soviet literature and its post-Soviet transformation; thirdly, Gulag literature; and finally, contemporary Russian literature.

Soviet modernism and socialist realism of the period 1930–1950

Since I have already discussed Soviet Arctic discourse of the 1930s, I begin by summarizing this point only briefly. Soviet literature of the 1930s was proclaimed not only as a means to depict and praise heroic deeds, but also as a major instrument of realization itself. To performatively realize the utopian vision was a programmatic aim of socialist realism, the aesthetic norm proclaimed in 1934.

Soviet literature and film of the 1930s focused again and again on the Soviet conquest of the Arctic. Central to all of these texts was the idea that the aim of all expeditions, of all polar and transpolar flights, was to totally integrate the Arctic into Soviet space by transforming it from an uninhabitable region of threatening cold into a friendly zone of human settlement. A corresponding image of heroism showed humans as making themselves completely independent of environmental conditions and simultaneously able to overcome them.

In Soviet modernism, as in modernism in general, we can locate four main symbolic meanings of "ice". Firstly, ice was a metaphor of death and a symbol of nature's power and hostility towards human life and therefore an important target of humanity's fight against nature. Ice is therefore a symbol of natural conditions that have to be overcome, as in the contemporary project of settling and warming the Arctic. Secondly, ice to the contrary was seen as a device for overcoming death by conserving life. In the context of Russian modernism, this idea has to be seen in connection with Russian cosmism represented by Nikolai Fedorov, who in his most important book *Philosophy of the Common Goal* (1906–13) developed a vision of overcoming mortality by freezing and raising our our forefathers from the dead when the time comes.[3] Thirdly, ice became a symbol of the presistance of the old and obsolete. Because ossification is a metaphorical meaning of freezing, ice as a medium hinders development, that is, modernization. The avant-garde writer Boris Pil'njak, for instance, used ice in his novel *Dvojniki* (Doubles) (1933–35) to symbolize stagnation: "The glaciers of tradition yet continue to exist in the frost of ages."[4] Fourthly, ice functioned as a metonym of the attempt to come first, to be the first to set one's feet on untrodden territory (or to detect and try to delete earlier traces – to make a tabula rasa).

Using ice as a metonym for the attempt to come first, to get somewhere where it is possible to start from zero, has much to do with modernism's tendency not to use memory as a means of self-legitimation, but rather to extinguish it as something that restrains freedom of action in the present. Statements by Nietzsche in "On the Genealogy of Morals" (*Genealogie der Moral* (1887): Erste Abhandlung. Chapter 11), Bertolt Brecht's poems "To townspeople" (*Lesebuch an die Städtebewohner*), where the refrain "Obliterate the traces!" ("Gedicht 1": "Verwisch die Spuren!") ties together reflections on life in a big city (Brecht 1930), and Walter Benjamin's reflections on "Experience and poverty" ("Erfahrung und Armut", 1932) illustrate this tendency very well. To begin reflections from point zero, regardless of whether the topic is as old as philosophy; to discard the old experiences and create a tabula rasa as Benjamin observes in a "new barbarism", (Benjamin 1980, 292)[5] and to declare memory to be useless, are all one gesture that corresponds to the adamistic mythopoetics of the avant-garde, to its pathos in destroying and creating language anew, to return to point zero of culture.

Accordingly, the new barbarians accepted only one memory, the memory of the initial act of transgressing the boundary of culture. It is exactly this central "attitude of modernism" (Menke 2000, 595) that, as Bettine Menke puts it in her article "Die Polargebiete der Bibliothek" [The polar regions of library], is visualized through the "Arctic regions": "the rejection and the clearing away of traces" (Menke 2000, 595: "das Abräumen oder Verwischen der Spuren").[6] The figure of coming first, of being the first to set foot on untrodden territory, as well as the corresponding figure of being late, of finding traces of earlier expeditions, of being unable to transcend the boundaries of already-known territory, are both crucial in this context. Significantly, ice is often addressed not as a matter that preserves traces and relicts, but rather – like glass[7] – as a substance that helps to blur traces and thus also memory. Just "in the ice crystal the separation from the burden of heritage becomes manifest" (Menke 2000, 52).

Multinational Soviet literature: Yuri Rytkheu

It was mostly the negative connotations of ice that became important in official Soviet modernism from the 1930s onward. In the context of this – as Terry Martin (2001) called it – "affirmative action empire", literary representations of the cultures of their Arctic home regions by authors from the so-called "small peoples of the North" did not develop alternative symbolic investments, but enacted exactly these meanings of ice and permafrost. Here I want to briefly elaborate on Yuri Rytkheu, the Soviet Chukchi author also discussed in the chapter by Audun J. Mørch in this volume.

Rytkheu is a typical representative of the so-called "multinational Soviet literature" that was first announced in 1934 at the First All-Union Congress of Soviet Writers. His career was typical of (non-Russian) Soviet writers from the periphery: all had to pass through educational institutions of the capital cities (usually the Gorki Institute of Literature in Moscow, but in Rytkheu's case the philological faculty of Leningrad State University). He wrote all his novels in Russian, but at the same time worked as a translator into Chukchi, thereby promoting its literary status. It was Rytkheu who realized the demand for socialist realist multinational literature by developing – in parallel with other non-Russian Soviet authors – a narrative of the Chukchi people becoming a modern nation that conformed to the socialist realist masterplot: the masterplot which told the story of the becoming of a Soviet hero as a passage from "natural giftedness" (stichijnost') to "conscience" (soznatel'nost') (cf. Clark 2000), that is, recognizing the right way.

In his early autobiographical trilogy *The Time of Melting Snows* (1958–67),[8] Rytkheu realizes the ideal-typical ethnic variant of Socialist realist masterplot that Yuri Slezkine has called the "long journey plot" (1997, 310ff.),[9] that is, the story of forging a Soviet hero out of a representative of a small and primitive people of the North. "Thawing" in this context becomes a topical metaphor for the process of enlightenment and modernization. This is also true of another novel by Rytkheu that he wrote a decade later: *The End of Perpetual Ice* (1977).[10] Here, Rytkheu transposes the metaphor from "snow" to "ice" or rather "permafrost", but retains the same meaning:

> For a long time the talents of the Chukchi people lay under a thick stratum of ignorance, darkness and superstition. Now great spring has come for our people – the time when snow thaws. Cold and heavy snow disappears, and free like flowers in the Tundra the talents of the people flourish, even

> more beautiful, happy, like a river, your life will spring up into real human happiness. (Rytkheu 1969, Chapter 6)[11]

> In the past, we didn't think about life: the search for food and concern about a warm home – that is what our thoughts circled around. Nowadays our life is changing, like earth in spring, when the snow is thawing. Many things aren't clear yet, but we are willing to understand, to learn how to manage our lives ourselves. (Rytkheu 1977, Chapter 26)[12]

Again we can observe the complementarity between the narrative of the Soviet conquest of the Arctic, where the active transformation of Arctic nature into a warm and inhabitable zone is a constitutive part of the narrative, and Rytkheu's use of "melting" as a metaphor to describe Sovietization and modernization as processes of enlightening the indigenous peoples.

In this novel, the complementarity becomes explicit, when Rytkheu writes: "Our next task will be to spread Soviet power over the territories of enduring ice, along the shores of icy seas and deep into the tundra" (Rytkheu 1977, 157).[13] At the same time, he uses the motif in the same way as in *The Time of Melting Snows*, for instance, in the following lines:

> Every sunny day grandmother Milyune came to the green hill. She passed the beautiful new community hall (dom kul'tury), the school, the television centre. To the left was the former regional government building, to the right – in place of the church, a drugstore, where Michail's father worked – big cars passed by, ahead the sea port roaring. The coast of Anadyr was full of vessels ... the time of enduring ice came to an end for the Chukchi people, disappeared into the deep past. (Rytkheu 1977, 188)[14]

There is one interesting innovation in *The End of Perpetual Ice*. Ice as a medium of preservation – and, metaphorically, as a medium for holding the past in the present – becomes itself an active element (actor) within the narrative. During the civil war, two revolutionaries were killed by counterrevolutionary, "capitalist elements" (Rytkheu 1977, 186)[15]. Thanks to permafrost soil, their bodies are preserved just as they were when they died: "In perpetual ice they will lie forever" (Rytkheu 1977, 186).[16]

Therefore, the beloved of one of these men can find him looking at her with the same gaze as ever even after some twenty years, when she is already living in a completely changed – modern, Soviet – world, together with their son. Due to this "revival" and the memorial site to the heroes of the revolution that has been erected now, the new era can finally begin.
In Rytkheu's prose, the emphasis on permafrost as a preservative that enables it to perpetuate someone's memory is a new way of dealing with the motif of ice, a way that finds its equivalent in how Gulag writers

develop the same motif. But – symptomatically – Rytkheu does not make any further use of it.

Rytkheu, like several of his fellow non-Russian Soviet writers (such as Chinghiz Aitmatov and Fazil Iskander), survived the end of Soviet Union as a writer, but – unlike the others – in a quite specific way. Although practically forgotten in post-Soviet Russia, his Chukchi plots that included shamanism met the esoteric interests of New-Age readers in Western Europe, especially in the German-speaking world. Almost all his novels written after 1990 were published in German translation. What about the motif of ice in these post-Soviet novels? Does the attempt to develop a new perspective on multi-ethnicity have any impact on the symbolic dimension of ice?

In Rytkheu's novel *Chukotskii anekdot* (*Chukchi Joke*) (2002), translated into German in 2006 as *The Gold of the Tundra*, the preserving function of permafrost raises false hopes, but precisely in this way makes empathic contact between local inhabitants and western travellers possible. A young American, whose grandfather was one of the gold diggers on the Chukchi Peninsula during the years of the October Revolution, comes to Chukotka in search of a gold treasure that her grandfather – as he told her before he died – had hidden under a Chukchi tent, a "yaranga". Only a few pages deal with the (secret and) unsuccessful search for the treasure. Most of the novel is dedicated to the description of the world with which the American traveller(s) are confronted: a world in hopeless decay, where corrupt cadres have taken over, and where there is no ray of hope for the local inhabitants. Whereas these impressions are likely to disillusion the young visitor, the fact that the gold has disappeared does not. On the contrary, the fact that somewhere something is left forever that belonged to her ancestor brings her into deep, empathic contact with this region. For some moments, this empathy is reflected in a closer relationship between the American woman and a Chukchi that gives hope for the possibility of a peaceful and friendly coexistence of different cultures as part of one humankind.

In his last book, which is entitled *Dorozhnii leksikon* (*Traveller's Encyclopaedia* or, in German translation: *The Alphabet of My Life*),[17] Rytkheu completely avoids new metaphors of ice. Instead, in a set of articles in alphabetical order, he confronts differences in language word by word – every Russian lemma is matched by the corresponding Chukchi word. The articles themselves mostly consist of reflections on the translatability of the words and their significant cultural implications. Emphasis is laid on comparisons of the meaning of words that denote central spheres of everyday Chukchi experience: e.g.: "Ice" in Chukchi:

"gilgil" – "salted ice" (Russian: "*соленый лед*") or "tintin" – "sweet" or "unsalted ice" (Russian: "*пресный лед*") (Rytkheu 2010, 180). These lexical comparisons give rise to reflections on cultural difference, its origins and its range, but also on the opportunities for translation and dialogue. Obviously, the description searches for empirical rather than metaphysical causes and aims to explain differences in language and cultural practice through environmental conditions – which are considered to be contingent – and not through differences of some ethno-cultural essence. With subtle irony, in the article on "Climate" (Russian: "Климат" – "*наргинэн*" or "*вагыргын*" – "literally – state of nature" (Russian: "*дословно — состояние природы*"), Rytkheu accentuates a symmetry of misunderstanding:

> The state of nature is what the Louravetlan is mostly interested in. In ancient times, shamans foretold the weather … Our vigilant hunters said e.g.: "The polar explorers brought a bulk of coal – ergo winter will be severe." At the same time, the polar explorers attentively observed how my compatriots heated their yarangas, how they prepared fur curtains, and, on their part concluded that winter is going to be severe…. (Rytkheu 2010, 140)[18]

The same kind of reflection can be found under the lemma "Cold" (Russian: "Холод") – "кивен" or: "кевягыргын", where the argument is that anthropological differentiation in terms of sensitivity to cold is nothing but a myth: Chukchi suffer from cold in exactly the same way as people from other regions. It is solely practical knowledge about how to handle it that makes a certain difference, from which on their part Arctic explorers like Amundsen or Nansen were able to profit (Rytcheu 2012, 321).

Rytkheu clearly pursues a universalistic strategy, as he did during the Soviet period. But this is a new kind of universalism. Whereas Soviet universalism aimed to include all cultures and nations along a single scale of social development into one, universal process of historical development, Rytkheu's post-Soviet universalism justifies mankind as a unity of different, but "translatable" and interactive cultures. Cultures, his essays tell us, are different only because of environmental conditions but not in essence. Taking a look at the articles that deal with "climate", "ice", "cold" and "permafrost", we cannot detect any metaphorical meaning – ice does not work as a metaphor for primitive, unenlightened society any more, neither is it ascribed any other symbolic meaning. Although Rytkheu elaborates extensively on the very specific aesthetic and aisthetic fascination and attraction experienced by the local inhabitants, he discusses ice here simply as a symptom of a certain climate and thereby neutralizes it symbolically.

Rytkheu's new universalism has yet another dimension: not only is humankind seen as a totality, but so is the earth in general: "No, we don't need any global warming!"[19] He rewrites and changes the globalist perspective of Soviet modernism as well. Now it is not conquest and power that motivate a global point of view, but the threat of climate change. And the perspective of transforming the whole earth into one technically civilized zone under human control is replaced by a vision of global and not really controllable warming. Now that man has lost and nature has regained the position of power, "warm" has radically changed its significance. The following sentence seems to underline this proposition: "In this landscape there was a strange feeling of eternity, of the imperishability of nature's laws."[20]

Gulag literature: Pavel Florenskii and Varlam Shalamov

Whereas all positions of modernism accentuated the quality of ice as a medium of memory, either negatively or instrumentally, authors who wrote during the same period from a radically different point of view – the point of view of those who had been persecuted, socially excluded and banished to the polar regions – interpreted this quality of ice in quite another way and used it for different symbolic purposes.

Pavel Florenskii's poem "Oro" (cf. Frank 2014) and Varlam Shalamov's "Kolyma tales" were extremely different in terms of poetics and literary quality, but in the context of our topic highly interesting and comparable pieces of literature on the topic of the Gulags. They significantly alter the Soviet modernist attitude towards the Arctic and its climatic conditions when writing about the polar "zone" of their banishment. On the one hand, they do not accept the project of mastering nature and changing its conditions, even though, as Gulag workers, they had been forced to actively participate in it. On the other hand, they do not simply continue to tell the story of the Arctic as a hostile zone per se, as a realm of death. By foregrounding the preserving quality of ice and permafrost they rather re-semanticize it. By presenting ice as a matter of nature that itself commands a certain agency, they provide the motif with new narrative significance.

Florenskii's and Shalamov's conceptualizations of ice and permafrost deviate in a threefold manner from modernist conceptualizations of Arctic ice. Firstly, they turn the concept of man upside down. The Gulag convict, who from the perspective of Soviet ideology was an essential part of the utopian project of "forging the new man" ("perekovka"), is shown by

Shalamov to be a "dokhodiaga", a nearly dead man. Man here is not the victor over harsh natural conditions, but its victim as much as he is the victim of the totalitarian regime. But insofar as climate conditions serve as an instrument in the hands of the tormentors, it is not nature itself – the cold and its symptoms of ice and permafrost – that is the enemy of Gulag convicts, but nature used by other human beings as an instrument of violence. In the case of Florenskii's poem, nature itself – from the narrator's perspective – becomes a victim of human violence insofar as the brutal methods of exploitation destroy it.

Secondly, Florenskii and Shalamov accordingly understand the human-nature relationship in a way that is completely different from that of other Soviet modernists. By means of their narratives, they ascribe a certain agency to nature; they demonstrate that, despite all human efforts, man has to face nature as an agent of history. In Florenskii's poem, nature (equating to permafrost, equating to the dying shamanic culture of the Orochons) withstands destruction through gold digging (equating to the new age, rationalism and materialism). In Shalamov's "Kolyma tales", ice is conceptualized as nature's weapon against oblivion. Nature itself "fights with might and main" against the "work of man", man's attempts to cover up his deeds: "Stone, North withstood with might and main the work of man, and didn't let the dead come inside."[21] Whereas Florenskii anticipates an ecological critique of modernism, Shalamov depicts the Gulag as the utmost consequence of radical Soviet modernism.

Thirdly, in interpreting the ice as an archive of human history – each, of course, in his own way – Florenskii and Shalamov not only formulate a sceptical perspective towards utopian modernism. Instead they – and especially Shalamov – insinuate a new critical theory of historical memory (under the conditions of modernity). In Shalamov's case, I would go so far as to see a parallel between his conceptualization of ice and Walter Benjamin's conceptualization of memory under the conditions of modernity. It was Benjamin who diagnosed a "new barbarianism", that is, a tendency to radically neglect cultural heritage. In his *Passagenwerk* and "On the notion of history", he criticized the simplified belief in progress, which (because of its blindness vis-à-vis past and present) he declared responsible for underrating fascism.

Benjamin's famous ekphrasis describes Paul Klee's "Angelus novus" as an allegory of this problem. Instead of epic narration, which has become impossible, Benjamin propagated a kind of memory that serves the present by presenting (condensed) fragments of the past. Is this not similar to what Shalamov attempts at the level of fiction (and also in his programmatic essay on "the new prose")? In Shalamov's opinion, the function of what he calls "new prose" is not to represent or to tell, but to give evidence by making present the most significant fragments of an unavailable past. Like Benjamin, Shalamov believes in the importance of "historical [or biographical] insight" (Shalamov 1998, vol. 1: "On prose") but considers it possible only if memory bursts the historical continuum, the temporal succession, and makes the past present in the form of a single, a singled out and condensed image. One of Shalamov's most convincing images here is the "glove" (cf. the story "Perchatka")[22]. For Benjamin, as for Shalamov, this is the only way to deal with the past in the context of a present that is not able to come to terms with its catastrophes, because it forgot to remember. When Benjamin compares the materialist historian to come with an archaeologist digging in a heap of ruins, and when Shalamov calls the ice a museum of cruelties, they both turn out to be archivists *avant la lèttre*.

Contemporary Russian literature: Sergei Lebedev and Vladimir Sorokin

In his first novel, *The Border of Oblivion* (*Predel zabveniia* 2012/ Eng. transl. 2016), Sergei Lebedev picks up not only the topic of the Gulag but also the way in which Shalamov has enacted the motif of ice. However, Lebedev, born in 1981, offspring of a family of geologists and a geologist himself, who grew up during the last years of the Soviet Union and therefore not only has no personal experience of the Gulags but also is not a direct descendant of the eyewitness generation, significantly transforms the ice motif in his novel. First of all, Lebedev's novel renews an established genre of Russian literature, the family novel, and gives it an innovative turn.

Lebedev's focus is on memory as a trans-generational family legacy. As one of the first ever in Russian post-Soviet literature, Lebedev elaborates on Gulag-memory by choosing the case of an offender's memory. The novel tells the story of a young geologist who, during a research trip, gets onto the track of his "second grandfather", a grandfather by choice, whose life was deliberately shrouded in mystery. Finally, by choosing to give blood to his grandson, he sacrificed his life for him and thereby became his real blood relative. It turns out that this grandfather was none other than the director of a Siberian Gulag whose conscience was burdened with thousands of lives. Here again it is the ice of the permafrost soil serving as an archive of nature that enables a revelation of these unpunished deeds. The ice gives evidence about them, holds them present and allows the grandson to become a witness, to confront the real face of his grandfather-by-choice and thereby to work out his inherited memory. Like Shalamov, Lebedev makes use of the fact that permafrost prevents the bodies of Gulag victims from decomposing. When the protagonist falls into a permafrost crater and, by accidentally hitting his head on the ground, gains insight into the gruesome truth, the ice turns out to be not only a medium but also an agent of memory that helps to reveal the truth about the past.

The scenery of the crater is described in the novel as grotesque in the primordial sense of the term. It is a grotto where vegetal, human and lithic forms become indistinguishable:

> I was falling into the hole and falling into the dark hole inside me […] my consciousness blinded, and I did not feel my fall to the bottom. What happened next did not take place on the border between reality and delirium […] I was in both states simultaneously […] Around me were the pecked-over bodies of animals – foxes, wolves – all those that had crossed the frozen river to find fortune in this gully and then had fallen inside […] there was order in how they lay, and I realized that someone had tried to create a dugout here […] my vision changed and in the black peaty protuberances, in the icy smears I saw the outlines of human bodies […] The funnel was filled with corpses, the permafrost had preserved them. The opening in the wall filled with grass turned out to be a mouth; a rounded bump was a head; mixed with dirt, dissolved in it, the dead seemed to be trying to step out, to break through the ice crust; what I had taken for tree roots were arms; the dead flesh had taken on the colour of the earth, and you could recognize it only by the form. […] I was in the belly of the earth; my brothers lay there, and their imperishability was not the incorruptibility of sainthood, but the absence of death. […] this incompletion, this everlasting moment turned to ice. (Lebedev 2016, 276-277)[23]

The narrator/protagonist identifies the funnel as "the eye of the earth" (2016, 239) or a "throat" (2016, 275). He explicitly hints at the psychoanalytical symbolism of this grotto: "I fell into the dark hole inside me" (2016, 276). But this double inner and outer meaning is not frozen into an equilibrium of undecidability. The real meaning concerns the relation between humans or living beings in general and inert matter, and reveals itself to the gaze of the now initiated viewer:

> The view of the quarry materialized the hopelessness of life, verging on despair, translated this sensation into a visual one, existing in the world as a thing, the quarry did not threaten to swallow up anything, did not beckon with fatality, it simply was – in the dull steadiness of a hole, the most immutable thing, and the fact that man is transient compared to its immutability was almost unbearable. (Lebedev 2016, 179)[24]

Lebedev takes up the thread first spun by Florenskii and Shalamov: in his interpretation, too, permafrost is the museum of human cruelties. But by comparing the immutability of nature (the hole) to human transience, he turns the screw further: as in the case of Florenskii and Shalamov, permafrost functions as a symbol of nature's agency. But in Lebedev's novel, it is not nature's agency to be on human's duty, it is not in service to the victims of totalitarianism and the Gulag. Lebedev's permafrost museum of cruelty collects, it gulps but it does not swallow anything, and when it allows revelation, its deep truth will not heal but horrify.

My second example of contemporary Russian literature is Vladimir Sorokin's so-called *Ice Trilogy* (2004), which develops a very special mythopoetics of ice. "Bro" (the Russian title is: "Put' Bro" – "Bro's way"), the first part of the trilogy, tells the story of the establishment of a group of chosen persons, the ritual of their initiation, the story of seeking and finding each other, the kind of insight they gain after being initiated, and the beginning of the act they are chosen to fulfil. After they have understood that the world they live in is nothing more than a "dreadful tumour, the cancer of the Universe" (Sorokin 2008, 79), they begin to free the world from it. The starting point of this esoteric plot is a trip by a geologist to the place where the famous Tunguska meteorite hit the earth in 1908. There, confronted with extra-terrestrial ice that bored deep into earth and has been preserved by permafrost soil, he experiences a certain awakening, a kind of initiation. He learns that his real name is Bro and that it is his mission to begin seeking and awakening his equals.

At first glance, this narrative looks as if it could hardly be taken as an example of recoding Nature by empowering it in its relation towards man. Is not ice used by humans as a simple, quite concrete, instrument to "awaken" Bro's equals, the chosen people? Yes, it is, but the very beginning of this process, the hero's initiation, is not an intended action. Bro wakes up by accident – and becomes recognizable as a prefiguration of Sergei Lebedev's protagonist: "The hummocks ended: ahead lay the smooth duckweed hollow. I took the one-hundred-and-first step. And sank up to my chest […] I was swimming in a bog as though swimming across a lake" (2004, 75). Later, his hands touch ice:

> I had found *the huge and intimate.* […] I pulled myself up and ran towards it, splashing through the water, trampling the sleepy, thousand-year-old duckweed. The Ice! The Ice sparkled white and blue! How pure! How powerful! How mine […] Running up to the patch, I slipped. / And fell, slamming my chest against the shining Ice. I lost consciousness. For a moment. / Then my heart began to resound from the blow of the Ice. And I *immediately* felt the entire MASS of the Ice. […] "Bro, Bro, Bro!" my

> heart trembled. / And the Ice answered my heart. […] The Ice was vibrating. It was older than everything alive on earth. The Music of Eternal Harmony sang in it. (2004, 76–77)

At least it is non-human agency here that makes the hero fall, and it is Ice itself that knocks him up.

Initiated – by chance or destiny – Bro gains a cosmological perspective and recognizes the meaning of Earth and human history in a wider context:

> The Earth alone violated the harmony of the Cosmos. For it was alive and developed on its own. The Earth became a dreadful tumour, the cancer of the Universe. […] the Universe that we created gradually began to scatter in the Emptiness. But a piece of the world of Harmony […] fell to the Earth. This was one of the largest meteorites that had ever fallen to Earth. A huge piece of Heavenly Ice, in which the Harmony of the Primordial Light sang […] The permafrost helped to preserve it. (2004, 79)[25]

This new insight helps him to bear even a scene of ultimate cruelty that he happens to witness somewhere nearby in the Siberian taiga, when two hungry polar bears kill a pregnant elk:

> This bloody scene in the midst of the dead forest graphically demonstrated to me the essence of earthly life: a creature that had not yet even been born became food for other creatures. The utter absurdity of earthly existence was here in this wheezing elk […] But the chaos of earthly existence wasn't frightening to me, since I had known the Harmony of the Primordial Light. (2004, 82)[26]

In the fictional world of Sorokin's novel, Nietzsche's "Übermensch" is enacted for one last time in order to deconstruct this myth which proved so fatal for the twentieth century. This deconstruction aims to remove the human from the centre of the "world" in a cosmological sense. Within the (ice-induced) vision of a cosmic order, human history no longer appears as the very aim, the telos of history – as Christianity-based modernism saw it – but rather as a deviation, an erroneous development that can be compared only to a "dreadful tumour" (2004, 79).

Salvation in this gnostic vision can come only from outside, and must finally bring this deviation to an end. But only some humans are able to realize this – and the only way to gain insight is to be knocked out by "vibrant matter", by cosmic ice brought to earth by the famous Tunguska meteorite. Interestingly, leading contemporary scientists who are looking for the origin of life on earth give Sorokin's story scientific underpinnings when they ascribe a cosmic origin to ice and argue that it must have come to earth as an ingredient of a comet. But what Sorokin develops in his fictional narration, however, implies weightier consequences. It is only at first glance that he tells an esoteric story of chosen people. On closer investigation, the cosmological frame of this story dramatically alters the cosmist worldview developed by Russian intellectuals of the early twentieth century, such as the previously mentioned Nikolai Fedorov. Whereas modernist cosmists, more radically than anybody before them, put man as the only sovereign at the very centre of the cosmos, Sorokin radically not only removes human beings from the centre, but, by declaring them aberrations, shifts them right to the edge of the cosmic order.

In my opinion, there is yet another scientific correspondence in Sorokin's novel, a rather unexpected and therefore all the more interesting correspondence to the geological approach of Anthropocene theoreticians. Is it not this same gesture of radically widening the frame of human history – a gesture which at the beginning of the twentieth century was intended to radically elevate the position of man – that at the beginning of the twenty-first century can only be meant to do the opposite?

Modernity and the Anthropocene

Finally, I come back to my observations at the beginning of this chapter, that is, to Vladimir Vernadsky and Paul Crutzen and their respective notions of the "noosphere" and the "Anthropocene". Paul Crutzen's definition of the term "Anthropocene" is in fact quite similar to what Vernadsky meant by the "noosphere". Both scientists are convinced that the driving force behind the designated era is human. But is it not exactly their very geological approach that subverts mere anthropocentrism? It is not only that the notion of the "Anthropocene" has an uncanny ("unheimlich") foundation (Horn 2014, 166) because it implies the possibility and even probability of an unforeseen impact and consequences of human intervention and therefore a loss of control. As geologists, they look upon human history from the perspective of a longer, non-human, timescale, that of geological time. Hence, and in a sense in contradiction to the term "Anthropocene", they contextualize the history of humankind in a way that dethrones humans from the position modernist thinking has ascribed to them.

Vernadsky went even further in this direction of recontextualizing and de-isolating human history, when he ascribed inert matter a certain vividness and activity on its own that uses living organisms as a kind of vehicle. Vernadsky thought of humankind as "a particularly potent mix of minerals" (Bennett 2010, 11), because people "redistribute and concentrate oxygen […] and other elements of Earth's crust into two-legged, upright forms that have an amazing propensity to wander across, dig into and in countless other ways alter Earth's surface. We are walking, talking minerals" (quoted in Bennett 2010, 11). These are ways of removing humans from the centre of history, and that is what makes these geologists the most relevant thinkers for the present, as their correspondence with Lebedev and Sorokin demonstrate quite convincingly.

In summary, what the examples of literature discussed here demonstrate, is, firstly, that the Arctic has been and still continues (but probably not for long) to be a crucial place for enacting and reflecting upon the modernist concept of humans in relation to nature. Secondly, literary texts throughout this period were and are a means of reflecting upon basic aspects of a scientific (and philosophical) worldview, and in this respect they are in permanent dialogue with scientific concepts and approaches. During the course of the twentieth and early twenty-first centuries, we can observe a significant shift in the conceptualization of the human-nature relationship, a shift from the human as the hoped-for sovereign of the Earth and the master of its fate, to the human as (a more or less insignificant) part of much larger geological processes. Thirdly, we can observe that during this process Arctic ice, due to its manifold material, medial and aesthetic qualities, has served and continues to serve as a symbol that was invested with very different, sometimes even contradictory, meanings. Yet, every time it was exactly ice that had to mirror the underlying anthropological point of view.

Obviously, the process of semantic shifting described here is not homogeneous and simply linear. We have seen that the point of view of radical modernism that was typical of official and mainstream discourse during the Soviet period became altered, firstly by authors who wrote from the point of view of the excluded and from positions that are deprived of power, as in the case of Gulag convicts from the intellectual elite. But, in tracing the motif of ice, we could also observe that not all positions deprived of power are equal. Indigenous Soviet authors from the periphery sometimes followed Soviet official discourse and norms even more insistently than authors from the centre. Interestingly, the example of Yuri Rytkheu indicates that the abandonment of Soviet ideology did not give rise to a fundamental shift in the human-nature relationship, but instead evoked a new type of universal humanism (at least in Rytkheu's post-Soviet writings).

Finally, the two examples of post-Soviet literature, the novels by Sorokin and Lebedev, in their way of enacting the motif of Arctic ice, display a strong tendency to empower nature, to return to it the agency of which modernism had deprived it. Whereas Lebedev, whose imagery is obviously inspired by Sorokin's earlier novel, focuses on ice as a symbol of nature's power over human will (to hide and forget), Sorokin – by using the symbolic layer of ice as the matter of the extra-terrestrial origins of life – gives the catastrophic view of human history a utopian vision of a way out, at least for the chosen few. Comparing these post-Soviet and very contemporary literary imaginations to scientific views of human history from an early twentieth-century perspective, we can finally detect that the seeds of the idea of placing human history in a wider context were sown by geologists, representatives of a discipline that principally thinks in wider, non-human timeframes. They therefore become all the more relevant in a context like ours. When faced with multiple eventualities of more or less self-caused extinction, one has to recontextualize human beings by removing them from the centre and demonstrating the entanglement of their existence, its inseparability from other living beings and from living and inert matter in general and to look for new historical narratives in accordance with this.

References

Benjamin, Walter. 1980. *Illuminationen*. Frankfurt: Suhrkamp.

Bennett, Jane. 2010. *Vibrant Matter: A Political Ecology of Things*. Durham, NC: Duke University Press.

Brecht, Bertolt. (1930) 1988. *Werke*, Vol. 11: *Gedichte 1*, 157. *Sammlungen 1918–1938*, Frankfurt: Suhrkamp Verlag.

Clark, Katerina. 2000. *Soviet Novel: History as Ritual*. Bloomington: Indiana University Press.

Crutzen, Paul. 2002. "Geology of Mankind." *Nature* 415 (23): 23.

Crutzen, Paul, and John R. McNeill. 2007. "The Anthropocene: Are Humans Now Overwhelming the Great Forces of Nature?" *AMBIO: A Journal of the Human Environment*, 36 (8): 614–621.

Fedorov, Nikolaj. 1906/1913. *Filosofija obščego dela*, Vo. 1: Vernyj/ Vo. 2: Moskva.

Florenskii, Pavel. 1998. *Oro: liričeskaja poėma, Zabajkal'e 1934, Solovki 1934–1937*, Moskva.

Fourier, Charles. (1808) 1966. "Théories des Quatre movement et des destinées générales" [The theory of the four movements]. *Oeuvres complètes de Charles Fourier*, Tome 1. Paris: Editions Anthropos.

Frank, Susi K. 2014. "Transformationsästhetiken der Moderne: Symbolismus – Sozrealismus. Am Beispiel der Symbolik von 'Weiß', 'Schnee' und 'Eis'." [Aesthetics of Transformation: Symbolism –

Socialist Realism. The Example of the symbolism of "white", "snow" and "ice"]. *Wiener Slawistischer Almanach* 73: 373–402.
Horkheimer, M., and Th. W. Adorno. (1947) 1968. *Dialektik der Aufklärung*. Amsterdam: Querido.
Horkheimer, M., and Th. W. Adorno. 2002. *Dialectic of Enlightenment: Philosophical Fragments*. Translated by Edmund Jephcott. Stanford, CA: Stanford University Press.
Horn, Eva. 2014. *Zukunft als Katastrophe*. [Future as Catastrophe]. Frankfurt: Fischer Verlag.
Huntington, Ellsworth. (1915) 2001. *Civilization and Climate*. Honolulu: University Press of the Pacific.
Latour, Bruno. 1993. *We Have Never Been Modern*. (*Nous n'avons jamais été modernes*, 1991.) Translated by Catherine Porter. Cambridge, MA.: Harvard University Press.
Lebedev, Sergei. 2012. *Predel zabveniia*, Moskva: Ėksmo.
—. 2016. *Oblivion*. Translated by Antonia W. Bouis. New York: New Vessel Press.
Martin, Terry. 2001. *Affirmative Action Empire*. Cambridge MA: Harvard University Press.
Menke, Bettine. 2000. "Die Polargebiete der Bibliothek. Über eine metapoetische Metapher" [The polar regions of the library: On a metapoetic metaphor]. *Deutsche Vierteljahrsschrift* 4: 545–599.
Nietzsche, Friedrich. (1887) 1988. *Zur Genealogie der Moral*. KSA 5. München: de Gruyter.
Pil'njak, Boris. 2003. *Dvojniki. Odinnadcat' glav klassičeskogo povestvovanija*. Moskva: Agraf.
Rytkheu, Yuri. 1977: *Konec vechnoi merzloty*. Moskva: Izdatel'stvo Novyj mir.
—. 1982. *Konec večnoj merzloty*, Moskva: Izdatel'stvo Sovremennik.
—. 2010. *Dorozhnyi leksikon*. Sanktpeterburg: izdatel'stvo zhurnala *Zvezda*.
Serres, Michel. 1995. *The Natural Contract*. Translated by Elizabeth McArthur and William Paulson. Ann Arbor, MI: The University of Michigan Press.
Shalamov, Varlam. 1994. *Kolyma Tales*. Translated by John Glad. Oxford: Penguin.
—. 1998. *Sobranie sochinenii v 4-kh tomakh*, Moskva: Vagrius.
Slezkine, Yuri. 1997. "Primitive Communism." In *Socialist Realism without Shores*, edited by Thomas Lahusen and Yevgeni Dobrenko, 315–336. Durham, NC: Duke University Press.
Sorokin, Vladimir. 2004. *Put' Bro*, Moskva: Zakharov.

—. 2008. *Ice Trilogy*. New York: New York Review of Books.
Vernadsky, Vladimir. (1945) 2005. "Some Words about the Noosphere." *21st Century* (spring): 16–21

Notes

[1] "Herrschaft und Besitz - das sind die Schlüsselwörter, wie sie von Descartes zu Beginn des wissenschaftlich technischen Zeitalters ausgesprochen wurden, als unsere abendländische Vernunft zur Eroberung des Universums aufbrach. (...) Unsere grundlegende Beziehung zu den Dingen lasst sich in Krieg und Besitz zusammenfassen." Serres 1994, 59

[2] "Die Weltherrschaft über die Natur wendet sich gegen das denkende Subjekt selbst" Horkheimer / Adorno (1947) 1968, 39.

[3] In *Philosophy of the common goal* (*Filosofija obščego dela*) Nikolai Fedorov, a Russian thinker of the late nineteenth century, developed his grandiose vision, firstly of the conservation of everything – and especially everyone – that has ever lived in a museum and, secondly, the resurrection of the "forefathers" in their totality as an act of final empowerment of man over nature, which Fedorov understood as a central task of Christianity. Cf. George M. Young, *The Russian Cosmists, The Esoteric Futurism of Nikolai Fedorov and His Followers*, Oxford: Oxford University Press 2012, 76–91.

[4] "Ледники традиций еще существуют в морозе веков", Pil'njak 2003, 201.

[5] "Barbarentum? In der Tat. Wir sagen es, um einen neuen, positiven Begriff des Barbarentums einzuführen. Denn wohin bringt die Armut an Erfahrung den Barbaren? Sie bringt ihn dahin, von vorn zu beginnen; von Neuem anzufangen; mit Wenigem auszukommen; aus Wenigem heraus zu konstruieren und dabei weder rechts noch links zu blicken. Unter den großen Schöpfern hat es immer die Unerbittlichen gegeben, die erst einmal reinen Tisch machten. Sie wollten nämlich einen Zeichentisch haben, sie sind Konstrukteure gewesen."

[6] "wird thematisch, was die Moderne macht und ausmacht, das Abweisen oder Abräumen der Spuren, um damit Nietzsche, Brecht und Benjamin zu zitieren" (Menke, Die Polargebiete der Bibliothek)

[7] Cf. Benjamin's notes on the quality of glass buildings to prevent traces in "Experience and poverty" 1980, 294.

[8] Rytkheu, Yuri 1969: *Vremja taiania snegov*. Moskva: Izdatel'stvo Izvestiia.

[9] Slezkine, Yuri 1997: "Primitive Communism", Lahusen, Thomas; Dobrenko, Yevgeni (ed.), *Socialist Realism without Shores*. Durham, London: Duke University Press, 315–336.

[10] Rytkheu,Yuri 1977: *Konets vechnoi merzloty*. Moskva: Izdatel'stvo Novyj mir.

[11] Долгое время таланты чукотского народа лежали под толстым слоем невежества, темноты и суеверия. Теперь для вашего народа и для вас наступила в жизни великая весна — время таяния снегов. Сойдет холодный, тяжелый снег, и свободно, как цветы в тундре, расцветут таланты вашего народа, еще краше, счастливее, полнее, как река, забурлит ваша жизнь для настоящего человеческого счастья! (my translation)

[12] Раньше мы не думали о жизни: забота о еде и теплом жилище — вот о чем были наши думы. Сейчас вся наша жизнь меняется, как земля весной, во время таяния снегов. Многое нам еще неясно, но мы хотим все понять и научиться сами разбираться, как и что в жизни. " (my translation)

[13] Наша ближайшая задача здесь, на просторах вечной мерзлоты, распространить нашу советскую власть на все побережье ледяных морей и в глубь тундры. (my translation)

[14] Каждый погожий день бабушка Милюнэ приходила на этот зеленый холмик. Она шла мимо нового красивого Дома культуры, школы, мимо телецентра. Слева оставалось бывшее здание уездного правления, справа — хозяйственный магазин на месте церкви, где служил отец Михаил.. .Но мимо проносились большие машины, впереди гремел морской порт. Анадырский лиман был полон кораблей.
... Кончилось время вечной мерзлоты для чукотского народа, ушло в далекое прошлое... (my translation)

[15] "Пали вы жертвой от рук капитала, Пали в кровавой борьбе." (my translation)

[16] "В мерзлоте они будут вечно лежать." (my translation)

[17] Rytkheu, Yuri 2002: *Chukotskii anekdot*, Moskva: Izdatel'stvo zhurnala Zvezda. Rytkheu, Yuri 2010: *Dorozhnyi leksikon*. Sanktpeterburg: izdatel'stvo zhurnala *Zvezda*. Rytcheu, Juri 2012: *Alphabet meines Lebens*. Zürich: Unionsverlag.

[18] "Состояние погоды становится главным интересом луоравэтлана. Обычно в старые годы шаманы предсказывали погоду ... Наши наблюдательные охотники говорили: 'Полярники завезли много угля, значит, зима будет суровой'. И в то же время полярники внимательно приглядывались к тому, как мои земляки утепляли яранги, обкладывали меховые пологи пучками сухой травы, и в свою очередь делали заключение в том, что, видимо, шаманы предсказали особенно суровую зиму...." (my translation)

[19] "Нет, глобального потепления нам не надо!" (my translation) Rytkheu, Jurii 2010: *Dorozhnyi leksikon*. Sanktpeterburg: izdatel'stvo zhurnala *Zvezda*, chapter "Climate".

[20] И в этом пейзаже было странное чувство вечности, нерушимости установленных природой закономерностей.

[21] "Камень, Север сопротивлялись всеми силами этой работе человека, не пуская мертвецов в свои недра." Shalamov, "Po lendlizu" In: Kolymskie rasskazy, Sobranie sochinenii v 4-kh tomakh, vol. 1, Mokva: Vagrius 1998, 350–357.

[22] cf. Shalamov's story "Perchatka", in which only permafrost soil, due to its ability to preserve, makes biographical continuity and therefore the healing of the trauma possible. Cf. Shalamov, Kolymskie rasskazy, Sobranie sochinenii v 4-kh tomakh, vol. 2, Mokva: Vagrius 1998, 279–307.

[23] Я падал в яму - и падал в темный провал внутри себя.... падения на дно я уже не ощутил. То, что было дальше, происходило не на грани яви и бреда, ... я как бы оказался в двух этих состояниях одновременно.... зрение мое изменилось, и в черных торфяных буграх, в ледяных наплывах я различил

очертания человеческих фигур. Воронка была полна мертвецов, вечная мерзлота сохранила их нетленными. Отверстие в стене, забитое смытой травой, оказалось ртом. Округлый выступ - головой. Смешанные с землей, растворенные в ней, мертвые словно тщились вышагнуть, сломать корку льда. То, что показалось мне древесными корнями, было руками. Мерзлая плоть обрела цвет земли, и узнать ее можно было лишь по форме. ... Я находился в желудке земли. Здесь лежали мои братья, и нетленность их была не нетленностью святости, а лишенностью смерти.... эта несавершенность, этот вечно длящийся момент, обратившийся в лед ... (Lebedev, Predel zabveniia, 394)

[24] Вид карьера материализовал ощущение безнадежности жизни, приходящее во время отчаяния, переводил это ощущение в разряд наглядного, существующего в мире как вещь. Карьер не грозил ничего поглотить, не манил гибельностью, он просто был - в тупом постоянстве дыры, самой неизменной вещи, и то, что человек преходящ относительно ее неизменности, было почти невыносимо - для чувства, для мужества, для твердости духа. (Lebedev, Predel zabveniia, 248-249.)

[25] Только Земля нарушала гармонию Космоса. Ибо она была живой и развивалась сама по себе. Земля стала уродливой опухолью, раком Вселенной. ... И Вселенная, созданная нами, стала
постепенно рассеиваться в Пустоте. Но на Землю упал кусок мира Гармонии ... Это был один из самых больших метеоритов, когда-либо падавших на Землю. Громадный кусок Небесного Льда, в котором пела гармония Света Изначального, миллиарды лет странствующий по Вселенной.... Вечная мерзлота помогла сохраниться. (Sorokin, Put' Bro, 2004, 104)

[26] Кочки кончались: Впереди лежала ровная, подернутая ряской мочажина. Я сделал сто первый шаг. И провалился по грудь.... Я плыл по болоту, как по озеру. (99) Мои руки коснулись Льда.... Я нашел *огромное* и *родное*.... Я приподнялся и побежал к нему, разбрызгивая воду, разгоняя сонную тысячелетнюю ряску. Лед! Лед сверкал белым и голубым! Какой чистый! Какой могучий! Какой родной! ... Подбежав, я посколзнулся. И упал, со всего маха ударившись грудью о сияющий Лед. Сознание покинуло меня. На мгновенье. Затем мое сердце словно зазвенело от упада об Лед. И я почувствовал сердцем *сразу* всю глыбу Льда.... - Бро! Бро! Бро! – трепетало сердце. И глыба Льда отвечала моему сердцу ... Лед вибрировал. Он был старше всего живого на земле. В нем пела Музыка Вечной Гармонии. (Sorokin, Put' Bro, 2004, 100–110)

ICEBERGS AND LIGHT: MODERNITY AND THE ARCTIC SUBLIME IN BARRY LOPEZ'S *ARCTIC DREAMS*

SIGFRID KJELDAAS

Can a melting and environmentally threatened Arctic still be sublime? Or must modern Western culture today let go of the century-old aesthetic of the natural sublime, and move towards a consideration of other aesthetic modes in its cultural and literary analyses of the Arctic?

The natural sublime was an important aesthetic mode for nineteenth-century literary romantics seeking unity and a fresh relationship with nature. It was also powerfully present in the more scientific narratives of nineteenth-century Arctic exploration. Because these explorations served to bring unknown regions like the Arctic into the sphere of modern Western culture's expanding systems of knowledge and commercial exploitation, the Arctic sublime may be read as part of a nineteenth-century modernity associated with heroic conquest of new lands and new natural resources. That these explorations and conquests potentially came at great cost to the individual explorer was inherent in the concept of the natural sublime, which evokes the threatening and overpowering forces of nature against which the human is ultimately helpless. Yet, at the same time, the aesthetic of the sublime rests on distinctly modern ways of relating to the natural world that presume and confirm human distance from and dominion over this world.

The literary scholar Benjamin Morgan has recently argued that in these present days of climate change, which signals the vulnerability of the Arctic's natural environment and its ecosystems, the Arctic sublime is historically and environmentally outmoded. Christopher Hitt, on the other hand, claims that the aesthetic of the sublime may still prove to be a useful tool in forming new and better ways of relating to the natural world. Through an investigation of Barry Lopez's *Arctic Dreams: Imagination and Desire in a Northern Landscape* (1986), this chapter engages with the discussion of whether or not an Arctic sublime can be part of contemporary modernity, or is still a useful concept at all.[1] It does so by presenting the way in which Lopez's late twentieth-century text reworks the concept of the Arctic sublime in a manner that responds to the criticisms of this aesthetic.[2]

The nineteenth-century Arctic sublime

Ian S. MacLaren has argued that British explorers charted the unknown landscapes of the Arctic aesthetically with much the same skill as they did astronomically, and that indeed this "description of new tracts [of land] as more or less sublime or picturesque" was as important in identifying where and how these lands related to the familiar landscapes of home as were the measurements of latitude and longitude identifying their position in relation to Greenwich (MacLaren 1985, 89). Whereas the taxonomic mappings and journal reports of nineteenth-century Arctic explorers bore witness that the Arctic, in certain places and at certain times of year, could "nourish much fauna", the reading public tended to ignore this information in favour of an image of the Arctic as a vast, powerful, terrifying and awe-inspiring lifelessness expressive of the cold and uncaring forces of nature (Loomis 1977, 104).

Unlike other natural sublimes, the nineteenth-century Arctic sublime ascribed sublimity to the Arctic *as a region* (Loomis 1977, 96). One reason for this was the fact that the Arctic existed quite literally beyond knowledge. With its radically unstable sea ice, its periods of utter darkness, and its weird visual phenomena, the Arctic by its very nature seemed to resist human conceptualization. As more and more reports of the explorers' many struggles became widely circulated, Arctic nature came to be perceived as "somehow vaster, more mysterious, and more terrible than elsewhere on the globe – a region in which natural phenomena could take strange, almost supernatural, forms, sometimes stunningly beautiful, sometimes terrifying, often both" (Loomis 1977, 96).

Written a century and a half after the great era of Arctic exploration, Barry Lopez's *Arctic Dreams* is a modern text of non-fictional American nature writing that employs the aesthetics of the sublime in its representation of Arctic landscapes. Due partly to its employment of this aesthetic, and partly to the way in which the text presents its readers with a poetic and ecological vision of the Arctic's natural environment, *Arctic Dreams* has been associated by (eco)critics with a distinctly American transcendentalist literary tradition (Paul 1992, 98). American transcendentalism has its roots in European romanticism, and shares several of its features. Because the criticism of romantic texts traditionally involves sympathetic readings attempting to respectfully seek out the "spirit" in which the individual author has created his (or her) text (Harland 1999, 72; Kittang 1976, 18; Phillips 2003, 231), Barry Lopez's open engagement with both old Arctic explorer narratives and scientific descriptions of Arctic natural phenomena have been more or less overlooked.

This engagement is exemplified in several references in *Arctic Dreams* to William Scoresby's *An Account of the Arctic Regions With a History and Description of the Northern Whale-Fishery* (1820) and William Edward Parry's *Journal of a Voyage for the Discovery of a North-West Passage from the Atlantic to the Pacific* (1821), both of which appear in the book's extensive bibliography. The unique combination of old and new, romantic and scientific ideas about nature (and human relationships with nature) employed in *Arctic Dreams* makes it a pertinent case study for a modern reinterpretation of the Arctic sublime. Lopez's sublime is of a different quality, and communicates different aesthetic, emotional and rational responses to the landscapes of the Arctic, than does the more traditional Arctic sublime theorized by Loomis. This chapter investigates the differences between the nineteenth-century sublime and Lopez's late twentieth-century version, focusing in particular on the way Lopez's sublime seems to anticipate and respond to recent critiques of the ideology implicated in the aesthetics of the sublime.

After the Arctic sublime?

Arguing for the outmodedness of the natural sublime in a distinctly Arctic setting, Benjamin Morgan is merely the latest in a long line of critics who have condemned the aesthetic of the sublime as both outmoded and ideologically unsound. From the late 1980s onwards, literary critics have focused upon the association between the aesthetic of the sublime and a more general romantic ethos of expansion that in both Europe and America worked to support the "heady imperialism" of the eighteenth and nineteenth centuries (Weiskel 1976, 6; cf. Novak 2007, 33).

One explanation for the effectiveness with which the aesthetic of the sublime could be applied in support of imperialist projects lies in its requirement for distance between the observer and the natural phenomenon observed (Kant 1953, 110; Spufford 2003, 30–31; Morgan 2016, 15). Whereas both Burke and Kant claim that physical and emotional distance are required in order for the viewer not to be overcome by fear in meeting with the sublime phenomenon, Kant's conviction that only transcendental reason can comprehend the infinitude of the sublime further distances the human mind from the sensual world (Burke [1759] 2008, 25; Kant 1953, 110).

Even the more positive and contemplative understanding of the natural sublime found in the romantic and transcendentalist literary traditions has had to face charges of "drastic egotism" (Weiskel 1976, 50). These charges have their basis in the way in which the all-powerful imagination of romantic poets works to transform the world's natural particulars into "substance of mind" (Weiskel 1976, 52; see also Hitt 1999). As the poet's imagination strips away the properties of these natural particulars and assigns significance and symbolic value to them according to its own making, it hides "the formal otherness of things" (Weiskel 1976, 59). The otherness of the natural world fades in the encounter with systems of human thought, emphasizing the human dominion over this world (Hitt 1999, 608).

However, the imperialist impulse of the sublime is not limited to the natural world. Among that "formal otherness of things" that the natural sublime works to hide is the otherness of living subjects: of animal and cultural Others, and of women. Literary critics of the past few decades have ascribed to the sublime "an expression of asymmetrical power relationships" within settings ranging from the colonial to the domestic and gendered spheres (Hitt 1999, 603). Based on these observations, the sublime may rightfully be accused of reducing the natural world to insignificance, of removing its otherness, and of transforming everything into the sameness of a transcendental human subject, whose power and authority is thereby accentuated. No wonder, then, that this aesthetic has fallen into disrepute among critics, and that ecocritics have been reluctant to engage with literary representations of the sublime (Weiskel 1976, 6; Hitt 1999, 604).

Literary criticism of nineteenth-century Arctic exploration narratives, on the other hand, has tended to be deeply involved with the aesthetics of the sublime. This is what provoked Morgan to direct the charge of outmodedness specifically against the Arctic sublime, arguing that modern Western culture is currently situated "After the Arctic Sublime" in more ways than one. According to Chauncey C. Loomis' periodization, this aesthetic category "reached its peak in the 1850s, following the disappearance of the British naval hero John Franklin's Arctic expedition, and […] faded by the century's end" (Morgan 2016, 3; cf. Loomis 1977, 112). This leaves us historically and scientifically "after" the Arctic sublime. Accepting this historical periodization of the Arctic sublime also means accepting Loomis' premise that a natural region, landscape or phenomenon, in order to retain its sublimity, "cannot be mapped" (Loomis 1977, 112).

This historical sense of "afterness", however, is not Morgan's main concern. The core of his argument is that a melting and environmentally threatened Arctic can no longer be considered sublime. The nineteenth-century Arctic sublime presented the Arctic in terms of a terrifying, awe-inspiring and lifeless power exposing the frailty of human life. With climate change, the modern Arctic has acquired new symbolic meaning as an expression of "the earth's profound vulnerability to collective human agency" (Morgan 2016, 3). This, Morgan argues, should cause a dismissal of the natural sublime because its very definition is founded on the recognition of nature's irresistible power; on the presumption of its permanence and inviolability (Morgan 2016; Hitt 1999, 618). In the present moment, the ontological premise of the aesthetic of the sublime seems outmoded due to the way in which it rests on the absolute Cartesian distinction between mind and matter, the human and the natural world, that arguably has brought about this state of ecological peril.

Morgan's approach thus constitutes one among a range of appeals from contemporary critics for new ways of understanding human interaction with the physical world. Such an understanding may be developed, Morgan argues, if literary critics working with texts like nineteenth-century Arctic exploration narratives would set aside the dominant aesthetic of the natural sublime in order to proceed with investigations of other forms of aesthetic judgment involved in these explorers' interactions with the physical reality under description. I agree with Morgan's claim that a change in focus is necessary in order to interpret the new place of the Arctic in the Western imagination. But does this also require that literary criticism relinquish the aesthetic of the sublime?

I find Morgan's "After the Arctic Sublime" invaluable for the way in which it addresses the question of how to define and use aesthetic concepts under today's different historical circumstances. This issue also reverberates throughout Lopez's *Arctic Dreams.* Morgan bases much of his discussion of the aesthetics of the Arctic on Scoresby's *Account*. *Arctic Dreams* also refers to Scoresby's *Account* but, as I will show, Lopez's reading of this text leads him to develop his own version of the Arctic sublime. Lopez's version seems, however, to incorporate much of what Morgan finds to be lacking in the traditional Arctic sublime, and might be regarded as an already existing answer to Morgan's call for a new, historically and environmentally updated aesthetic framework for the modern Arctic.

"After the Arctic Sublime" involves a timely critique of literary scholarship's evasive treatment of the scientific aspects of old explorer narratives, which mirrors literary critics' neglect of the scientific parts of *Arctic Dreams*. Morgan interprets this neglect as a sign that scholars have forgotten the very significant manner in which "descriptions of the Arctic environment were, literally, descriptions of the Arctic environment" (Morgan 2016, 10).

To reverse such neglect, Morgan proposes a double strategy of reading intersections: "the intersections of the lyrical and the technical in the Arctic voyage narrative", and between the "moods or affects" that frame the scientific observations presented (Morgan 2016, 11, 12). This denotes reading for an aesthetic more in line with the original Greek term *aesthesis*, whose meaning refers to "the whole region of human perception and sensation, in contrast to the more rarefied domain of conceptual thought" (Eagleton 1990, 11). Morgan argues that this sense of the aesthetic is indeed present in nineteenth-century exploration narratives but has been neglected in favour of the aesthetic of the sublime (2016, 14).

I believe that a new materialist ecocritical approach would be extremely helpful in bringing out the somatic dimension of the aesthetic that Morgan wishes to promote. Taking as its starting point the proposition that "matter possesses agency", new materialism focuses on concrete physical matter, the bodily dimension of experience and "non-binary object-subject relations" in a reality of complex and interacting natural-cultural phenomena (Iovino and Oppermann 2012, 77, 76). By focusing on these material aspects of Arctic narratives, a new materialist approach might also aid in an analysis of the kind of effects that different forms of representation (like the scientific and the lyrical) have on our understanding of Arctic natural phenomena.

In analysing Barry Lopez's *Arctic Dreams*, I include scientific descriptions of Arctic animals and the Arctic's natural environment that form vital parts of the text. In this, my approach differs from that of earlier critical work on *Arctic Dreams*, which (as already mentioned) has focused primarily on the text's romantic and transcendentalist qualities. Critics like Nicholas O'Connell and John Gatta have emphasized the spiritual dimension and "sacramental" qualities of *Arctic Dreams* (Gatta 2004, 183), while Scott Slovic has pointed out Lopez's "self-conscious interest in the psychology of awareness" and treated the environmental details of the text in terms of their symbolic value (Slovic 1992, 141). In more recent critical work engaging with the ecological aspects of *Arctic Dreams* (like that of Coles 1993; Buell 2001; Phillips 2003; Browne 2007; Chou 2013), the text's representations of Arctic animals and Arctic natural phenomena have tended to be neglected or selectively employed in the support of new ecological and/or literary models of the world. Neither Buell nor Browne, who argue for the centrality of the text's descriptions of the natural environment by noting how the Arctic ecosystem "is made to revolve around the concentrated portraits" of large Arctic mammals (Buell 2001, 220), and how "the extreme otherness of the [physical] Arctic" that Lopez presents "forces us to look at ourselves differently" (Browne 2007, 149), analyse in any detail these animal and landscape "portraits" or their function in the text.

My discussion of Lopez's scientific descriptions of ice and icebergs takes seriously the materiality of the setting and the objects presented. This approach engenders precisely that awareness of "the messy entanglements" of the human and the geological – and, I would add, of the human and the biological – that Morgan finds the aesthetic of the sublime to prohibit (Morgan 2016, 22). Although presented in narrative form rather than in the tables and lists included in nineteenth-century exploration narratives (with the exception of the indexes at the back of *Arctic Dreams*), Lopez's descriptions may nevertheless be considered scientific catalogues. However, this narrative form of presentation allows Lopez to blend the lyrical with the factual and objective. The effect of this blending is an aesthetic much like that called for by Morgan, in constant interchange between information and sensation (Morgan 2016, 12–14). It is not one that relinquishes the aesthetic of the sublime. As I show in the following, Lopez actively employs this old aesthetic in order to rework it in a manner that highlights rather than suppresses the reality and otherness of the Arctic's natural environment.

Less distant versions of the natural sublime

Morgan argues that we are located "after" the Arctic sublime both because this aesthetic category belonged to a historically limited time period and cultural sentiment, and because what is needed today are analyses of embodied human interactions with the materiality of the world, which violate the sublime's requirement for distance (whether physical, emotional or intellectual). There are two reasons why I do not agree with Morgan on this. The first resonates with Hitt's frustration over ecocriticism's "scholarly neglect" of the aesthetics of the sublime; a neglect he finds peculiar and problematic considering the central role played by nature in both Burke's and Kant's theorizations of the sublime (Hitt 1999, 605).

For all its problematic valorization of the (male, Western) human, "humility before nature" has always been an "elementary part" of, and indeed "a prerequisite" for, the natural sublime (Hitt 1999, 606, 607). This humility, in turn, is what generates in the observer the admiration and respect for nature that supplement the emotions of wonder and awe in experiences of the natural sublime. Hitt's valorization of the natural sublime finds support in environmental historian William Cronon's account of how this aesthetic helped transform the idea of wilderness from "a place of spiritual danger and moral temptation" into a place in which "the supernatural [lies] just beneath the surface" and the human might enjoy a more or less original and innocent state of being (Cronon 1996, 73). Hitt argues that, because it engenders these positive attitudes towards the natural world, "the concept of the sublime [still] offers a unique opportunity for the realization of a new, more responsible perspective on our relationship with the natural environment" (1999, 605). As an alternative to the former transcendental or egotistical sublime, he proposes a new *ecological* sublime. The ecological sublime is one in which the encounter with the untamed forces and radical otherness of nature "jolt[s] us momentarily out of a perspective constructed by reason and language, a perspective that, in modern Western culture, has rendered nature mute" (Hitt 1999, 617). The observer's mind does not transcend the physical world. In a reversal of power relationships, the powers of the physical world transcend the human *logos*: our language and reason, our categories and judgement (Hitt 1999, 616).

The second reason why I am critical of the idea that modern Western culture is situated "after" the Arctic sublime is connected to Francis Spufford's explanation of how the Arctic sublime came to allow for and include threats much closer to the observer than was the case for earlier and more original versions of the natural sublime. This implies that the demand for distance in the Arctic sublime did not remain unchanged or absolute. Spufford turns to the work of late eighteenth-century female writers of gothic novels for evidence supporting his claim. Well versed in "Burke's menu of sublime effects", these women played in their fiction with the opposition between male activity and female passivity, employing the aesthetic categories of the sublime and the beautiful to do so (Spufford 2003, 33). As long as the threat faced by the heroines of this fiction conformed to the "sublime pattern of dangers" and was obscure enough not to identify the specific ways in which self-preservation was threatened, they could "feel the sublimity *of the threat*" even while under this threat (Spufford 2003, 36). These gothic novels furthermore invited their readers to identify with something stronger than the self and "bent on menacing" the self (Spufford 2003, 36). In this manner, writers of gothic novels undermined the sublime's requirement for distance.

For Spufford, this fictionally reworked idea of the sublime functions better than the original to explain real-life polar explorers' attraction towards a natural environment that is constantly threatening to extinguish their lives. The obscure but immediate and ubiquitous threats of intense cold, sudden violent storms and the movement of ice confirmed the authority of the sublime Arctic, and the explorers, Spufford claims, "responded by identifying themselves with its sublimity, glorying in the place even as it thwarted or even hurt them" (2003, 37).

To elicit a sense of identification with this environment without diminishing its power to thwart and overturn human plans, aspirations and orders, is precisely the aim of Lopez's *Arctic Dreams*. This environmentally concerned text seems to respond to Hitt's call for the recuperation of the natural sublime, without reiterating what Morgan and other critics have identified as the problematic aspects of this aesthetic. Lopez's choice to apply the natural sublime as a means of promoting his own scientifically updated modern ecological vision of the North American Arctic seems both peculiar and perfectly natural to this text, which throughout employs both postmodern and romantic modes of representation. Having so far presented some of the problems this aesthetic causes for an eco-centric perspective on the world, the following analysis will demonstrate how *Arctic Dreams* reveals that it may also promote such a perspective.

Terror and distance in Lopez's Arctic

In both Lopez's late twentieth-century account of the Arctic and early nineteenth-century ones, we find aesthetic as well as scientific descriptions of ice. Narratives like Scoresby's 1820 *Account* and Parry's 1821 *Journal* include scientific descriptions of the physical characteristics of different kinds of ice, as well as the various shapes and movements of ice that they observed. Whether presented in the form of lists of physical shapes or technical terms, these meticulous records of ice in "technical language" are precisely what Morgan scorns literary critics for having dismissed on the formalist presumption that they are "practical and essentially unliterary" (2016, 8). In these texts, the natural sublime tends to be reserved for the presentation of icebergs and mountainous Arctic landscapes, as exemplified in Scoresby's descriptions of Spitsbergen (Scoresby [1820] 1969, 104, 110).

Lopez uses Scoresby's *Account* to give an idea of the traditional natural sublime in an Arctic setting. But instead of selecting one of the more obvious examples of Scoresby's use of the sublime to do so, he chooses a scene in which Scoresby has had to abandon his ship, which is caught in enormous fields of moving ice off the east coast of Greenland. As Lopez reports, Scoresby is "mesmerized" by the "sheer power" and "daunting scale" of the ice that is threatening to slowly crush his ship (214; cf. Scoresby 1969 [1820] 250). Although neither author explicitly identifies this as a sublime moment, Scoresby's emotional response and his positioning in relation to the scene reveal it to be one. His response arises when he is some distance from the ship, and at a point at which he has "lost the sense of plight that spurred him", and become a mere "careless spectator" of the ordeal (247, 250; cf. Lopez 214, 215).

Lopez's selection of this particular scene emphasizes the paradox involved in the application of an aesthetic of distance and transcendence in circumstances like these. Whatever emotions are evoked, and whatever response the mind has in this sublime moment, Scoresby will soon lose his role of "careless spectator". Because he and his men are absolutely dependent on their ship to protect them from the obscure but ubiquitous threats of an environment too cold, too unsheltered, and (in most cases) too barren to sustain them, any threat to this cocoon of safety becomes an immediate reminder of what Kant described as man's "physical helplessness" in the meeting with "the irresistibility of the might of nature" (Kant 1953, 111). Unlike more southern mountaineers facing terrible mountain vistas or the raving destructions of distant thunderstorms, Scoresby cannot merely walk away from the sublime scene. Instead he must walk back into the life-theatening chaos and fear caused by the moving ice fields. As a result, his mind will soon find itself neither completely stunned by fear nor free to disengage from concrete reality and slip into transcendental contemplations of its own dominion over nature. Rather, it will be deeply engaged in the practical consequences of the very real and direct contact with the agency of ice, currents and weather. Hence, in Arctic scenes like this one, the aesthetic distance evoked is exposed as a constructed, imagined or merely temporary distance.

Arctic Dreams makes it very clear that most travellers to the Arctic could not afford the luxury of distance required by the traditional aesthetic of the sublime. In the Arctic, Lopez writes, "ice at rest one minute was moving the next" (215). "Damage was routine, some of it serious" (216). In this place of water and moving ice, of overpowering meteorological forces and life-threatening cold, the threat to self-preservation was ubiquitously present and often so physically invasive as to make all aesthetic considerations impossible. Instead, the narratives of Arctic travel to which *Arctic Dreams* refers are replete with images of how meetings with icebergs and ice-cold water left the travellers physically exposed and battling against the overpowering forces of a dangerous environment. In this manner, Lopez evokes the idea of the Arctic sublime while simultaneously displaying the constructed nature of distance involved in this aesthetic.

The modern traveller and the deterioration of the sublime

So far, the discussion has revealed the extent to which the idea of the sublime is associated with the positioning of the observer in relation to the natural phenomenon observed, both physically and mentally. Since the early nineteenth century, the modern geographical and natural sciences have done much to demystify the Arctic and bring it closer to us. Modern technology has further made it possible for men and women less heroic and more reliant on modern comforts to enter the Arctic as visitors. On the basis of access to updated scientific knowledge of its physical and ecological qualities, and because its direct threat to human life has been alleviated, a modern traveller in the Arctic can represent the region in a different manner than could the daring explorers a century ago.

The significance of this difference is evident in the way Lopez's first personal encounter with icebergs is presented. The encounter takes place aboard the *MV Soodoc* in the northern Labrador Sea. It is dark, and a storm is approaching, carrying with it "a picket line of ice the size of cathedrals" (204). Under these circumstances, the ship's radar is unable to distinguish the icebergs from the surrounding sea, and Lopez and the crew face a very real danger of having to suddenly abandon ship and lower away "into ice and darkness in 20-foot seas", "brought down to the raw edge of life" (205). Although it reportedly left "some of us" sleepless, the terror of the scene is reported only in calm retrospect, and only as the theoretically possible outcome of a storm that never hit (205). It seems as though Lopez purposefully refrains from associating icebergs with the Burkean sublime of terror, choosing instead to present the scene in the neutral and objective language of most explorer narratives.

In watching the icebergs from the bridge of the *MV Soodoc*, Lopez is half protected, half exposed to the elements: "I rested my forearms on the sill, feeling the warmth of the bridge heaters around my legs and a slipstream of cool air past my face" (206). Lopez here draws attention to how differences in modes of transportation influence the conception of landscape at different points in historical time. The modern "machinery" that supports Lopez's experiences in the Arctic "compresses time and space, and comforts [...] because of the authority with which it keeps danger at bay" (218). Lopez seems to realize that the differences between his own depictions of the Arctic's natural environment and those found in the historical narratives to which he refers result in part from this change in physical positioning, and from an associated change in the perception of threat. "[F]rom these quarters, its scale reduced", he writes, "we appraise the landscape very differently" (218).

One might suspect that this difference in physical positioning would create an even stronger sense of the Kantian sublime in Lopez's text, as it seems to testify to the power of the rational mind to produce the technology that secures our simultaneous protection from and dominion over the natural world. However, this does not happen because the text activates a historical and textual in-between position that parallels the narrator's partly exposed physical one, and presents two different viewpoints on Arctic natural phenomena: that of the safe and well-protected contemporary traveller *and* that of historical whalers and explorers whose very lives were at stake in physical encounters with these phenomena.

This dual point of view leads to Lopez giving several paradoxical depictions of ice. One of the most striking of these is of the sea ice as simultaneously life-generating and "the great silencer" of heroic explorers like Frobisher, Davis and Baffin (205). Another example is found in the passage below, in which icebergs are figured both in terms of victims of violence and sites of tremendous power. Although close by, they seem somehow remote, and due to the fact that "four-fifths of [their] height and seven-eighths of [their] mass" exists beneath the water line, they are both visible and invisible (208). Through all of this, as in the way they both reflect and trap light, icebergs embody the paradoxical qualities that both Burke and Kant associated with the sublime.

Following the night of the anticipated storm, in daylight and calm Lopez proceeds to give us the kind of detailed description of icebergs that the relative safety and comfort of the *MV Soodoc* allows:

> The first icebergs we had seen, just north of the Strait of Belle Isle, listing and guttered by the ocean, seemed immensely sad, exhausted by some unknown calamity. We sailed past them. Farther north they began to seem like stragglers fallen behind an army, drifting, self-absorbed, in the water, bleak and immense. [...]
>
> Farther to the north they stood their journeys with greater strength. They were monolithic; their walls, towering and abrupt, suggested Potala Palace at Lasha in Tibet, a mountainous architecture of ascetic contemplation. [...] I would walk from one side of the ship to the other, wondering how something so imposing in its suggestion of life could be approached so closely, and yet still seem so remote. It was like standing in a dirigible off Annapurna and Everest in the Himalayas. (206)

Lopez here uses personification to portray icebergs as having life. Personification is an efficient way of ascribing a sense of life and/or agency to non-living matter, and was a literary tool much in vogue among romantic writers. A more contemporary characteristic of Lopez's icebergs is the way they change according to context. In the southernmost part of their range, they seem like victims of violence, affected by a battle between north and south that reverses the traditional aesthetic association between violence and latitude. Further north, they become stronger and more majestic, until their mountainous forms bring associations with the most sublime of all mountain landscapes: the Himalayas. This association confirms the affective power of Lopez's icebergs, and ties his representation to the romantic tradition of the aesthetic contemplation of alpine glaciers (Wilson 2003). However, the reference to Potala Palace, rather than to the Alps of Europe, subtly expands the romantic search for the world's underlying unity from a conventionally presumed Western origin to include the eastern philosophy of Buddhism.[3]

Lopez makes use of the fact that several human cultures associate mountains with spirituality to re-direct the idea of spirituality away from the Western Christian tradition. Rather than reading the icebergs as manifestations of God, in a manner resembling the way in which American nature writers before him read mountains, Lopez turns them into symbols of a cross-cultural human longing for spirituality; a form of spirituality that does not depend on an external authoritative force but must be sought individually – and internally. This is why Lopez claims that "when I saw them [the icebergs], it was as though I had been waiting quietly for a very long time, as if for an audience with the Dalai Lama" (205–6).

Lopez's references to Buddhism and the Himalayas signal that a modern understanding of icebergs should not be restricted to one cultural or historical point of view. Thus, in the same way as the first half of *Arctic Dreams* allows the reader to envision the Arctic's natural environment from the point of view of various of its animals, Lopez's portrayal of icebergs activates several physical perspectives and hints at other possible cultural ones. While the phenomenon of icebergs, and the scientific study of them, is by necessity dependent on the physical context, knowledge production should aim to transcend established categories while still maintaining an awareness of the very cultural and historical context of this transcendence.

Another characteristic feature of the icebergs described in *Arctic Dreams*, and one that clearly distinguishes them from the icebergs of Scoresby's *Account*, is the scientific description of how they generate rather than threaten life. "The suggestion of life" around the icebergs, writes Lopez, "was not an illusion. Harp seals and flocks of seabirds were drawn to fish schooling in the nutrient-rich waters at their base – an upwelling driven by fresh-water runoff from the iceberg" (206). His portrayal of icebergs allows life and agency to these structures, while in Scoresby's text they remain the passive reflection of outside forces.

In Scoresby's description, icebergs:

> differ a little in colour, according to their solidity and distance, or state of the atmosphere. A very general appearance is that of cliffs of chalk, or of white or grey marble. The sun's rays reflected from them, sometimes give a glistening appearance to their surfaces. Different shades of colour occur in the precipitous parts, accordingly as the ice is more or less solid, and accordingly as it contains strata of earth, gravel, or sand, or is free from any impurity. In the fresh fracture, greenish-grey, approaching to emerald-green, is the prevailing colour. (Scoresby [1820] 1969, 254–55)

One hundred and sixty years later, Lopez provides us with an almost parallel description of icebergs:

> Where the walls entered the water, the surf pounded them, creating caverns, grottoes, and ice bridges, strengthening an impression of sea cliffs. […] Where meltwater had filled cracks or made ponds, the pools and veins were milk-blue, or shaded to brighter marine blues, depending on the thickness of the ice. If the iceberg had recently fractured, its new face glistened greenish blue – the greens in the older, weathered faces were grayer. In twilight the ice took on the colors of the sun. […]
>
> The burden of rocks, gravel, silt, and sand that icebergs carry within them streaks their sides; as they melt, they rise higher in the water and the debris in their shoal water creates a series of waterline marks. As they fracture and tilt, the patterns of water-line marks cross at odd angles and slant skyward. (207)

Lopez presents more or less exactly the same iceberg features as does Scoresby. Yet Lopez's modern perspective evokes a completely different conception of these natural phenomena. Through their interaction with air and ocean, light and rocks, Lopez's icebergs undergo constant processes of change. These processes become inscribed into the icebergs' very structure, enabling us to read their lives from their physiognomy. Like sea ice, icebergs are features of the landscape through which the world's creative forces manifest themselves on a shorter time-scale than the geological one on which glaciers operate. This relative time compression is another reason why Lopez's portrayal of the Arctic supersedes romantic portrayals of the agentic quality of matter. Icebergs may look like mountains but, made up as they are of travelling, more easily morphing matter, they are more life-like and more unpredictable than the glaciated mountain faces of the Alps.

In Lopez's representation of icebergs, human form and agency are superimposed upon matter through personification. However, this matter already has an agency of its own, and the human is not the only form the ice takes. In one particularly interesting iceberg, we find "a human forehead against the sky", "the pattern of a sperm whale's lacerated tun", and "[f]loating, orographic landscapes", some of which have surfaces of "raw jade" and "abraded obsidian" (207). Expressive of the larger poetic project of *Arctic Dreams*, through its imagery this iceberg unites humans, animals, ice and mineral landscapes into a living, evolving unity. Combining lifelike characteristics with qualities and dimensions unknown, icebergs are to Lopez "creatures I have never seen before" (208).

Changing the iceberg metaphor

As "creatures" combining life-threatening and life-generating qualities, the icebergs of *Arctic Dreams* begin to subtly evoke a new sense of sublimity. "[T]he behaviour of light around the icebergs" – how they take "their dimensions from the light" and "their color from the sun, and from the clouds and the water" – makes Lopez marvel (206). Signalled through this emotional response, one could imagine this new sense of sublimity to be the result of changes in the observer's relative positioning to the natural world, and his recourse to modern scientific knowledge about it. But this is only part of the explanation. Equally important is the text's combination of scientific fact and religious symbolism.

Icebergs are the most symbolic of all the physical particularities of Lopez's Arctic. Despite the general emphasis in *Arctic Dreams* on empirical, scientifically accurate and updated knowledge about the Arctic, Lopez simply cannot resist relating to the Western tradition of symbolic representation of the Arctic. This would have created a narrative as culturally inadequate and historically de-contextualized as those earlier exploration narratives from which *Arctic Dreams* aims to distance itself. Instead, the reader finds Lopez, in his encounters with icebergs, "grappling with language, […] grasping for metaphors" in order to "get the images just right" (Slovic 1992, 161). Looking at the number of culturally determined images that Lopez projects onto his icebergs, in the following I will bring out the intentionality and cultural critique implicit in this act of "grasping", and the way in which it contributes to Lopez's reworked idea of the natural sublime.

As Lopez points out, "[a] comparison with cathedrals has come to many Western minds in searching for a metaphor for icebergs" (248). His own descriptions of the icebergs of Davis Strait combine scientific measurement with aesthetic descriptions, mixing topographical imagery with the architectural imagery of cathedrals; of ramparts, spires and columns (207–8). But why does Lopez insist upon projecting onto his icebergs the culturally loaded imagery of mountains and cathedrals? Why does he continue to grasp at such old and seemingly out-dated metaphors when he seems to have found new and better ones? This question becomes especially pertinent when one considers that elsewhere Lopez explicitly warns his readers that "[t]he risk we take" in applying metaphors as tools for understanding new places or stretches of land "is of finding our final authority in the metaphors rather than in the land" (247).

The answer lies in in how Lopez uses this process to display the range of metaphors with which icebergs may be described, while simultaneously examining a few of those most commonly used by outsiders travelling in the Arctic. By reflecting on the origin and function of a few of these metaphors, Lopez aims to enhance his readers' awareness of the aesthetic concepts that have historically shaped Western culture's perception of the Arctic. By further testing these metaphors against an agentic and authoritative Arctic natural environment that at times proves their constructed nature and distortion of reality, *Arctic Dreams* emphasizes how the land itself contributes to the construction of meaning. Indeed, Lopez argues that the land "urges us to come around to an understanding of ourselves" (247). What creates this understanding in *Arctic Dreams* is the text's exposure of how "we bring our own worlds", our own cultural metaphors and symbolic meanings, "to bear in foreign landscapes" (247).

Lopez's subsequent discussion of the cathedral metaphor provides an example of the transposition of symbolic meaning structures from the home context to the Arctic. In its application to the mountain-like structures of icebergs, the cathedral metaphor generates associations with an American Christian tradition of reading mountains as cathedrals (Muir [1914] 1988; Cronon 1996, 478), as well as with the more Burkean mountain sublime that would later influence perceptions of the Arctic sublime. Lopez argues that the "appropriateness" of the cathedral metaphor holds a more profound significance than obvious similarities of "line and scale" (248). It has to do with Western culture's "passion for" and symbolic interpretation of light (248). Lopez finds the gothic cathedrals of Europe, "with their broad bays of sunshine, [their] flying buttresses that let windows rise where once there had been stone in the walls" to express a cultural longing for light, and for a "God" who "is light" and mind (248).

The striving for a relationship with something larger than the self is the aim of Lopez's exposition of the Western symbolism of cathedrals, and what causes him to juxtapose this exposition with its critique. To Lopez, "the erection of the cathedrals" represents "the last wild stride European man made before falling back into the confines of his intellect"; a confinement from which nothing except "a more complicated manipulation of materials, a more astounding … grasp of the physical principles of matter" has been accomplished (250). To get away from this intellectual confinement and engage once more in striving for a relationship with something outside the self, necessitates a symbolic re-interpretation of light, and of God. Based on other scenes from his travel narrative, in which he alternately delights in or despairs of the physical influence of light on his body, one might be tempted to conclude that Lopez is searching for a new materialist understanding of light. But what his text combines is both this physically concrete and sensory impression of light and a more romantic sense of its interconnectivity.

Through his detailed and scientific depictions of Arctic ecosystems, Lopez represents light as the physical energy that binds all things together: living things that rely on light's energy to provide them with open waters, green fields and other living things for nourishment, as well as non-living but agentic things like icebergs that attract, distort and take their "dimensions" and "color" from the light (206). As a complement to this, he creates an aesthetic and symbolic sense of light through references to nineteenth-century landscape painting. In this respect, his discussion of the difference between European and American traditions is significant. Whereas the European tradition, Lopez claims, presented pastoral landscapes to its viewers, "the world viewed from a carriage window", North-American visual artists had to "struggle with light and space" in facing the enormous open spaces of the new continent (245). Through this struggle, American luminists developed paintings in which the atmosphere was "silent and contemplative", and in which the light was "like a creature, a living, integral part" of landscapes represented as "numinous, imposing, real" (245). Lopez thus implies that North-American visual artists have managed to evoke a different, more sensory interpretation of light, without robbing it of its spirituality. Lopez reads this symbolic re-definition of light, which moves it away from traditional Christian interpretations, as running parallel to a re-conceptualization of landscape away from the "merely symbolic" European landscape painting tradition

(245). Both of these aesthetic movements progressed towards a distinctly American understanding of the land as a concrete and determining presence in human life.

Following this re-definition of light, Lopez proceeds to re-define one of the central concepts of Christian spirituality: *agape*. Rightfully defining *agape* (or *charity*) as "the love of another for the sake of God" (250), Lopez does not include in his definition the way in which *agape* is understood to be the manifestation of "the reciprocal love between God and man" ("charity"). Instead, he expands the concept away from its anthropocentric origin and towards the experience of "spiritual affinity with the mystery [it] is 'to be sharing life with other life'" (250). *Agape* becomes to him:

> a humble, impassioned embrace of something outside the self, in the name of that which we refer to as God, but which also includes the self and is God. We are clearly indebted as a species to the play of our intelligence; we trust our future to it; but we do not know whether intelligence is reason or whether intelligence is this desire to embrace and be embraced in the pattern that both theologians and physicists call God. Whether intelligence, in other words, is love. (250)

Lopez here superimposes a relationality based on love upon the material relationality of Arctic phenomena presented throughout the text, suggesting that the two come together in a new understanding of the nature of God. While Christian love is re-defined in relational, ecological terms, God is re-defined as the interconnectivity of all things. As a result, the barriers between God and the self, and between the self and 'other' non-human life outside the self, are diminished.[4] In this manner, the dimension of love and care extends along with the human's sense of social and material relationship into the relational networks of the natural environment.

Only after having thus been associated with the 'sharing of life with other life' – with the 'embracing of something outside the self' – do the icebergs of *Arctic Dreams* become truly sublime. As he watches them one last time from the *MV Soodoc*, Lopez now finds the icebergs "so beautiful" that they make him "afraid" (251). Although his recent allusions to spirituality may tempt readers to interpret the effect of the icebergs on Lopez in terms of a romantic poet's attunement to the workings of a transcendental or universal mind, an awareness of the setting in which the scene takes place disrupts this interpretation. Lopez's final experience of icebergs is not that of a lone traveller intuiting the overwhelming forces of the natural elements. Rather, it is a physically comfortable experience shared with the ship's second engineer, who has drifted into the Arctic from Guyana. It ends with Lopez gazing over "the extraordinary fluidity of [the geometry of ice] on the calm waters of Melville Bay" (251). The "extraordinary fluidity" of the ice signals the way in which absolute certainties in Lopez's portrayal of the Arctic have disappeared before the physical and cultural relationality and creative "play" of the world. There is order to the Arctic's physical environment, but this order is neither universal nor unchanging.

Lopez's icebergs bring together ice and light. They are at one and the same time real and embodied parts of the land and carriers of symbolic meaning. In his final description of them, Lopez reflects on how they "so embod[y] the land. Austere. Implacable. Harsh but not antagonistic. Creatures of pale light" (251). Thus ends a chapter in which the nature of both ice and light have been explained in great scientific detail. The "grasping" for metaphors that Lopez has enacted in his depictions of icebergs has allowed a juxtaposition of old cultural and symbolic with modern scientific (and ecological) metaphors that works to critique Western culture's overly simplistic and overly symbolic understanding of light, and of icebergs. Through this juxtaposition, the nature of both symbols changes. The icebergs are no longer antagonistic, threatening structures, but part of the land in its all-inclusive relationality: parts that – if you look closely – make you aware of your relationship with that which exists outside yourself. The imagery signals this change in signification by the way in which the light of the icebergs has faded slightly, perhaps as a result of the way their new ontological position blocks them from being interpreted as mere physical structures reflecting God's spirituality. Lopez's icebergs are no longer mere physical structures, but creatures engaged in the absorbing, reflecting and transforming of light, as well as in the giving, taking and shaping of life. Simultaneously beautiful and evoking fear, icebergs become the hallmark of his revised form of the Arctic sublime.

It should be noted that the depiction of icebergs is among the very few instances in which Lopez digresses from his generally empirical presentation of physical Arctic phenomena. He does so in order to endow the material world with spirit in a way reminiscent of nineteenth-century romantic poets. The evocation of this spiritual dimension allows Lopez to engage in a critical dialogue with the well-established tradition of aesthetic representation of Arctic landscapes in general, and icebergs in particular. It also engages us on a more profound and/or emotional level than the text's earlier and more scientific presentations have managed to do. The activation of this more affective response is crucial to the text's establishment of alternative value judgments and to its criticism of our culture's focus on rationality and enlightenment.

A modern Arctic sublime

The modern or re-defined sublime in *Arctic Dreams* has its basis in other forms of mystery than the geographical. I have already associated these mysteries with the recognition of otherness, in both its animate and inanimate forms. Mystery is also inherent in the sense of how modern Westerners know the Arctic – particularly in the recognition of the limits to what modern science may identify and quantify. These mysteries are tacitly but ubiquitously present in *Arctic Dreams* through Lopez's evocation of the imperceptible forms of interconnectivity of the Arctic's natural environment. Similarly, the text's many references to quantum physics remind contemporary readers that the interconnectivity of all matter – even at its most basic, atomic level – remains beyond modern science's most advanced forms of knowledge.

The text's continuous oscillation between microscopic and macroscopic imagery also engenders a sense of sublimity by reminding us of those non-local relationships that quantum physics claims govern the way in which cosmic and local forces interact in the world's continuous act of creativity. Accordingly, one may argue that the modern or re-defined aesthetic of the sublime in *Arctic Dreams* emerges in part as a *result* of the text's positivistic use of science to explain Arctic phenomena. Whereas science's quest for certainty has unveiled much of the mystery of the Arctic as a geographical region, other and more intricate mysteries have presented themselves through this very quest.

In place of the geographical vastness and tremendous inorganic forces of the previous aesthetic, Lopez's Arctic sublime presents us with another form of vastness: the vastness of life. The first half of *Arctic Dreams*, which describes different Arctic ecosystems and their animals in great detail, firmly deconstructs the idea of the Arctic as a lifeless, blank space. The Arctic's plethora of life is, however, not limitless. Whereas the text presents the relationality of life as limitless, life itself, in all its myriad expressions, persists on the very brink of existence. In addition to the strain of long, unabated periods of cold and darkness, life in the Arctic also has to deal with abrupt changes in environmental parameters vital to existence.

Thus Lopez presents landscapes that possess paradoxical qualities resembling those of the traditional natural sublime: "horror within magnificence, absurdity within intelligibility, suffering within joy" (411).[5] Their dark aspects are further darkened by the continuous underlying threat of modern human activities and their "ability to alter the land" (411). This particular form of environmental concern, in combination with the text's ecological focus and repeated evocation of feelings of respect and awe in encounters with environmental others, justifies adding the label "ecological" to Lopez's re-defined and relational Arctic sublime. Like Hitt's, Lopez's ecological sublime reminds modern readers that "[t]here will always be limits to our knowledge, and nature will always be, finally, impenetrable" (Hitt 1999, 620). It does so precisely by presenting the ecological sublime's characteristic sense, not of fear and trembling, but of *wonder* in the encounter with concrete, physical nature (Hitt 1999, 620).

Aiming for a modern vision of the Arctic and its landscapes, *Arctic Dreams* employs the aesthetic of the sublime, but in a way that brings this aesthetic back to the material. Arguably, in relation to the Arctic, more extensive knowledge of natural and cultural particulars is still necessary in order to counter longstanding and deeply rooted misconceptions. This explains why Lopez's predominant orientation in this text is one of non-interference, of letting facts speak for themselves to reveal the simultaneous otherness and sameness of environmental others. In this way, he avoids the romantics' tendency to "charge [...] biological occurrences with human significance" (Wilson 2003, 179–180). Yet, as is evident in his discussion of cathedrals, to Lopez the development of Western society over the past few centuries represents a crisis of the human imagination that is associated with modern scientific society. Thus, he reiterates the romantic trope that sees rationality and science as stifling the imagination. In *Arctic Dreams*, however, he simultaneously argues that science *need not* do this. Not only does the technologically advanced science of today allow insights into animal lifeworlds and migratory ways of life that were previously unknown to us; it has further yielded knowledge about the very boundaries of what we *may* know. Thus, if other, more humanist, lines of inquiry are allowed to complement it, science may generate new sources of wonder and awe – and of the sublime.

Arctic Dreams was written before climate change made its ubiquitous and game-changing impact on modern Western conceptions of the natural world. Although it therefore does not deal with issues related to climate change, I would argue that its way of re-thinking our relationship with the natural environment is still useful in the present day. As the Arctic itself melts and changes, the material changes to this region are generating incomprehensible but radical and potentially life-threatening changes to natural environments worldwide, as well as to planetary biogeochemical systems. Even technocratic modern science has no ultimate overview of the intricate web of relationships these systems constitute, and no reliable means of predicting how they are about to change. The Arctic, which like the Antarctic plays a key role in stabilizing these planetary systems, thus represents the failure of the modern scientific enterprise to control and mould nature for the comfort and benefit of humankind.

For this reason, the Arctic might still be thought of in Loomis's terms as a region "somehow vaster, more mysterious, and more terrible" than other regions in its imperceptible yet ubiquitous power (Loomis 1977, 96). In this sense, climate change ultimately confirms that nature's power over human lives remains unchanging and inviolable, even at the very moment when the Arctic itself is proving to be malleable and vulnerable. The idea that the human mind might transcend the limitations of the physical world is severely challenged by the early effects of climate change, which leave little doubt that in practice our lives do indeed exist within these limitations.

Ecologically informed, the new natural sublime outlined in *Arctic Dreams* combines aesthetic reflection with an insistence on material reality, and recognizes the simultaneous power and vulnerability of the Arctic. Similar to the natural sublime identified by Spufford in nineteenth-century exploration narratives, this more contemporary Arctic sublime invokes a sense of identification with the natural environment, but one of a different and perhaps more complex character. Once the biological life and life-generating natural forces of the Arctic are recognized, the region becomes simultaneously life enhancing and life threatening. Lopez's Arctic sublime thus causes different emotional and rational responses from its predecessors. Because the present physical and cultural climate recognizes that the cause of its life-threatening power is ultimately to be found in the imperceptible and unintended forces set in motion by human exploitation of the natural world, when modern Westerners now identify with the sublime Arctic, our sense of its vulnerability is at the same time the sense of our own vulnerability. The new and daunting threats associated with the melting Arctic might have been initiated by us, but they represent the power, not of the human, but of the tremendous forces of nature over and against the human. Retaining some version of the sublime in contemporary portrayals of the Arctic may act as an efficient reminder of Western culture's long-time recognition of this power.

References

Browne, Neil W. 2007. *The World in Which We Occur: John Dewey, Pragmatist Ecology, and American Ecological Writing in the Twentieth Century.* Tuscaloosa: University of Alabama Press. Accessed 6 May, 2015. ProQuest Ebook Central.

Buell, Lawrence. 2001. *Writing for an Endangered World: Literature, Culture, and Environment in the U.S. and Beyond*. Cambridge, MA: Belknap.

Burke, Edmund. (1759) 2008. *A Philosophical Enquiry into the Origin of our Ideas of the Sublime and Beautiful*. Mineola, NY: Dover.

Capra, Fritjof. 1983. *The Tao of Physics: An Exploration of the Parallels between Modern Physics and Eastern Mysticism*. Glasgow: Flamingo.

"charity." *Encyclopædia Britannica. Encyclopædia Britannica Online*. Encyclopædia Britannica Inc., 2017. Web. 14 Jan, 2017.

Chou, Siuh-huah Serena. 2013. "Barry Lopez's Arctic Dreams: Organicism and the Relocation of the Wild." *Foreign Literary Studies* 35 (1): 24–34.

Cicovacki, Predrag, ed. 2009. *Albert Schweitzer's Ethical Vision: A Sourcebook*. Oxford: Oxford University Press.

Coles, Romand. 1993. "Ecotones and Environmental Ethics: Adorno and Lopez." In *In The Nature of Things: Language, Politics, and the Environment*, edited by Jane Bennett and William Chaloupka, 226–249. Minneapolis, MN: University of Minnesota Press.

Cronon, William. 1996. "The Trouble with Wilderness; or, Getting Back to the Wrong Nature." In *Uncommon Ground: Rethinking the Human Place in Nature*, edited by William Cronon, 69–90. New York: Norton.

Eagleton, Terry. 1990. *The Ideology of the Aesthetic*. Oxford: Basil Blackwell.

Evernden, Neil. 1992. *The Social Creation of Nature*. Baltimore, MD: Johns Hopkins University Press.

Gatta, John. 2004. *Making Nature Sacred: Literature, Religion, and Environment in America from the Puritans to the Present*. Oxford: Oxford University Press.

Harland, Richard. 1999. *Literary Theory from Plato to Barthes: An Introductory History*. New York: St. Martin's Press

Hitt, Christopher. 1999. "Toward an Ecological Sublime." *New Literary History* 30 (3): 603–623. Accessed 6 May, 2015. http://www.jstor.org/stable/20057557.

Iovino, Serenella, and Serpil Oppermann. 2012. "Material Ecocriticism: Materiality, Agency, and Models of Narrativity." *Ecozon@* 3 (1): 75–91.

Kant, Immanuel. 1953. *The Critique of Judgement*. Translated by James Creed Meredith. Oxford: Clarendon.

Kittang, Atle. 1976. *Litteraturkritiske problem: teori og analyse*. Bergen: Universitetsforlaget.

Loomis, Chauncey C. 1977. "The Arctic Sublime." In *Nature and the Victorian Imagination*, edited by U. C. Knoepflmacher and G. B. Tennyson, 95–112. Berkeley, CA: University of California Press.

Lopez, Barry. 2001. *Arctic Dreams: Imagination and Desire in a Northern Landscape*. New York: Vintage.

MacLaren, I. S. 1985. "The Aesthetic Map of the North, 1845-1859." *Arctic* 38 (2): 89–103.

Morgan, Benjamin. 2016. "After the Arctic Sublime." *New Literary History* 47 (1): 1–26. Accessed Jul. 25, 2016. https://muse.jhu.edu/article/626111.

Muir, John. (1914) 1988. *The Yosemite*. San Francisco, CA: Sierra Club.

Novak, Barbara. 2007. *Nature and Culture: American Landscape Painting, 1825–1875.* Cary, NC: Oxford University Press.

O'Connel, Nicholas. 2000. "At One With the Natural World: Barry Lopez's Adventure With the Word & the Wild." *Commonweal* 127 (6): 11–17.

Paul, Sherman. 1992. "Making the Turn: Rereading Barry Lopez." In *For Love of the World: Essays on Nature Writers*, 67–107. Iowa City, IA: University of Iowa Press.

Phillips, Dana. 2003. *The Truth of Ecology*. Cary, NC: Oxford University Press.

Scoresby, William. (1820) 1969. *An Account of the Arctic Regions with a History and Description of the Northern Whale-Fishery*. New Abbot Devon: David & Charles Reprints.

Slovic, Scott. 1992. *Seeking Awareness in American Nature Writing*. Salt Lake City, UT: University of Utah Press.

Spufford, Francis. 2003. *I May be Some Time: Ice and the English Imagination*. London: Faber and Faber.

Weiskel, Thomas. 1976. *The Romantic Sublime: Studies in the Structure and Psychology of Transcendence*. Baltimore, MD: Johns Hopkins University Press.

Wilson, Eric G. 2003. *The Spiritual History of Ice: Romanticism, Science, and the Imagination*. New York: Palgrave Macmillan.

Notes

[1] All further references to *Arctic Dreams* are given parenthetically.

[2] In the following discussions of the Arctic natural sublime and contemporary Western cultural attitudes towards the natural world, the term *modern* will refer to contemporary modernity, as distinct from the nineteenth-century "heroic" and clearly humanist form of modernity.

[3] In a brief introduction to Buddhism, Fritjof Capra writes: "As always in Eastern mysticism, the intellect is seen merely as a means to clear the way for the direct mystical experience, which Buddhists call the 'awakening'. The essence of this experience is to pass beyond the world of intellectual distinctions and opposites to reach the world of *acintya*, the unthinkable, where reality appears as undivided and undifferentiated 'suchness'" (Capra 1983, 106).

[4] Additionally, Neil W. Browne has also commented on how in *Arctic Dreams* "[t]he impassioned *relationship* between self and world replaces a traditional God" (2007, 156).

[5] Lopez's translation of Albert Schweitzer. In Pedrag Cicovacki's edition, the quotation reads: "In the world as it is, we see horror mingled with magnificence, absurdity with logic, and suffering with joy" (2009, 179).

RE-ENACTING MODERNITY IN THE WILDERNESS OF THE CANADIAN NORTH: AN ECOCRITICAL EXAMINATION OF HELGE INGSTAD'S *THE LAND OF FEAST AND FAMINE*

FREDRIK CHR. BRØGGER

If any one book is to be regarded as the progenitor of Norwegian wilderness literature of the twentieth and twenty-first centuries, it must be Helge Ingstad's *The Land of Feast and Famine* (1933), originally published in Norwegian two years before as *Pelsjegerliv blant Nord-Canadas indianere* (literally translated, "Fur-hunter life among North Canada's Indians"). Ingstad's tale of his excursions into the Canadian wilderness towards the end of the 1920s has inspired many subsequent Norwegian adventurers of the late twentieth and early twenty-first centuries to travel to the northern regions of North America and venture forth into this Arctic and Subarctic Promised Land of allegedly unspoiled natural splendour and wildness – and to write about it afterwards. Due to the prominent position of *The Land of Feast and Famine* as a quintessential exemplar of Norwegian wilderness literature, it is in my opinion high time to subject it to close critical analysis.

In contradistinction to for instance the rich and diverse tradition of wilderness prose in North America (epitomized by writers such as Henry David Thoreau, John Muir, Gary Snyder, Barry Lopez and Rick Bass), the Norwegian wilderness adventure narrative as exemplified by Ingstad's book tends to view nature from a more limited perspective.[1] The focus of *The Land of Feast and Famine* is not so much on the diversity of natural life in the wilds as on the trials and tribulations of the narrator-adventurer himself, face-to-face with the dangers of an environment situated far away from the comforts and shelter of civilization. Ingstad's story may thus be argued to be primarily preoccupied with the narrator's own strength, practical skills, and self-reliance in his struggle *against* nature. Such a perspective seems, almost by definition, to entail a limited awareness of the impact of human activities upon the wilderness, including those of the protagonist himself.

My subsequent critical comments on Ingstad's *The Land of Feast and Famine* are in no way meant to belittle his stature as a Norwegian explorer. One indication of Ingstad's international renown is the fact that two geographical landscape features in North America have been officially named after him. During the last winter chronicled in *The Land of Feast and Famine*, Ingstad lived on the tundra in a dell with spruce trees, a little lake, and a small river in Canada's Northwest Territories; the latter has later been officially given the name of *Ingstad Creek* (the groundwater in the area has apparently sunk in recent times, reducing it to tiny brooklet). And in the Brooks Range in Alaska we find *Ingstad Mountain*, 1461 metres high, approved by U.S. Board on Geographic Names in 2006 and originally proposed by the Nunamiut tribe in gratitude for his efforts on their behalf after his wintering with them in 1949/50, chronicled in his book *Nunamiut: Among Alaska's Inland Eskimos* (1954).

Ingstad combined a zest for adventure with strong interests in anthropology and archaeology and may be said to figure in the Norwegian popular imagination as a blend between a latter-day Fridtjof Nansen and an arctic Thor Heyerdahl. Together with his wife Anne Stina, an archaeologist, Ingstad was undoubtedly most famous for pursuing and proving their theory of a Norse presence in North America, which culminated in their discovery in 1960 of the remains of a Viking settlement on the coast of Newfoundland, confirming old Norse tales of a *Vinland* in the West and verifying conclusively that Greenlandic Norsemen had crossed the Atlantic to North America some five hundred years before Christopher Columbus. It is no wonder that, to mid twentieth-century Norwegians, Ingstad was a household name.

As a teenager in Oslo in the 1950s, I found Ingstad's tale about hunting and trapping in the Canadian Northwest Territories, which had become a bestseller in Norway. I read Ingstad's narrative with all the fervour and longing of an inbred city boy for stories of woods and mountain plateaus unaffected by civilization. I can still remember my own intense wish fulfilment evoked by a passage in the beginning of Ingstad's narrative when he and his companion have started paddling across Great Slave Lake with a fishing line trailing behind the canoe (here cited from the English translation, *The Land of Feast and Famine*):

> But it was not until after we had left the islands behind that the trout began to bite in earnest. Then there was really something doing! Great big fellows, weighing from fifteen to twenty pounds, were on the line in a flash. Ingenuous creatures were these, too, lacking utterly the *kultur* which

> distinguishes our trout back home in Norway. The finer points of discrimination between the divers types of flies were utterly unknown to them. Nor did they insist upon gut leaders and such; they would snap at a tinfish fastened directly to a thick line, and if they failed to hook themselves the first time, they would try it again just as eagerly as before. (Ingstad 1992, 38–39)

Think of coming across *uncultured* trout! Weighing between seven and ten kilos! In wildernesses far beyond the reach of civilization, teeming with wildlife!

Even today there is no shortage of true-story wilderness accounts of the far North in Norwegian literature. We may thus trace a direct line from Fridtjof Nansen's *The First Crossing of Greenland* (1892) and *Farthest North* (1897) through Helge Ingstad's *The Land of Feast and Famine* (1933) and *East of the Great Glacier* (1937) to the present-day Norwegian adventure narratives by Lars Monsen, for instance *Til fots gjennom Alaska* [On foot through Alaska, 1996] and *Nådeløs villmark. Canada på tvers* [Ruthless Wilderness: Across Canada, 2002]. Among Scandinavians and Finns of the late twentieth and early twenty-first century, northern Canada and Alaska have seemed to yield a particularly strong fascination as the last remaining wildernesses of the North.

Ingstad's *The Land of Feast and Famine* is first and foremost the vivid record of an individual struggling to survive for several years in the northern Canadian wilds, without access to help from the outside. As the Norwegian philosopher Peter Wessel Zapffe put in his occasional poem celebrating Ingstad's seventy-fifth birthday:

> You came on the heroic era's closing-day,
> When the glow of a fearless one-man's plight
> Paled like the waning sunset's final ray
> And nothing more seemed really worth the fight. (Zapffe 1977, 87)[2]

The Land of Feast and Famine represents Ingstad's dramatization of the fearless endeavours of the single individual fighting to survive in the wilderness of the North. As Frode Skarstein notes in his biography of Ingstad, the narrative "has in the close to eighty years since its publication acquired a cult status. The book is talked about as a bible not only by dogsledders and serious wilderness enthusiasts, but also by ordinary Norwegians interested in outdoor life" (Skarstein 2010, 32).[3] The renown of Ingstad's wilderness chronicle, however, also extends beyond the borders of Norway. It has been published in many European countries including Russia, and has been translated into quite a few languages. The re-publication of *The Land of Feast and Famine* by McGill-Queen's University Press in 1992 testifies to its continued importance also for Canadian cultural history.

Quite a few Norwegian wilderness devotees have paid tribute to Ingstad's name by quite *literally* walking in his footsteps or by having been inspired by him to undertake wilderness adventures of their own in northernmost America. In the first-mentioned category is for instance an expedition in 1998 undertaken by six men, most of them well-known Norwegian sportsmen and adventurers,[4] in order to find the lake that Ingstad in *The Land of Feast and Famine* called "Moose Lake" and to hopefully discover the remnants of his cabin there, where he spent his first year together with Hjalmar Dale; this also resulted in the production of a TV film broadcast in 1999.

In fact, in the early twenty-first century this part of the Northwest Territories seems to have functioned as a magnet for Norwegian wilderness enthusiasts on their pilgrimage to the regions in which Ingstad fished and hunted. After an unsuccessful search in 2010 and several years of preparations with repeated, meticulous readings of Ingstad's book and advice from locals in the Great Slave Lake region, a group of Norwegians carried out an expedition in 2013 to Ingstad Creek where they claim to have found the "tundra camp" in which Ingstad spent his last winter hunting and trapping ("Sandy Lake Expedition 2013," www.helge ingstad.net).[5] The heritage of Ingstad is thus kept very much alive even at the present time, although such chronicles may also be said to confirm Zapffe's assertions that they no longer have the same lustre as the one described in *The Land of Feast and Famine*.

Recently published adventure books of the Far North also pay homage to Ingstad's influence, for instance Sverre Hovind's *I Ingstads fotspor: På villmarkstur i Canada* [In Ingstad's footsteps: A wilderness hike in Canada, 2009] and Lars Monsen's *Nådeløs villmark: Canada på tvers* [Ruthless Wilderness: Across Canada, 2002]. The latter was even made into a six-episode TV series. Hovind reveals that Ingstad's *The Land of Feast and Fortune* spell-bound him – "a boy's dream incarnate!"[6] – and triggered him to set out on his month-and-a-half trip in Ingstad's footsteps from Old Snowdrift and Great Slave Lake to the area around Ingstad Creek and down Snowdrift River. And in the chapter entitled "Drømmen om Canada" [The dream of Canada] that introduces the narrative about crossing its "ruthless" wilderness, Monsen too emphasizes that Ingstad's book about hunting and trapping helped sharpen his lust for adventure. Ingstad's stature is also illustrated by the fact that he surfaces in Monsen's thoughts several times during the latter's thirty-one months of crossing

Canada, for instance when four hundred caribou are in full flight before his own dogsled on the ice of a lake: "He [Ingstad] has made this classical Canadian wilderness scene immortal in the minds of many Norwegians. Now I stand here, in the middle of the same adventure, hearing the same sounds, seeing the same animals and sucking in the same fresh winter air!"[7] In Monsen's description the world of perception becomes suffused with, and indistinguishable from, intertextual discourse. Most moving, perhaps, are Monsen's comment when he belatedly receives the news of Ingstad's death while he is camped by Great Slave Lake:

> All we can do is to thank him, deeply and wholeheartedly, for all the inspiration that he gave, not to speak of his scientific contributions. Now it is left to us who remain to carry on his wilderness traditions.
>
> The 101-year-old wilderness king and science champion Helge Ingstad [...] made tracks for the happy hunting grounds on the 28th of March 2001, where he almost certainly is hunting caribou in his renowned style and filling his canoe with fish.[8]

Wilderness adventure as a modernity project

In view of these eulogies to Ingstad and the unquestioned position of *The Land of Feast and Famine* as the apotheosis and exemplary model of Norwegian wilderness writing, I find it important to submit the text to a more scrupulous, detailed analysis. Its repute notwithstanding, Ingstad's chronicle evinces in my opinion relatively little interest in nature as such. There is little description of the diversity of life in this northern Canadian wilderness (except, of course, when it comes to the animals that provide valuable fur), and the impact of the activities of the hunters and trappers on this northern environment is furthermore not an issue of serious deliberation. Ingstad's *The Land of Feast and Famine* (as well as the wilderness narratives of his present-day Norwegian heirs) focuses instead on celebrating the skills and tenacity of an archetypally male adventurer – which is to say extolling his anthropocentric endeavours to hold sway over the forces of nature.

Maybe the dearth of ecological perspectives in the wilderness literature by Ingstad and his successors may be argued to result from the demands of the genre itself, premised as it is on the idea of a pristine nature. The genre requirement that the adventurer of the far North must confront an *unspoiled* nature may be said to situate it, by definition so to speak, outside of the realm of human history. At the same time, of course, the wilderness writer is nonetheless a product of his own culture. My central contention in this connection is that Norwegian wilderness literature in general, and Ingstad's *The Land of Feast and Famine* in particular, are propelled by what I would call the ideology of modernity, which serves as an effective barrier to genuine environmental concern.

"Modernity" is a term that has been used in many different ways in philosophical, sociological, political, and literary contexts, to name only a few. However, when contending that Ingstad's *The Land of Feast and Famine* is first and foremost expressive of the ideology of modernity, I have in mind the specific worldview that broke with classicism and medievalism and that emerged with the Renaissance and the Enlightenment era in Europe, coming to full fruition during the age of nineteenth century capitalist industrialism. It represented the ideological superstructure of a purely anthropocentric project directed at an ever-increasing domination of nature. Its basic tenets were the beliefs in rationalism and continuous social progress combined with the principle of freedom from restraint (economically, politically, and personally). In the words of Peter Wagner,

> Modernity is the belief in the freedom of the human being – natural and inalienable, as many philosophers presumed – and in the human capacity to reason, combined with the intelligibility of the world, that is, its amenability to human reason. In a first step towards concreteness, this basic commitment translates into the principles of individual and collective self-determination and in the expectation of ever-increasing mastery of nature and ever more reasonable interaction between human beings. (Wagner 2012, 4)

The ideology of modernity thus reflected a strong adherence to the idea of individual autonomy and, one may add, colonial conquest. This European ethnocentric orientation also included, as an article of faith, the exploitation of both animate and inanimate nature for human (Western) advancement and profit. Modernity was thus thoroughly permeated by the Cartesian dualisms of culture vs nature, mind vs body, human vs animal, man vs woman, and us vs them. The relationship between man and nature constituted a one-way connection, dictated by man.

The central paradox of Ingstad's narrator-protagonist is that, having escaped from industrial, urban, bourgeois society into "untouched" nature, he ends up affirming the core ideology of modernity that helped create that society in the first place, that is, the pioneering and entrepreneurial values of the self-reliant individual attempting to master nature. Historically, of course, modernity was quite literally predicated upon the exploitation of wilderness for profit, regardless of whether the spoils were fur, cotton, minerals, or oil. At the end of his preface to the Norwegian 1977 edition of his book, Ingstad indeed notes that his chronicle represents "one of the last leaves in Canada's engrossing pioneer history about fur hunters who ventured forth into virgin wilderness and paved the way for the so-called civilization" (1977, 11).[9] However much propelled by the desire to *escape* civilized life, the wilderness adventure of the fur-hunter is, paradoxically enough, a way of *partaking* in the commencement that self-same civilization, this time in Canada.

The tendency in Ingstad's narrative to exaggerate the pristine and primeval character of the natural environment of northern Canada may thus be linked to his discontent with his own present-day society that entailed the erosion of the traditional culture of modernity in the first quarter of the twentieth century. Ingstad's reverence for unspoilt nature is accompanied by his celebration, intensified in compensatory fashion, of the hard and rugged values of a pioneer-type modernity. My ecocritical examination of *The Land of Feast and Famine* therefore involves an analysis of the text's predominantly utilitarian approach to nature; its exaggeration of the aboriginal state of the Canadian wilderness and of its First Nations peoples; its colonialist and capitalist attitude to the environment and to its native inhabitants; and its exaltation of masculine values, individualism, and man's mastery of nature.

A utilitarian approach to nature

The attitude to nature that we find in Ingstad's book is first and foremost a utilitarian one. That subsequent adventurers and hunters have read his narrative as a sort of instruction manual is not surprising; Ingstad's very survival depended on his rapid acquisition of practical wilderness skills. *The Land of Feast and Famine* informs the reader of how to best carry a packing board, how to arrange the dog-pack, how to treat the sled dogs, how to build a log cabin and a cache for storing food, how to partition a caribou, and so on. The following passage from Ingstad's description of the ample rewards of hunting at the onset of the last winter is indicative of the pragmatic attitude that informs his relationship to nature throughout the narrative:

> With dried meat hanging in thick rows in front of the tent, our camp had assumed a homelike atmosphere. Our spirits brightened and the wilderness, which before had seemed so bleak and oppressive, now wreathed itself in smiles for our benefit. (Ingstad 1992, 281)

The book indeed suggests that it is first and foremost your own food supply that determines your view of nature. That the wilderness is said to "wreathe itself in smiles for our benefit" is a direct expression of Ingstad's anthropocentric orientation towards his environment. An unruly and unpredictable animate wilderness has finally been ordered by man into "thick rows" of dried meat. The passage also reminds the reader that the English-language title of Ingstad's book is highly appropriate: the northern Canadian wilderness is indeed evoked as the land of feast and famine, where an unsentimental attitude to nature is the prerequisite for getting enough to eat. At the same time, the overriding concern with his own welfare is precisely the attitude that seems to make it hard for the narrator-protagonist of this wilderness story to make room for reflections on the wider consequences of his own and other people's activities in the wilderness.

The exaggeration of the aboriginal state of the wilderness and of its First Nations peoples

Ingstad's promotion of a pristine, unspoiled Canadian wilderness is at the same time repeatedly contradicted by information that he himself provides. In his 1977 Preface he notes approvingly that "animal life was as nature had intended it to be" (Ingstad 1977, 8).[10] At the same time, however, his narrative promotes for instance the hunting and killing of wolf by the trappers (at the end of his chapter on the caribou). In this connection Ingstad also contrasts the attitude of the some of the First Nations peoples with that of the whites. The former refrain from killing wolves (Ingstad 1992, 266), something which Ingstad regards as the result of superstition, whereas the trappers who "live on the flesh of the caribou are simultaneously waging war upon its arch-enemy, the wolf. In order to gain some impression of the havoc wreaked by wolves, one would have to witness with his own eyes their wasteful slaughter" (Ingstad 1992, 165). The dualisms of culture/nature and human/animal are quite evident in such passages. Caribou is civilized into "flesh" for human consumption, whereas wolves are linked to "wastefulness"; the humans consequently engage in the rational and systematic response of "war", whereas wolves simply wreak "havoc". Ingstad's story of wildlife "as nature had intended it to be" is repeatedly undercut by his own dualistic and anthropocentric ethos.

My objective with this analysis is certainly not to anachronistically criticize Ingstad for lacking a modern ecological consciousness in the narrative of *The Land of Feast and Famine*. My discussion points to its disregard of environmental perspectives in general – perspectives that had been brought into the public domain long before Ingstad's own adventure in the Northwest Territories. A deep concern for the North American wilderness certainly existed in the United States in the nineteenth and early twentieth centuries with writers of wilderness preservation such as

Henry David Thoreau, John Burroughs, and John Muir (and subsequently for instance Gene Stratton Porter and William J. Long). In Canada in the late nineteenth and early twentieth centuries there were advocates of wilderness protection such as Charles G.D. Roberts and Ernest Thompson Seton, who were also famous far outside Canada's borders for their animal stories.[11] In her book *Survival: A Thematic Guide to Canadian Literature*, Margaret Atwood writes that "[t]hose looking for something 'distinctly Canadian' in literature might well start right here" (Atwood 1972, 73).

According to Atwood, the Canadian animal story represents a genre unique for its approach to the living world. Whereas stories of animals in the United States are told from the point of view of the successful hunter who "wishes to match himself against them" and "conquer them", reflecting, as Atwood puts it, "the general imperialism of the American cast of mind", the Canadian genre involves tragic or pathetic stories "*told from the point of view of the animal*". In Atwood's succinct summary, "English animal stories are about 'social relations,' American ones are about people killing animals; Canadian ones are about animals *being* killed, as felt emotionally from inside the fur and feathers" (Atwood 1972, 74). Ingstad's book may be said to belong to the American rather than the Canadian genre, whereas the Norwegian writer Mikkjel Fønhus may certainly be regarded as a prominent representative of the latter type of fictional animal stories, publishing his first important wilderness books in the first half of the 1920s. No matter whether Ingstad had read these particular authors or not, apprehensions concerning animal population decimations were certainly widespread in Western culture by the advent of the twentieth century.

Specific details that occasionally come to the surface in Ingstad's own text reveal that the Canadian wilderness was considerably less unspoiled than his dominant discourse makes it out to be. Early in *The Land of Feast and Famine* we are told that the buffalo has been decimated west of the Slave River and in fact must be reintroduced there (Ingstad 1992, 7); the chapter on the caribou informs us that in recent years these animals have disappeared from the areas "along the western extremity of the Arctic coast" due to the Eskimos' recent acquisition of fire-arms" and that indeed "west of the Coppermine River, tame reindeer have been imported from Alaska in order to compensate this loss" (Ingstad 1992, 162). Likewise, the chapter on "the Barren Ground Indians" informs us that the musk-ox is "a dying race" because of the natives' use of modern weaponry, and that the musk ox therefore presently enjoys "complete protection" (Ingstad 1992, 254). On the other hand, everything is relative: there is no doubt that the Canadian wilderness at Ingstad's time still held great numbers of much-desired animal species such as beaver, otter, marten, mink and arctic fox whose fur was coveted for profit. Ingstad's book, however, seems devoid of a self-reflexive awareness of his own paradoxical situation: that he is in fact involved in a trade that is in the process of decimating the wildlife whose primeval quality is the very prerequisite for the appeal of his own narrative.

The decimation of animal life of the Northwest Territories, more pervasive than Ingstad's occasional observations suggest, was part of a systematic *colonization* of these areas by Western modernity. Beaver – an animal that Ingstad is very much intent on hunting – was for instance already becoming scarce in most areas of North America by the beginning of the twentieth century, partly because of hunting by the natives themselves. As Greame Wynn points out, many native people had at this time been "left with little alternative but to increase their take of beaver to survive on increasing quantities of store-bought or traded goods. All of this fit well with the views of the fur traders, who considered native peoples better employed hunting furs than food" (Wynn 2007, 71; see also 395). And as far as the Arctic fox was concerned, both the Inuit and the First Nations tribes had traditionally shown little interest in hunting them until the beginning of the twentieth century, when the fox became fashionable in Europe as a garment accessory. By 1930, therefore, the Hudson Bay Company had fur trading posts "strung along the coastline of the Canadian mainland, from Hudson Bay to the Yukon, as well as on most of the major islands to the north" (Hamilton 2008, 177, 179). The Hudson Bay Company "reported buying 30,000 arctic foxes from its post in the Canadian Arctic in 1943 alone" (Hamilton 2008, 181).

What actually saved parts of the wildlife in Canada from large-scale decimation was, paradoxically enough, fur farming. Already in the course of the 1930s, a decade or so after Ingstad's Canadian adventure, fur farming began to provide steadily larger portions of the supply of animal furs, and in the years following World War II mink and chinchilla farming had drastically reduced the need for wild furs (Newman 1989, 194, 196). Ironically, one might say that Canadian wildlife was saved by capitalism itself, finding cheaper ways to procure pelts, although hunting not only for subsistence but profit is still an important livelihood in present-day Northern Canada.

Ingstad's propensity for making the wilderness excessively pristine is also reflected in his tendency to exaggerate the unspoiled character of First Nations tribes of the Northwest Territories at this time. There is thus a double primevalization of northern Canada at work in *The Land of Feast and Famine*, of the land as well as its native peoples. With regard to the latter, his sympathy for indigenous tribes seems to fluctuate according to their degree of indigeneity. His highest praise is reserved for the Caribou-Eaters, a small Chippewyan tribe with whom he spent a whole winter by the Upper Thelon River, as they existed almost exclusively on caribou and lived "a more isolated life than the other tribes" (Ingstad 1992, 167). His portrayal of them is characterized by considerable understanding and empathy and points forward to his inclusive and detailed depiction twenty years later of the Nunamiut Eskimo tribe in the Brooks Range in northern Alaska, with whom he wintered 1949/50.

In *The Land of Feast and Famine*, however, Ingstad exaggerates the purely aboriginal character of the Caribou-Eaters' way of life, insisting that they "lived essentially like their ancestors, except for the fact that they now had firearms [...]" (Ingstad 1977, 8).[12] The off-hand remark on the natives' acquisition of modern weapons is, as suggested above, actually a fact of far-reaching importance. In addition we learn from his own account that the Caribou-Eaters have been in contact with Roman Catholic missionaries (Ingstad 1992, 226); that members of the tribe also regularly visit the trading post further south; and that the young seventeen-year old girl that courts him (the main reason, it seems, for his fleeing their camp in the spring) has "spent some time in the Roman Catholic mission in the south" and indeed prays every evening to the Virgin Mary, although "she was still far from being tamed" (Ingstad 1992, 236).

Ingstad's celebration of aboriginality is accompanied by his misgivings about cultural hybridity. These are attitudes that he certainly shared with a great many Western ethnologists of the time. As he bids "little band of human beings on the shore of Nonacho Lake" farewell, his reflections are deeply tainted by Western paternalism: "As they stood there in a little circle about me, there was something pathetic about these children of the wilderness who were clinging to the old life of their forefathers there in the forest, but who would soon have to go down beneath the wheels of fate, inexorably approaching from the land of the white man" (Ingstad 1992, 240). His cultural pessimism on their behalf is at the same time a strong testimony to the fact that the Canadian wilderness of the time was being aggressively colonized and changed.

A colonialist and capitalist attitude to the environment and its native inhabitants

This process of colonization is more strikingly described in the situation of the First Nations peoples that Ingstad meets elsewhere during his three-year sojourn, and his depictions of these natives are correspondingly less positive than of the Caribou-Eaters. The lives of "the Barren Ground Indians" have clearly been radically changed:

> The advent of the fur-traders opened new possibilities to these natives. A rough choice of the goods of civilization was then accessible to them, and the price for these was pelts. The Indians no longer found it necessary to keep wandering about the country in quest of game in order to remain alive. The doors of the trading stores stood open to anyone who had a beaver, fox, lynx, or marten to offer in exchange. There were not only food and tools to be had, but weapons which were more effective than the old. (Ingstad 1992, 246)

Ingstad also notes that "In spite of their splendid harvests of pelts, the majority of the Indians are in debt to the traders" (Ingstad 1992, 250).

The First Nations peoples have thus become part of the capitalist modernity from whose presence in Norway Ingstad flies, but with which he is still interacting, albeit in a new context and a new manner. His own narrative seems largely oblivious to his own complicity in the advance of modernity; he shows little awareness of being an intruder onto the land of somebody else's resources for subsistence. If the fate of the First Nations entails exploitation, Ingstad seems largely to blame the natives themselves for this – in itself a prototypical colonialist attitude to indigenous peoples. A case in point is his description of how the stores of the two fur-trading posts at Snowdrift at the Great Slave Lake are packed to the doors with people from the First Nations when spring has arrived:

> Otter-, fox-, beaver-, marten-, mink-, and muskrat-skins find their way over the counter until there is a mighty mound of pelts. In front of each Indian there collects an ever-mounting pile of conglomerate trash and fripperies, until the last skin has been handed over. Whatever they may chance to desire *must* be theirs [...] (Ingstad 1992, 136)

This scene seems strongly reminiscent of descriptions of white traders exchanging goods with Africans in the previous centuries.[13]

Here it may be pertinent to note that the greatest stumbling block to environmental stewardship in the history of our globe is precisely what Garrett Hardin in 1968 called "the tragedy of the commons" – the prioritization of the exclusive and exclusionary short-term interests of the single individual or the single group or the single enterprise or the single community over the long-term, collective welfare of the ecosystem itself. This tragedy may be argued to be closely connected to the doctrines of individualism that represent the ideological cornerstone of Western modernity, which in Ingstad's book are also closely linked and entangled with the arch-masculine features attributed to the world of the wilderness.

The foregrounding of masculine values

In the 1977 preface Ingstad endows Hjalmar Dale (with whom he built a cabin and spent a winter hunting and trapping) with ideal masculine qualities: “He [Hjalmar] had the reputation across the Northwest Territories as one of the most outstanding long-distance trekkers by dogsled and canoe and as an exceptionally tough guy. […] In his strong blue eyes I never saw fear, and he had a relentless determination never to give up” (Ingstad 1977, 8).[14] In this preface Ingstad also speaks of “the anticipation of unknown land emerging and not least the excitement during the hunting. Occasionally it could certainly happen, when everything went awry, that I said to myself: What in the world are you doing in this freezing country instead of sitting cozily in a city by a table with everything dished out and where there are women!” (Ingstad 1977, 9).[15] The arena of the far North and the Arctic emerges as a singular sanctuary in which the purely masculine can be nostalgically re-experienced. At the same time, this setting becomes synonymous with freedom from social conventions and restraints.

Ingstad’s book is male-centred not only because of the scarcity of (white) women in the wilderness, but because his narrative is a tribute to male comradeship with an emphasis on the ideals of ruggedness, endurance, and physical strength. Masculinity seems in Ingstad’s story to be called for by the natural environment of the wilderness itself. Living “face to face with the wilderness” means to Ingstad that a man “must rely entirely upon his dogs and his gun” (Ingstad 1992, 125). Whether one is lucky or unlucky, writes Ingstad, “why, great Heaven, it is the gamble that makes the life of a trapper such an interesting adventure. One must always take the bitter with the sweet” (Ingstad 1992, 222). The last sentence about taking the bitter with the sweet is too freely translated in the English edition; the original Norwegian version is far more pertinent here: “Men det skal menn til å møte den slags med et smil” (Ingstad 1957, 155) (“But it takes men to meet such stuff with a smile”).

In fact, the only feminine entity in *The Land of Feast and Famine*, aside from native women and female animals, is indeed the wilderness itself. As Ingstad describes his journey east of Great Slave Lake in his 1977 preface: “Here I had come to a wilderness just as I wanted it – practically virginal. There was something so enthralling about this untouched nature and the way of life here […]” (Ingstad 1977, 8).[16] In his narrative Ingstad returns constantly to the pleasures of enjoying a nature that is untouched. As Annette Kolodny has amply demonstrated in her ground-breaking work *The Lay of the Land*, beneath such gendered, "virgin land" references lurks the idea of male conquest by force.

The trappers themselves – their "spirit of swashbuckling independence" notwithstanding – also owe their soul to the company store. As Ingstad notes about the trapper, he must "know something about his business [of hunting], if he expects to have anything left for himself after he has paid his debt to 'Hudson Bay' [Company]" (Ingstad 1992, 80). The capitalist connotations was well as the dualism of masculinity vs femininity are particularly strong in Ingstad's description of his own hunting booty during his last wintering in a spruce hollow at the edge of the Barrens, near a little lake:

> Along one wall of the tent hang my white-fox pelts – delicate masses of long silken white hair. I let my eyes linger upon them for a time and feel like a veritable capitalist, though the majority of these furs are doomed to find their traditional way into the warehouse of "Hudson Bay" to balance my indebtedness.
>
> One fox-skin in particular stands out from the others; it is larger and more luxuriant, a perfect cataract of white. Where will it finally come to rest? In Paris, perhaps? I permit my fancy to roam. A gleaming limousine draws up in front of l'Opéra in Paris; a young lady steps out and glides up the steps to the door. About her bare throat and shoulders floats a billowy mist of white fox … mine! Lucky white fox …" (Ingstad 1992, 300)

His erotically tinted fantasy is more than understandable in his solitary existence in the middle of the wilderness, but the passage also pinpoints, quite directly, a compelling conjunction between the exploitation of the Canadian wildlife and the world of European high fashion. The sensuous vision of a southern *lady's* delicate and luxuriant femininity brings us literally and figuratively far away from the protagonist's *rugged* masculinity of the far North. At the same time, however, the passage speaks eloquently not only of contrast but *connection*. It is, ironically enough, precisely the capitalist civilization – the culture of modernity – that enables Ingstad to enact his traditional masculinity in the northern wilds in the first place.

The exaltation of individualism and man's mastery of nature

In the world of Ingstad's narrative, values of masculinity and individualism go hand in hand. As Ingstad declares in the third paragraph in the opening of the book about the men who congregate in Edmonton at the entrance to the northern Canadian wilderness: "Their varieties are many, but their type is one. All express the same spirit of swashbuckling independence, the same unshakable feeling of self-confidence, found in men accustomed to doing for themselves" (Ingstad 1992, 3). Or as he declares about his wilderness adventure in the 1977 preface: "A free and strong life, about as primitive as you can get it without tools of stone. With canoe and dogsled into the wilderness, gun and axe the important instruments, everything was to be wrenched from the land, everything hinged on your own effort" (Ingstad 1977, 9).[17] The idea that everything hinges on the individual's self-reliance, required for survival in the wilderness, seems to be intimately connected with the idea of conquest and mastery – with the attitude that one's living has to be *wrenched* from nature.

This mind-set and type of experiences could just as readily have been those of Fridtjof Nansen a generation earlier. Nonetheless, it is important to take into account the general transformation of society that occurred in the time lapse between Nansen's early Arctic adventures and those of Ingstad. In an address in 1926, Nansen himself put his finger on some of the cultural changes that in his opinion had begun to take hold at this time, and which served to undermine the "spirit of adventure" of the late nineteenth and early twentieth century. Nansen's address connects his own former achievements of Arctic exploration with values such as persistence ("the spirit of adventure will not give in […] till the work is done and done well"), "patience and industry", and, most importantly, self-reliance ("for 'self-trust is the first secret of success'" (Nansen 1927, 25, 26). In short, Nansen points to ideas that typified the culture of modernity of the previous century. In 1926, however, Nansen's warns that these values are becoming outmoded in a rapidly changing world, threatened by what he sees as a drift towards uniformity:

> Increasing urbanization, uniform education, the rapidly improving means of transport and communication tend to abolish distance, and to wipe out those characteristic differences between peoples, nations, and cultures which have made life interesting and beautiful, and acted as an important stimulus to new thought.
>
> There are several ideals in vogue nowadays, which, if realized, would lead us towards a dangerous monotony, a uniform greyness, in which it would be difficult to develop one's own personality. (Nansen 1957, 17)

Intimations of the kind of cultural change that worries Nansen here – the increasing uniformity and monotony of modern life – could in fact be said to have emerged already at the end of the nineteenth and the beginning of the twentieth century in many Western urban cultures. As T.J. Jackson

Lears, for instance, demonstrates in his study *No Place of Grace: Antimodernism and the Transformation of American Culture 1880–1920*, a great many people in this period gave voice to being afflicted by a blend of anxiety and ennui, experiencing "a spreading sense of moral impotence and spiritual sterility – a feeling that life had become not only overcivilized but also curiously unreal" (Lears 1981, 4–5).

We also find this sensation of unreality and alienation reflected in the literature and art of this period, in various movements such as neo-romanticism, decadence, and naturalism – and dramatized for instance in the Norwegian artist Edvard Munch's painting *The Scream* from 1893. A sense of the strangeness and fragmentation of everyday life seemed to gain further momentum in American and Western European culture in the 1910s and 1920s, culminating in the modernist movement in literature (with writers such as T.S. Eliot, Virginia Woof, and James Joyce) and in various expressionist, cubist and surrealist movements in art (with painters such as Pablo Picasso, John Marin, and Marcel Duchamp). The sense of disorientation and unreality expressed in modernist art has been linked to the rapid social change instigated by urbanization and the creation of the modern mass society, which was accompanied by the erosion of traditional values, the loss of a sense of individual autonomy, and a growing sensation of anonymity – precisely the cultural development that Nansen bemoans in his speech.

In Ingstad's 1977 preface we are told that he as a young man had established a thriving business as lawyer in the Norwegian city of Levanger whose expansion made him "afraid of becoming so prosperous and bourgeois as to abandon what I really wanted. Then I cut and ran; in 1926 I sold my business and travelled to Canada. I hankered north" (Ingstad 1977, 7).[18] There is considerable historical irony in the fact that Ingstad felt he could escape the culture of upper middle-class comfort and conformity in Norway by journeying into the wilderness of a foreign land where he then could *reinvent* himself (for a limited time) as a heroic individual by *re-enacting* the "hard" and rugged ideology of modernity in a natural environment so pristine as to appear premodern.

The untenability of the dualisms of modernity

Ingstad's journey back into premodernity may also be seen as a declaration of anti-modernism, which in Elizabeth C. Child's definition involves "the ultimate individualist quest to recuperate the self in a rejection of a modern urban world perceived as overly industrialized, bureaucratic, and rational" (Childs 2001, 50). From such a perspective, Ingstad's narrative of hunting and trapping in the

Canadian wilderness may be interpreted as an expression of cultural backlash – a pastoral, nostalgic yearning for the way of life of a previous era (cf. Wærp's chapter in this volume). His tale of adventure is, of course, that *as well.*

In my view, however, it is more fruitful to regard *The Land of Feast and Fortune* as an embodiment of heterogeneous and partly disparate beliefs that existed *contemporaneously*, side by side, in the culture at large at Ingstad's time. Ingstad's narrative may be said to reflect the complexity of the Western culture of the first half the twentieth century in which he lived – with its web of interconnected and conflicting patterns of beliefs and values, some of them inherited from the past and some of them engendered by present-day developments, all of them in flux, entangled with and predicated upon each other.

The mishmash of the values of *premodernity* + classic *modernity* + *antimodernity* that *The Land of Feast and Famine* gives expression to, could therefore be said to confirm Bruno Latour's dictum in *We have Never Been Modern* (1993). At least not in the classic, unadulterated sense of the term we have obviously never been modern. According to Latour, modernity overlooked, when giving birth to humanism, "the simultaneous birth of 'nonhumanity' – things, or objects, or beasts" (Latour 1993, 13). What Latour calls "the constitution of modernity" erected instead an absolute divide between nature and society, separating the laws of nature from social convention and abolishing any conception of their intricate interconnectedness. In a similar manner, the predominant discourse of Ingstad's book operates with modernity's radical dualisms of nature/culture and wilderness/civilization, but what the narrative itself may be argued to end up with is, to use Latour's words, "the exploration and proliferation of hybrids" (144), a potpourri of pre- and pro- and anti-modernisms.

The discursive heterogeneity that I have thus dredged up from *The Land of Feast and Famine* is, however, a result of my reading of Ingstad's narrative against its own grain, as it were. The great paradox of the text is that, in the course of his anti-modernist act of entering the premodern and aboriginal wilderness, Ingstad nonetheless ends up pursuing the core spirit of a pioneering modernity. On closer thought, this may perhaps be said to be inevitable. There was certainly no way the fur-hunter Ingstad could *become* premodern, no way he could escape his own socio-cultural identity. This is not, however, an issue that is problematized by the narrator himself. Instead Ingstad makes his story exist in a cultural vacuum. The silence with which *The Land of Feast and Famine* attempts to situate itself outside of time is overwhelming. Ingstad's wilderness adventures took place at the very end of the 1920s, yet there is no mention of the dire socio-economic crisis that radically affected Europe and

America at this time, at the very threshold of the Great Depression – not even as a contrast or backdrop to his own wilderness chronicle. History and society are, as it were, refused admission into this tale of a seemingly premodern and pristine natural environment.

Again, this may be linked to the demands characteristic of the wilderness-adventure genre itself. In his famous essay "The Adventure" from 1919, the German sociologist Georg Simmel argues that the adventure as such is predicated on the idea of doing something extraordinary, something that is not part of the familiar continuity of everyday existence. He sees it as a type of experience or story that is "independent of the 'before' and 'after'" of conventional social life; it constitutes instead "an island in life which determines its beginning and end according to its own formative powers [...]". According to Simmel, this makes the adventurer "the extreme example of the ahistorical individual, of the man who lives in the present. On the one hand, he is not determined by any past [...] nor, on the other hand, does the future exist for him" (Simmel [1919] 1997, 223). These features are in my opinion astonishingly apt as characterizations of Ingstad's wilderness narrative and those of his Norwegian heirs. Their northern adventures tend to be insulated – physically, socially, and temporally – from ordinary life; their stories are sealed off at both ends, disconnected both from the historical world they left and from the self-same world to which they finally return.

The great paradox of Ingstad's aboriginal wilderness adventure

At the end of *The Land of Feast and Famine*, however, Ingstad's description of his return to civilization at least acknowledges the general economic-industrial forces that will turn his adventure into an insulated interlude of the past. In the conclusion to his story he draws a sharp line of distinction between the life he led during his three-four years in the Northwest Territories, symbolized by the steamer that took him upriver at the beginning of his adventure, and the life of modernity that will eventually conquer the North, symbolized by the plane that flies him out of Fort Resolution at the end his narrative. As in the poems discussed by Renée Hulan in her chapter, this latter scene of the plane heralds to Ingstad a future when civilization will prevail with its "factories, smoke-darkened cities, milling humanity" (Ingstad 1992, 331).

Ingstad's assertion that 1930 represents a radical boundary in the history of the Canadian wilderness is in line with his dualistic discourse throughout *The Land of Feast and Famine* of nature vs culture, pristine vs spoilt, aboriginal vs civilized. As my chapter has tried to show, however, the tensions and contradictions in Ingstad's narrative disclose that such radical dualisms are exaggerated and ultimately untenable. In reality, of course, the historical transition from the pre-modern to the modern Far North had been going on for a protracted period of time. The capitalist exploitation of the North American wilds had been initiated not least by the trappers and hunters themselves long before Ingstad's own trip. A perspective excluded from Ingstad's hegemonic discourse of the wilderness (but which may be weaned from the plot of his narrative) is that the fur trade was a commercial business among several profitable enterprises that for centuries had been bent on extracting riches from the "wild" lands of Canada, or more precisely, from the homelands of its indigenous peoples. As Adriana Craciun notes in a newspaper article on the recent location of the wreck of one of the Franklin expedition ships, "earlier Arctic exploration had been entangled with imperial concerns and commodity extraction, including furs, coal, whale oil and gold" (Craciun 2014).

In the references to his role as a hunter selling his bounty to the Hudson Bay Company, Ingstad's narrative – inadvertently, as it were – shows that he has not escaped from civilization at all; he serves instead as its willing instrument. And it is here that Ingstad's homage to a pristine nature assumes an extremely paradoxical and deeply ironic tinge. The primitive life of fur-hunting that he celebrates is not disconnected from the colonialist, capitalist, and anthropocentric world whose ethos accompanied the exploitation of northern Canada; quite the opposite. Ingstad's eulogy to the wilderness (less untamed than his dominant discourse constructs it to be) is first and foremost the product of his antimodernist yearning for a premodern arena on which he can enact and affirm, however temporarily, the discourse of an individualistic, masculinist, and pioneering modernity.

References

Atwood, Margaret. 1972. *Survival: A Thematic Guide to Canadian Literature*. Toronto: Anansi Press.

Buell, Lawrence. 2005. *The Future of Environmental Criticism: Environmental Crisis and Literary Imagination.* Malden, MA: Blackwell.

Childs, Elizabeth C. 2001. "The Colonial Lens: Gaugin, Primitivism, and Photography in the Fin de siècle." In Lynda Jessup, ed. *Antimodernism and Artistic Experience: Policing the Boundaries of Modernity.* Toronto: University of Toronto Press.

Craciun, Adriana. 2014. "Franklin's Errors should inform Canada's map for Northwest Passage." Winnipeg Free Press, Sep. 12, 2014. http://www.winnipegfreepress.com/opinion/analysis/franklins-errors-should-inform-canadas-map-for-northwest-passage-274860051.html?cx_navSource=d-tiles-1. Accessed Sep. 12, 2014.

Hamilton, Garry. 2008. *Arctic Fox: Life at the Top of the World*. Richmond Hill, Ontario: Firefly Books.

Hardin, Garrett. 1968. "The Tragedy of the Commons." *Science* 162, no. 3859: 1243–48.

"Helge Ingstads Vinterkvarter" [Winter Quarters] 1929–30". www.helgeingstad.net. Accessed Oct. 19, 2016.

"I Helge Ingstad's fotspor gjennom Canada" [In Helge Ingstad's footsteps through Canada]. NRK1 [Norwegian Broadcasting Corporation, TV1], Jan. 2, 1999, 18: 55; http://arkiv.nrk.no/programoversikt/avansert/index83f8.html. Accessed Oct. 19, 2016.

Ingstad, Helge. 1937. *East of the Great Glacier*. New York: Alfred A. Knopf.

—. 1977. "Forord" [Preface]. In *Pelsjegerliv blant Nord-Canadas indianere* [A fur-hunter's life among North Canada's Indians], 7–11. Oslo: Den Norske Bokklubben.

—. (1933) 1992. *The Land of Feast and Famine*. Trans. Eugene Gay-Tifft. Montreal: McGill-Queen's University Press, 1992.

—. 1954. *Nunamiut: Among Alaska's Inland Eskimos*. London: Allen & Unwin.

—. 1957. *Pelsjegerliv blant Nord-Canadas indianere* [A Fur-hunter's life among North Canada's Indians]. Oslo: Gyldendal.

Kolodny, Annette. 1975. *The Lay of the Land: Metaphor as Experience and History in American Life and Letters*. Chapel Hill, NC: University of North Carolina Press.

Latour, Bruno. 1993. *We Have Never Been Modern*. (*Nous n'avons jamais été modernes*, 1991.) Translated by Catherine Porter. Hemel Hempstead: Harvester Wheatsheaf.

Lears, T. J. Jackson. 1981. *No Place of Grace: Antimodernism and the Transformation of American Culture 1880s–1920*. New York: Pantheon

Lyon, Thomas J. 1989. "A Taxonomy of Nature Writing." In Thomas J. Lyon, ed. *This Incomperable Lande: A Book of American Nature Writing*. New York: Penguin Books, 1991.

Monsen, Lars. (1996) 2009. *Til fots gjennom Alaska* [On foot through Alaska]. Oslo: Larsforlaget.

—. 2002. *Nådeløs villmark. Canada på tvers* [Ruthless wilderness: Across Canada]. Oslo: Lars Monsen Outdoors.

Nansen, Fridtjof. 1892. *The First Crossing of Greenland.* London: Longmans, Green.

—. 1897. *Farthest North*. London: Macmillan.

—. 1927. "Adventure." In *Adventure and Other Papers,* 7–42. London: Hogarth Press.

Newman, Peter C. 1989. *Empire of the Bay: An Illustrated History of the Hudson Bay Company*. Toronto: Madison Press Books.

"Sandy Lake Expedition 2013!" htttp://www.ingstadcreek.net/. Accessed Feb. 21, 2014.

Simmel, Georg. 1919. "The Adventure." Translated by David Kettler. In *Simmel on Culture: Selected Writings*, edited by David Frisby and Mike Featherstone, 221–232. London: Sage Publications, 1997.

Skarstein, Frode. 2010. *Helge Ingstad. En biografi* [Helge Ingstad: A biography]. Oslo: Spartacus Forlag.

Wagner, Peter. 2012. *Modernity: Understanding the Present*. Cambridge: Polity Press.

Wynn, Greame. 2007. *Canada and Arctic North America: An Environmental History*. Santa Barbara, CA: ABC-CLIO Inc.

Zapffe, Peter Wessel. 1977. "Helge Ingstad 75 år" [Helge Ingstad 75 Years]. In *Spøk og alvor. Epistler og leilighetsvers* [In jest and in earnest: Epistles and occasional verse], 86-90. Oslo: Gyldendal.

Notes

---

[1] *Wilderness literature* may be described as writing that has pristine and uncultivated nature as its central subject matter and/or setting. Some brands of wilderness literature may therefore be included as a sub-category within the broad and heterogeneous genre of *nature writing*. Some scholars have defined nature writing so eclectically as to include all publications that have nature as a subject, including for instance field guides, natural history articles, travel books, adventure literature, essays on farm life, and general writings on man's role in nature (cf. Thomas J. Lyon's classic piece "A Taxonomy of Nature Writing", 1991). Quite often, however, nature writing is defined more specifically as literature that portrays the environment of a particular place or area by combining detailed descriptions of the diversity of its natural life with personal, philosophical, and often autobiographical reflection. The archetypical work of such a genre would be Henry David Thoreau's *Walden* (1854). By focusing on the endeavors of the protagonist rather than the diversity of natural life, however, many Norwegian wilderness stories can hardly be subsumed under nature writing thus defined. Nature writing may be furthermore distinguished from *environmental literature*, a genre of academic writing encompassing studies of ecology and biology and work

within what may be called "the environmental humanities" (Buell 2005, vi), the latter including the field within literary studies called *ecocriticism*, which is to say the analysis of literature from an environmental perspective (the academic field to which this chapter belongs).

[2] "Du kom i heltetidens siste time,
da glansen fra en fryktløs énmands-færd
falmet som solnedgangens siste strime
og intet mere syntes møien værd." All translations from Norwegian are mine.

[3] "Den [*Pelsjegerliv*] har i de snart 80 årene som har gått siden utgivelsen opparbeidet seg kult status. Boken omtales som en bibel ikke bare av hundekjørere og seriøse villmarkslivsentusiaster, men også av den jevne friluftsinteresserte nordmann."

[4] The expedition members were Nils Lund, Stein P. Aasheim, Jan Fasting, Erik Sandberg Ingstad (Helge Ingstad's grandson), Thorbjørn Storskjær and Vegard Ulvang.

[5] The expedition members were Roy Kvamme, Christian Sverdrup Engelschiøn, Peter Sverdrup Engelschiøn, Harald Sollund, Erling Sagneskar, Åge Svinøy, Thomas Pedersen, and Jan M. Akre. See also the more recent entry, "Helge Ingstads Vinterkvarter [Winter Quarters] 1929-30," http://www.helgeingstad.net.

[6] "Skildringene var jo reineguttedrømmen!" (Hovind, 8).

[7] "Han har gjort denne klassiske, canadiske villmarksscenen udødelig i mange nordmenns bevissthet. Nå står jeg her, midt i det selvsamme eventyret, hører de samme lydene, ser de samme dyrene og suger inn den samme friske vinterlufta!" (*Nådeløs villmark*, 123).

[8] "Det er bare å takke dypt og inderlig for all inspirasjon han ga, for ikke å snakke om den forskningsmessige innsatsen. Så er det opp til oss som er igjen å føre de norske villmarkstradisjonene videre. Den 101 år gamle villmarkskongen og forskerkjempen Helge Ingstad [...] satte kursen mot de evige jaktmarker den 28. mars 2001, der han nå temmelig sikkert i kjent stil jakter caribou og fyller kanoen med fisk" (*Nådeløs villmark*, 163). My note: Encyclopedias and Internet entries that I have checked, however, list the day of Ingstad's death as March 29; cf. Skarstein's biography of Ingstad: "Da han døde den 29. mars 2001[...] var det forsidestoff i alle landets aviser" (295) ("When he died March 29, 2001 [...] it was front page news in all the Norwegian newspapers").

[9] "[...] her er et av de siste blad i Canadas fengslende pionér-historie om pelsjegere som tok seg frem i jomfruelig villmark og viste vei for den såkalte sivilisasjonen."

[10] "Dyrelivet var som naturen hadde villet det."

[11] There is even a reference to Ernest Thomson Seton in *The Land of Feast and Famine*, who in an expedition in 1907 estimated the number of caribou "at thirty million" (Ingstad 1992, 160).

[12] "Indianerne ['Villrein-eterne'] levet i det vesentlige som sine forfedre, bortsett fra at de nå hadde gevær" [...]"

[13] Ingstad refrains from considering the serious consequences of making the First Nations people dependent on the modern economic relations in which he himself also takes part. The catastrophic result is perhaps best seen a couple of decades

after Ingstad's own adventure, when the trapping of arctic fox was no longer profitable. It became clear that the real losers in this venture were above all many Inuit in the Arctic, and to some extent also the peoples of the First Nations a little further south: They had grasped the opportunity of going from general subsistence hunting to a specialized fur trade with white fox that was, for a time, extremely profitable but unfortunately also a very precarious and unstable business, making the native hunters wholly dependent on the fluctuating prices of the world market. Arctic fox prices crashed both during the Great Depression and after World War II; when fur farming of animals in small cages rose rapidly in the 1950s and 1960s, a great many Inuit were ruined and impoverished, at which time the Canadian government began to encourage them "to move to permanent settlements and adopt western-style lifestyles based on salaried employment" (Hamilton, 186, 187, 188-89).

[14] "Han hadde ry over Nordvest Territoriet som en av de ypperste langveisfarende med hundeslede og kano og som en hardhaus av de sjeldne. [...] I hans sterke blå øyne så jeg aldri frykt, og han hadde en innbitt vilje til aldri å gi opp."

[15] "forventningen når ukjent land steg frem og ikke minst spenningen under jakten. Det kunne jo hende når alt slo seg vrangt at jeg sa til meg selv: Hva i allverden skal du i dette iskalde landet istedenfor å sitte varmt i en by ved dekket bord og hvor det er kvinner!"

[16] "Her var jeg kommet til en villmark som jeg ville ha den – så å si jomfruelig. Det var noe så besettende ved denne urørte natur og tilværelsen her [...]"

[17] "Et fritt og sterkt liv, omtrent så primitivt som en kan få det uten steinredskaper. Med kano og hundeslede inn gjennom villmarken, gevær og øks de viktige hjelpemidler, alt skulle vristes av landet, alt stod og falt med egen innsats."

[18] "Den [advokatforretningen] gikk aldeles glimrende og jeg ble redd – det var fare for at jeg skulle bli borgerlig velhavende og svikte det jeg egentlig ville. Så brøt jeg overtvert, i 1926 solgte jeg forretningen og drog til Canada. Jeg ville mot nord."

"THE POETRY OF THE AEROPLANE": ARCTIC FLIGHT IN TWENTIETH-CENTURY CANADIAN POETRY

RENÉE HULAN

In the 1980s, the Innu people of Nitassinan, also known as Labrador, organized to protest low-level flights by NATO fighter jets over the Innu homeland. Their resistance exposed the powerful intersection of military, industrial, and political interests informing Canada's engagement with northern regions and highlighted how easily those interests could pit national sovereignty and economic development against the need to protect renewable resources and sustainable ways of life.[1] Aviation has played a central role in generating ideas and images of the Arctic by creating an array of aerial images from patrolling fighter jets to vast, pristine Arctic landscapes. With climate change, the dominant image of the Arctic as a land of permanent ice and snow is being replaced by new images of a melting, endangered Arctic. As it melts, the Arctic is increasingly seen as a victim of the excesses of wasteful, industrial modern life. However, many works of literary and visual culture reveal a different picture of the Arctic, not as an empty stage where the consequences of Modernity are played out, but as an actor in the unfolding story of Modernity.

In this chapter, I argue that the apparently rival discourses of economic development and environmental protection flow from common sources in the twentieth century. The history of Canada's expansion north illustrates what Zygmunt Bauman describes as the era of "heavy modernity" when exploration of the "blank spaces" propelled empires to "spread to fill every nook and cranny of the globe" (Bauman 2000, 114). Bauman's allusion to Conrad's Marlow, who yearns to explore all the parts of the map that appear empty, is deliberate and evocative: modern empires were built by finding, mapping, and acquiring territory when "[w]ealth and power was firmly rooted or deposited deep inside the land – bulky, ponderous and immovable like the beds of iron ore and deposits of coal" (Bauman 2000, 114), or, as in Canada, the fields of minerals, oil, and gas that continue to draw capital north. The process that Bauman analyses in *Liquid Modernity*, the movement from heavy forms of technology tethered to the ground to technology that moves at the speed of light, did not occur overnight, or even with the rapid development of electronics; rather, it unfolded simultaneously with the industrialization that defines the modern age. The technology of the airplane makes this explicit: with air travel, the speed with which the globe could be explored accelerated, collapsing time and space as it did. According to Bauman, a lighter and increasingly "liquid" form of Modernity has replaced the noise and dirt of machinery and hardware with the movement of information by electronic signal and, with it, the mastery of time. With time no longer "a factor of the event" or part of "the calculation of value" (Bauman 2000, 117), the formerly "blank spaces" like the Arctic are no longer remote, and distance becomes no obstacle to human mastery. This modern future was anticipated in the early history of aviation when the irresistible chance to transcend earth-bound reality inspired the poets who first experienced it.

The poetry of flight celebrates the thrill of danger, the immensity of space, and the experience of transcendence over physical and spiritual limitations. "Oh, I have slipped the surly bonds of earth", says the speaker in John Gillespie Magee's "High Flight", whose flight climbs ever higher until its stirring last lines:

> And, while with silent, lifting mind I've trod
> The high untrespassed sanctity of space,
> Put out my hand, and touched the face of God.
> (quoted in Vance 2002, 268–9)

The popularity of "High Flight" which serves as the title of Jonathan Vance's cultural history of aviation in Canada, owes something to the circumstances of Magee's death after ejecting from his plane over Lincolnshire. Born in Shanghai to American missionary parents, Magee was only nineteen years old and a cadet in the Royal Canadian Air Force (RCAF) for less than a year before he died. But the poem, which captured the feelings of freedom with its "footless halls of air" and wonder at "the long delirious, burning blue", was a Canadian standard and the official poem of the RCAF (Vance 2002, 269) long before Ronald Reagan used it in response to the Challenger disaster in 1986. There is perhaps no better example of the symbolic transcendence achieved by the poetic equation of flight and divinity, or of the illusion of omniscience achieved in flight. Yet, every poetic flight is made possible by the right material conditions. The transcendent position of the airborne can only be achieved by faith in the mechanical means to transcendence – the friction and lift, nuts and bolts, steel and glass, ignition and combustion. As the viewer transcends the mundane, material world over distances that could once only be imagined, the imagination resolves or represses this paradox as the poetic eye surveys the view.

For many early travellers to the Arctic, the exhilarating experience of flying was augmented by the excitement of encountering the Arctic for the first time. Arctic aviation gradually transformed the way the North was imagined in the twentieth century with Arctic landscapes viewed from the air. In the nineteenth century, Arctic exploration and exploitation had advanced both the technological and conceptual innovations of Bauman's "solid" or "heavy" Modernity, beginning with the rudimentary cameras that travelled with the British explorer Sir John Franklin and the experimental panoramas and early moving pictures circulating through the century. As Russell Potter shows in *Arctic Spectacles*, the representation of Arctic landscapes developed along with new media in the nineteenth century, and "it was principally through the technologies of *vision* that the Arctic was most keenly and energetically sought" (Potter 2007, 4).[2] Indeed, "vision", in both its literal and figurative senses, is one of the most frequently used terms in the discussion of the Canadian North.

Arctic aviation had an immediate impact on Canadian culture as access to government-sponsored travel by airplane increased and not only in the spheres of science and technology: "The changes created by the interaction between the aircraft and the multilayered conditions of the Mackenzie region rippled out from the North, shaping Canada's economic, political, and cultural development in the 1920s, 1930s, and beyond" (Cronin 2006, 41). One of these ripples coursed through Canadian poetry. As Janine Rogers points out, in 1922, the poet Duncan Campbell Scott observed that a "poetry of the aeroplane" had yet to be written (Rogers 2014, 6). In his presidential address of that year to the Royal Society of Canada, D. C. Scott had praised the group for giving equal space to science and literature, two intellectual enterprises that he subsequently differentiated by identifying one as concerned with the material world and the other with the imagination (Scott 1922, 299–301). The speech, titled "Progress and Poetry", put forth Scott's idea that poetry gives expression to the new and the unknown through imagination, though in a "groping way" (Scott 1922, 316). Though Scott found the poetry of his time "wayward and discomforting" in its devotion to experimentation, and looked back to the classics for his models, he also looked forward to a "poetry of the aeroplane" that would eventually "pass beyond the expressions of bird-flight in the older poets and will awaken images foreign to their states of feeling": "The aeroplane has a beauty and daring all its own, and the future poet may associate it with some transcendent flower to heighten its world-taking beauty" (Scott 1922, 316). Progress was an important value to Scott whose career with Indian Affairs was characterized by support for assimiliationist policies and for his inaction when the appalling conditions in residential schools were brought to his attention. Progress would relegate the Native to the past as Scott and others looked to the future.

It is not clear whether the Canadian legal scholar and poet, F. R. Scott consciously "took up Duncan Campbell Scott's indirect challenge for a 'poetry of the aeroplane'" with his poem "Trans Canada" (Rogers 2014, 13), yet produce a poetry of the aeroplane he did. Appearing in *F.R. Scott: Selected Poems* in 1966, the poem "Trans Canada", which was also the name of one of Canada's first commercial airlines, describes the novel experience of flying by contrasting air and land and calling on natural images to describe this new form of travel:

> Pulled from our ruts by the made-to-order gale
> We sprang upward into a wider prairie
> And dropped Regina below like a pile of bones.
>
> Sky tumbled upon us in waterfalls,
> But we were smarter than a Skeena salmon
> And shot our silver body over the lip of air
> To rest in a pool of space. (Scott 1966, 34)

The air is a new system of roads "that are not seen or laid", a new frontier that predicts "Only the start of space, the road to suns" (Scott 1966, 35). Although Jonathan Vance cites this poem as evidence of how the "growing sophistication of aircraft engendered admiration of the human ingenuity" (Vance 2002, 277), "Trans Canada" alternates between fascination with the new perspective on air, clouds, and space that being airborne offers, and scepticism towards human conquest of nature and a future in the embrace of the "everlasting arms of science" (Scott 1966, 35). For Scott, flying did not create community, instead "every country below is an I land", and there is a strong sense of human hubris in the image of "Man, the lofty worm" tunneling through space (Scott 1966, 35). As chair of the Committee for Medical Aid to Spain, and working closely with Norman Bethune, Scott helped organize the visit of a delegation of Spanish republicans eager to raise money to support their government (Djwa 1987, 171). According to his biographer Sandra Djwa, "the failure of the Western powers to intervene on behalf of Spain, a democratic nation, infuriated and sickened Scott" (Djwa 1987, 176). In "Spain 1937",

Scott responded to the growing disillusionment with the Loyalist cause, which sees democratic "vision thwarted", diplomatic neutrality masking "the will to kill", and Guernica appears as a synecdoche for human suffering that "is more real than our daily bread" (Scott 1945, 11) That the bombs rained down from airplanes does not seem to concern Scott as much as holding on to the ideals of the republicans in this poem which proclaims: "This issue is not ended in defeat" (Scott 1945, 11).

Public enthusiasm for the airplane was dampened by the massacre at Guernica when the airplane "so recently seen as the pinnacle of human achievement, was evolving into a sinister killing machine capable of spreading destruction" (Vance 2002, 221), yet Canadians remained interested in the potential of aviation. If the airplane could be the means of mass destruction, it could also be a useful tool to understanding. In the summer of 1956, F. R. Scott wrote to Maurice Lamontagne, "then economic adviser to the Privy Council in St. Laurent's government", and was put in touch with "officials of Eldorado Mining and Refining Company, who agreed to transport him down the Mackenzie to Norman Wells during the last trip of the season before freeze-up, about mid-August" (Djwa 1987, 318). Scott was preparing to teach Canada-American relations as a visiting professor at Michigan State University, and he decided he needed first-hand experience of the North before he left. Perhaps best known for the impact the trip had on Scott's travelling companion, the young Pierre Elliott Trudeau – even though, as Djwa notes, Scott was reluctant to let Trudeau tag along – the trip had a profound impact on the two men who would eventually form a lifelong friendship that had a lasting influence on the ideology of the Liberal party and the policies of the governments Trudeau would lead. Although Scott wanted to experience the North for himself prior to writing about it, many of the poems inspired by the trip are written in the abstract, impersonal style associated with the cosmopolitan Modernism Scott and his fellow Canadian poets admired. In Europe, the avant-garde enthusiastically embraced the airplane which seemed "proof that the old narrow world of the nineteenth century was doomed to extinction and that another world of iron and steel was about to take its place" (Wohl 1994, 268). Like their counterparts in Europe, the Canadian Modernists wished to shake off the vestiges of the poetry that continued "to paint the native maple", which they considered parochial, outmoded, and stodgily Victorian, as the speaker laments in the satirical poem "The Canadian Authors Meet" (Scott 1966, 70). In its place, a Modernist and cosmopolitan Canadian poetry would create the world anew, and the poet, as the frustrated speaker in A.M. Klein's "Portrait of the Artist as Landscape" wishes, would be the "nth Adam taking a green inventory / in a world but scarcely uttered" (quoted in Bentley 1997, 254).[3] Through Scott's poetic renderings of his northern travels, Canadian Modernism itself, it can be said, was shaped by the technological changes that made Arctic exploration more accessible.

With the development of aviation, the Arctic grew paradoxically closer and more remote: a landscape that could be viewed from far above or a remote and vast airspace to be defended with words and weapons. As Marionne Cronin explains, the heroic narratives of polar exploration had created "the illusion that the Arctic existed as a space outside the modern world"; now the airplane introduced "a symbol of the modern technological world in the heart of the wilderness" (Cronin 2013, 61). By lifting the explorer above the harsh and dangerous environment, flight offered an advantage that some saw as the end of the heroic age of the Arctic, particularly polar, exploration pursued so vigorously by British, American, and Scandinavian explorers (Cronin 2010, 100). Some explorers, Roald Amundsen among them, signalled new dangers imposed by this mode of travel, especially in the extreme Arctic conditions while others, Matthew Henson included, regarded exploration using aircraft as less risky than travel overland (Cronin 2010, 114). During Amundsen's 1925 attempted flight to the North Pole, engine failure forced the crew to abandon one of the planes. As Cronin documents, the "newspaper coverage of the expedition emphasized the magnitude of the hazards faced, the risks encountered, and the suffering endured", and "Amundsen [was] described as a Viking of the air: brave, indomitable, cheerful in the face of danger, and able to make light of his suffering", thus inscribing the modern technology within the heroic narrative mode as Amundsen became "the embodiment of the modern Norwegian Viking character and spirit" (Cronin 2010, 111–112). While the heroic exploits of Arctic aviation were celebrated as national achievements in many countries and appropriated to support a collective identity in the Soviet Union (see McCannon 2010), the intersections of national, commercial, and artistic interests that provided the rationale for northern expansion were rarely made as explicit in Canada.

In contrast, Canadian exploration involved aerial surveying and mapping driven by "an overarching concern with practical utility and the collection of useful information" (Cronin 2007, 329). The growing reliance on aviation fundamentally altered northern science, transforming field methods and social networks with them (Bocking 2009, 290; Zaslow 1975, 344). As Stephen Bocking shows, the "objective perspective of the disembodied, disinterested observer high above the earth" that airborne geographers and other scientists experienced became a viewpoint shared by prominent figures moving through "university, government, corporations" as decision makers throughout Canadian society eyed the region's economic and military value (Bocking 2009, 290). Like adventurers and explorers, individual scientists no longer needed to travel through the land, overcoming the challenges posed by climate and terrain, nor did they rely as much on local indigenous people and knowledge, except when lost or broken down. In addition to changing the relationship between scientists and ndigenous people, aviation changed both the means and method of conducting research. At the same time, the increased availability of flights opened the Arctic to more outsiders and eventually to settlement. From the air, the presence of indigenous people could be rendered invisible, and as Bocking shows, the shift of scientific attention from renewable to non-renewable resources would further marginalize indigenous people. Through the age of heavy Modernity, in which machinery and new technologies of transportation made possible the acquisition of wilderness, aviation seemed to bring proof of the triumph of progress over tradition, feeding attitudes of superiority and perpetuating colonialism. By 1970, the myth of the North as a "cornucopia of riches" was firmly attached to policy decisions. "In Ottawa, the priority was now on pursuing a vision of the North as a hinterland supplying mineral wealth to the southern economy. In contrast, renewable resources such as caribou were increasingly seen as mere vestiges of a soon to be displaced native economy" (Bocking 2007, 878). The government would take an active role in this displacement by relocating northern populations and by removing indigenous children to residential schools often aided by air travel. Designed "to kill the Indian in the child", the long term consequences of these schools are only beginning to be acknowledged today, and the legacy of Canada's colonial administration of the North continues to be felt profoundly in indigenous communities.

In the early twentieth century, aviation further transformed the way the Arctic was imagined as aerial photography became more common. The view of Arctic landscapes, typically taken from patrolling ships or from a high point of land, changed with the advent of aviation. The Arctic landscape could now be represented, not as distance to be traversed with hidden obstacles to surprise and challenge the traveller, but as a surface to glide over. Avant-garde painters and sculptors in Europe experimenting with spatial representation embraced this "new way of seeing the world and the objects in it" and "abandoned traditional laws of perspective" (Wohl 1994, 272). As a Modernist aesthetic was taking shape through the 1920s, 30s, and 40s, new techniques in photography and the new medium of film were enlisted in the Canadian government's administration of northern regions, as Peter Geller explains: "The all-seeing eye of the motion picture camera, floating through Arctic waters as it captures images of people and place, is a vivid metaphor for the perceived extension of government control in the Canadian north" (Geller 2004, 49).[4]

As D. M. R. Bentley observes, Canadian Modernists sought "to gain distance from 'fallen' society", a "distance that cures enchantment with the merely local and permits the perception of simple patterns and universal archetypes" (Bentley 1997, 255). To achieve this, Bentley writes, "the high Modernists enacted in their work various strategies of removal in the horizontal and vertical planes" (Bentley 1997, 255). Flight was the perfect device for the Modernist sensibility seeking to detach itself from the particulars of place. For Bentley, Scott's poetic technique describing the flight north satisfied the Modernist desire to be cosmopolitan and therefore part of the Modern age. For, although Scott was a socialist who spurned bourgeois values, a leader of the CCF party and tireless defender of civil liberties in his home Quebec, his poetry was written in "a cosmopolitan mode that co-operated, not least in its progressivist emphasis on new techniques, new ideas, and new names, with the expansion of Anglo-American culture on an international and multi-national scale" (Bentley 1997, 277). Various critics, including Bentley, Wanda Campbell, and D. G. Jones, have pointed to the tensions and ambiguities in Scott's thought, and in these poems, there seems to be a very deep ambivalence.

Canadian poets sharpened their Modernist aesthetic on northern landscapes and enlisted the Arctic in the project of Canadianizing high Modernism, polishing the *vers libre* on remote northern landscapes viewed from above or afar. In "Laurentian Shield" first published in 1954 and later included in *F. R. Scott: Selected Poems*, Canada is "[i]narticulate, arctic" waiting to be voiced, its mute, unnamed remoteness serving both aesthetic and moral objectives (Scott 1954, 16; Scott 1966, 38). It is therefore difficult to agree with D. G. Jones that though the land "appears mute, empty as paper, the land has its own identity" and is "not merely raw material to be exploited, though Scott's diction allows for that possibility" (Jones 1983, 47), except for this acknowledgement of the ambiguity in Scott's work. The poem "North Stream" also renders the northern environment mute when the speaker takes on the river's voice, saying "[i]ce mothers me", […] "My bed is rock / Over sand I move silently" (Scott 1966, 18).

For Modernist poets and artists throughout America and Europe, the airplane "represented clearness of function and the daring required to break with the past" (Wohl 2005, 294), as they were "moved not only by the appearance of the airplane, but also by the principles it represents – a mastery of mass and weight, a victory over the force of gravity through motion" (Goodyear 2003, 225). The fame of World War I flying aces and aviator pioneers like Charles Lindbergh in popular culture forged a link between the airplane as a fascinating modern technological innovation and a cultural phenomenon. The nationalism that tinged the representation of aviation throughout Europe drew on the futurism of the Modernist art forms which were sometimes appropriated in political propaganda. The relationship between the Futurists and the emerging fascist parties, and the "connections between the aerial perspective and [their] aesthetic sensibilities are well established": Italian air painters Marinetti and Ambrosi used bold lines and vertical angles to create "the aesthetics of speed and violence associated with the aeroplane" that conveyed the futurist vision of the modern world (Adey, Whitehead, and Williams 2013, 9). "Mussolini understood, with his journalist's flair for the marketable image, that airplanes could be used symbolically and politically to enhance his popularity and to invest him and his newly formed Fascist movement with the cachet of modernity and glamour that aviation and aviators had acquired during the war" (Wohl 2005, 60). Poets were also drawn to the bold simplicity of this aesthetic although futurism never took off as a movement in Canada where nationalism continued, and continues, in the romantic mode.

Within this Modernist literature, F. R. Scott's northern poems could easily find themselves assimilated to a rhetoric of technological and industrial progress animating the policy of northern development contemporary with his voyage to the Arctic, and his poetic flights greatly influenced how Canadians imagined it. According to Bentley, "the simplification, devaluation, and decontextualization of reality that is widely evident in high Modern Canadian poetry participates in the industrial and capitalistic enterprise whose most obvious effects in the present century have included the degradation and homogenization in Canada and elsewhere of distinctive natural and social environments" (Bentley 1997, 264–5). In other words, the poetic expression of the Modernists was part of a nascent and liquefying Modernity in Canada.

The poems "Flying to Fort Smith" and "Mackenzie River", which also feature a speaker seeing the North from an aerial perspective, first appeared in the 1966 collection *F. R. Scott: Selected Poems*. In the 1973 collection *The Dance is One*, Scott gathered the suite of poems inspired by the journey under the title "Letters from the Mackenzie River" with "Flying to Fort Smith" and "Mackenzie River" reprinted as the first and last poems enclosing the rest. In the first poem, the Mackenzie Delta is a vast, silent territory of undeveloped potential awaiting the arrival of Modernity. The speaker in "Flying to Fort Smith" is as enchanted by the potential in the land below as by the novel experience of travelling by air:

> The spread of silver wing
> Gathers us into long lanes of space.
> We peer through panes of glass.
>
> The plain of lakes below
> Is bound with bands of green
> Fringed by darker green
> Pocked with drops of ponds.
>
> Everywhere
> A huge nowhere,
> Underlined by a shy railway.
>
> Snaking brown streams
> At every islanded corner
> Widen their reaches
> Leaving blue pools behind. (Scott 1973, 55)

This image of the vast, untapped "nowhere" of a wilderness of undeveloped potential animated government policy to develop Arctic regions from the early days of Arctic flight when the northern regions were imagined to hold unlimited riches. In the next stanza, Scott harkens back to that image:

> An arena
> Large as Europe
> Silent
> Waiting the contest.
>
> Underground
> In the coins of rock
> Cities sleep like seeds. (Scott 1973: 55)

As early as 1927, Hudson's Bay employee Duncan MacLaren had proclaimed to readers of the *Northern Miner*: "In a rush of air and a roar of sound, [the modern prospector] rapidly surveys its salient features, plans his campaign, he and his machine a perfect combination of efficiency and mobility. Into this great arena throw the youth of this country and who, then, will deny Canada's claim to ownership of the treasure trove of the world?" (quoted in Vance 2002, 134). High above the earth, Scott's speaker summons images of emptiness and metaphors of progress: the rocks produce the raw minerals and generate the value of "coins" that will bring future wealth and settlement, while the comforting, soft sibilance of the alliteration in "cities", "sleep" and "seeds", contrasts the hard consonance of "coins" and "rocks", softening the tone and lulling the reader into a tranquil acceptance of this harsh future.

The land is abstract and remote in this poem, but when the speaker touches down in the other poems in "Letters from the Mackenzie River", the local landscape is far from enchanting. Although Scott's enthusiasm for potential growth and development seems nostalgic for an earlier time, his socialism led him to see it from other perspectives as well. In 1940, Scott had been invited by the miners to serve on a commission investigating a dispute between the Dominion Coal Company of Nova Scotia and the United Mine Workers of America (Djwa 1987, 188). Visiting the mine, watching the miners cut into each new seam as the pit props collapsed behind them made a lasting impression on Scott. His fascination with their skill and his support for the miners as working men shaped his attitude towards the mining industry that had made his northern journey possible. Landing in "The Camp at Bell Rock" where the trees "leaned in broken ranks":

> The plane was ringed in dust
> As we whisked the sandy runway [...]
> Bull-dozers were widening
> A strip of weeds and grass. (Scott 1973, 56)

As in the sequence's closing poem, "Mackenzie River," which is "[a] river so Canadian / it turns its back / on America" (Scott 1973, 69), the Slave river in "The Camp at Bell Rock" is "[r]unning away from America / yet bringing America with it" in the barges and diesel tugs and the rough language of the site boss (Scott 1973, 56). While observing the industrialization marking the landscape with the detachment that characterizes the sequence of poems, the speaker also sees a beauty that elevates the mundane world:

> The cargoes speak the language of life:
> Muckers for mines, acids for ores,
> Barrels of oil and gas, timber and pipe,
> Bull-dozers and cranes, fork-lifts and drills,
> Tough hardware, and the fuels of power. (Scott 1973, 57)

The assortment of equipment and chemicals becomes a metaphor, "the language of life", to the speaker who seems resigned to the hard facts of this solid Modernity. In other poems such as "Norman Wells", Scott records the reality of resource extraction: the land is littered with the debris of human activity, the untidiness of the "civil development" of the region undertaken by the Canadian government and its commercial agents. But, even in the poems that dwell on the material conditions "on the ground", the speaker maintains distance. The activities, such as mining, that opened up the Arctic and sub-Arctic regions to travel from the south are explored with the same detachment as the descriptions of the reality on the ground. If turning the materials of capital into the subject of poetry seems at odds with his commitment to socialism, the poem "Norman Wells" clarifies his position with a sharp critique of capitalist exploitation. The "tanks of oil" seen as they round the bend in the river represent "[p]ower and light and heat / For whatever the uses of man" but "PRIVATE PROPERTY" prevents their use for a greater good (Scott 1973, 64). The sign prohibiting trespassers unleashes the speaker's fury in lines whose repetitive structure intensifies the emotion:

> Man is the absent fact
> Man is the aim and need
> Man is the source of wealth
> But Property keeps him out. (Scott 1973, 64)

What Scott sees here is that the wealth generated by the oil that "lies so close to the soil / It films the pools in the woods" (Scott 1973, 64) is of no benefit to the people:

> And the Indians wonder, who first
> Lived off this very soil
> And now are outcast and dying
> As their substance is drained away. (Scott 1973, 64)

The choice of the Mackenzie River as destination was no accident. As Marionne Cronin's extensive research shows, Scott was tracing the path taken by prospectors and geologists from the 1920s on. In the absence of other opportunities, such as passenger and freight, and government funding focussed on aerial photography and forest fire patrols almost exclusively, the aviators of the 1920s had cultivated a market in the resource industries, particularly mining. In Canada, Arctic aviation developed as science and technology adapted to facilitate rapid expansion northward to pursue the resource extraction and development that was the *raison d'être* of Arctic exploration in Canada. The use of aircraft in prospecting "started early and developed rapidly" (Dunbar and Greenaway 1956, 488) virtually opening the Mackenzie district to outsiders in the 1920s (Zaslow 1975, 335). The Imperial Oil Company "pioneered flying in the Northwest Territories", flying to Norman Wells in 1920 (Zaslow 1975, 359), and by 1930, there were six commercial air companies operating in the Mackenzie Delta alone (Dunbar and Greenaway 1956, 486, 494). From 1919 to 1928, Canadian surveyors embraced aviation because "the difficulty of overland travel and the sheer immensity of the territory made existing methods of mapping and surveying inappropriate" (Cronin 2007, 303). In 1923, the world's largest airport was at Goose Bay, Labrador (the location that would later be chosen for NATO testing), and the number and capacity of air bases expanded during the Second World War to support Arctic supply routes.

During the Second World War, future Prime Minister Lester B. Pearson championed Arctic aviation as a means to prosperity. Pearson had been a fighter pilot in Britain's Royal Flying Corps (RFC) during the First World War (English 1992, 37), and after being injured, served out the end of the war as an aviation navigation instructor in Toronto (English 1992, 52). Flying was a lifelong passion, and in his vision of northern expansion, aviation would play a vital role. He even got his nickname "Mike" from his commanding officer in the RFC (English 1992, 43). "Like so many young men of that day", his biographer John English observes, "Pearson was fascinated with speed and flight" (English 1992, 42) and like many of his generation, he cultivated lifelong "airmindedness" from his early experiences as a fighter pilot and instructor. "We now know," Pearson wrote in *Foreign Affairs* in 1946, "that there is great wealth in the Land of the Midnight Sun, and that there are ways of making that wealth available. That, again is primarily due to the conquest of the air. […] The northern skies are humming with activity, smoke is coming from northern chimneys; adventurous settlers are moving in" (Pearson 1946, 645).

Resource development would focus Canadian northern policy as the literary and visual representation of the Arctic reflected its newfound value as both territory and airspace to be defended by military flights over the Arctic. British studies of aviation such as *From Above: War, Violence and Verticality* understandably focus on the military history of airpower, showing the complex relationships between technologies of flight and technologies of war as aviation advanced military capability, and military objectives propelled aerotech research and development because "the affective effects of airpower have both enchanted and terrified entire populations" since the air raids over England in the Second World War (Anderson 2013, 248).

In Canada, Jonathan Vance has found evidence that the romantic figure of the pilot actually served as "the ideal antidote to the frightening uncertainties of modern war" (Vance 2002, 71). In a war that reduced individual lives to "cannon fodder" or staggering numbers of dead, the aviators flying above the misery of the trenches provided a heroic image: "Free of the suffocating constraints of the trenches, he sailed aloft in a realm of clarity, cleanliness and order" (Vance 2002, 71). After the First World War, there was increasing "airmindedness" (see Vance 2002 and Wohl 2005) in Canada as elsewhere, though the scientific exploration necessary for resource development was as important an incentive to aviation as military ambitions. Because of the country's sheer size, "aviation, in its early days, meant more to Canada than to other countries" (Vance 2002, 287). In a country of vast spaces, flight could reduce the time needed to traverse these distances, unifying the nation in ways the railroad had been intended to do, and aviation seemed to bind together the regions by accelerating communication and travel between them. The Cold War would further encourage the development of Arctic aviation as part of military training and testing. According to Whitney Lackenbauer and Matthew Farish, military activity during the Cold War was "not just a physical engagement with soil, muskeg, permafrost, water, and ice; it was critical to the formation of a new imaginative geography of the Arctic" (Lackenbauer and Farish 2007, 927).

The mineral-rich Mackenzie region was the prime area for these activities (Cronin 2006, 57, 61). Surveyors, and the mining industry in particular, pushed development as they adopted aerial surveying, and their work enabled the resource development that led to the settlement of the Mackenzie region and the disruption of Dene lifeways that accompanied it. Scott's trip in 1956 courtesy of Eldorado Mining and Refining allowed him to document the detrimental effects of development and lament the conditions in which he saw the indigenous inhabitants without making the connection to his own presence.

If injustice and inequality are created by capitalism, Scott's "Norman Wells" suggests that these consequences were arriving from elsewhere. Arctic modernity, in short, was already becoming liquid as borders gave way to the transnational flow of ideas and investment. As "America's overspill / Invades the tundra and lakes / Extracting, draining away, / Leaving a slum behind" (Scott 1973, 67) in the poem "A New City: E 3", the "Indian and Eskimo watch / The slow, inescapable death / of this land which has waited so long / For the sentence already pronounced" (Scott

1973, 67). Throughout the sequence, the urban landscape is dirty with a litter of oil drums or cans, populated by the displaced whites, who are given names, and the bewildered indigenous people who are not. These "broken people" watch silently in "Fort Providence" where the residential school offers no "library or workshop" just a "firetrap mental gaol" festooned with religious imagery and "Silk-screen prints of the Group of Seven" (Scott 1973, 62). Scott seems to be on the verge of naming the dispossession of the Dene in this poem, but falls short of blaming those responsible or offering any solution. Certainly, he was aware of the colonial relationship between the Canadian state and the North – he even refers to it as a "colony" – yet there is a tinge of fatalism in the description of the people as simply "broken", as if in a natural state.

In *Technology and Empire*, George Grant had identified North America's "conquering relation to place" (Grant 1969, 17). Grant observed that Canada, a modern nation emerging from the contest of European empires, had become part of the new technological empire led by the United States. The resistance to "America" in literary culture, therefore, articulated a critique of Canada's role in advancing the technological society it desired to be even as the solid forms of national authority were beginning to dissolve. Resistance to American culture also blinded Canadians to their own internal colonialism and imperialist ambitions. In "Canada's Own Dark Heart: F. R. Scott's 'Letters from the Mackenzie River'", Neil Querengesser highlights the "ironic contrast" between the observer's conception and the geographic reality that makes Scott's river journey strikingly similar to Conrad's classic, and also susceptible to Chinua Achebe's insight that Conrad was able to see and condemn "the evil of imperial exploitation but was strangely unaware of the racism on which it sharpened its iron tooth" (Querengesser 1992, 262).[5] Scott's anti-imperialism reaches a similar limit: he cast a critical eye on the conditions in which the indigenous people live and on the constellation of capitalist interests draining the land without ever giving up the dream of future development. Querengesser traces the speaker's attempts to grasp the immensity of this unknown place to Conrad's distanced view of the enigmatic African continent, demonstrating parallels in the way the outsider enters the wilderness and regards the presence of the other within it. The aerial view of the landscape creates a "strong sense of separation between the observer and the observed" in Scott's poems that "reveals that the observer is attempting to overcome, by means of language, the anxiety created by that very separation" (Querengesser 1992, 92). Alongside Conrad's dying Africans, Querengesser places Scott's description of the Dene and the future cities that are a "hidden presence" throughout. Like

Conrad, Scott sees the human toll of northern expansion into "Canada's colonies", and like Conrad, he presents an ironic critique that "leaves us with an ambiguous vision of Canada's dark heart" (Querengesser 1992, 102).

In "Fort Smith", the local Anglican priest takes Scott and friends "[t]o explore Canada's colony" in a Volkswagen that the speaker dubs "the Nazi car" (Scott 1973, 58). The group must head out of Fort Smith, with its Bank of Commerce, Hudson's Bay store and hotel, RCMP headquarters, government offices, churches, restaurant, barber shop, liquor store, pool room and "Imperial Oil Compound" (Scott 1973, 58), to find the Slave river where "lies the pre-Cambrian rock" and the rapids roar:

> Pierre, suddenly challenged,
> Stripped and walked into the rapids,
> Firming his feet against rock,
> Standing white, in white water,
> Leaning south up the current
> To stem the downward rush,
> A man testing his strength
> Against the strength of his country. (Scott 1973, 59)

The image of the individual man pitting himself against nature represents the conflict of human industry and natural environment, imbuing this encounter with retrospective heroic and national importance. If there is an answer to the disastrous effects of resource extraction, Scott suggests, it will be found in human contact with nature while Pierre Trudeau would develop his attitude towards wilderness, especially in the North, from such experiences.

In "Mackenzie River", the last poem in the sequence, the land is again viewed from a distance, and although the speaker's location is not described, an aerial view of the land is once again evoked in images such as the "arteries of streams" and "maze of ponds and dikes" and "servient valleys" reaching to "far hills" (Scott 1973, 69). The distance is also rhetorical as the river is personified and given abstract, symbolic meaning:

> A river so Canadian
>   it turns its back
>   on America (Scott 1973, 69)

It is as if the gritty reality of Norman Wells and Fort Smith had to be envisioned before it could be transcended, bringing the poem and the sequence to its inexorable concluding stanza:

In land so bleak and bare
a single plume of smoke
is a scroll of history. (Scott 1973, 70)

Noting the frequent shift to the global in several of Scott's poems, D. G. Jones remarks a tension between "technological adventure" and "human community" throughout the sequence (Jones 1983, 49). In "Mackenzie River", the profound feeling for the natural world "gives the lie to the extreme optimism of a liberal technological society, that vision of progress and of man's ever-expanding empire over nature" (Jones 1983, 52) and the "clear recognition of the fragility and integrity of earth, of the ecological interdependence of all things, of limits and of entropy, that tempers the expansive optimism of other poems and gives to his overall vision of the collective project its weight and conviction" (Jones 1983, 52).

If Modernist poets, including Scott, had used the Arctic landscape to Canadianize Modernism, they also created an aesthetic based on the aerial perspective, a "poetry of the aeroplane" that the environmentally-minded poets of the 1960s and 1980s would continue as they travelled northward. Flying to Baffin Island in 1965 was as transformative for the poet Al Purdy as it had been for Scott and Trudeau. Remembering the flight north in his autobiography *Reaching for the Beaufort Sea*, Purdy claims, perhaps disingenuously, to be unaware of poetic precursors like Scott, to whom he had actually dedicated the collection *North of Summer*:

> Baffin Island in 1965 was a beginning. Leaving Montreal in early June, flying over Ungava and Hudson Straits to Frobisher Bay and snow-splashed hills of Baffin like recumbent Jersey cows—that was exciting. I rushed from one side of the Nordair aircraft to the other feeling delighted with myself that I had chosen to come here; nursing a big bottle of booze under my HBC parka in case of shortages; the strangeness and probably discomfort ahead all part of my fascination. No other poet I knew of had ever gone to the Arctic (except Robert Service, and he didn't count), it was virgin territory for me, untouched except for the mundane prose of explorers and scientists. (Purdy 1993, 190)

In the Arctic, however, Purdy trades the aerial for an eye-level view of the ground in poems like "Arctic Rhododendrons" and "Trees at the Arctic Circle", and he would later describe feeling curious about everything he saw and experienced on Baffin Island. As D. M. R. Bentley explains, "Purdy's well-known response to the 'vast lonely' spaces of the Arctic was to use a short-lined, open-ended free verse and to concentrate, by his own admission, on details and on people" (Bentley 1980, 15). Purdy adopted the free verse forms that characterized Modernist poetry to explore different perspectives and themes such as the rapacious effects of modernity on the northern environment.

In "Trees at the Arctic Circle", for example, Purdy explores the challenge that this new environment poses for the outsider laden with attitudes and assumptions nurtured elsewhere. At first, the speaker chastises the "Coward trees" for "crawling under rocks / groveling among the lichens / bending and curling to escape / making themselves small / finding new ways to hide", comparing the Arctic ground willow to magnificent Douglas fir trees, maples and oaks, and even the "dwarf shrubs of Ontario" that "mock them / Coward trees" (Purdy 1986, 84). To signify the speaker's gradual acquisition of new knowledge, there is a clear turning point at the twenty-sixth line:

> And yet – and
> yet – their seed
> pods glow
> like delicate grey earrings
>
> their leaves are veined and intricate
> like tiny parkas. (Purdy 1986, 85)

With a new appreciation for how the ground willow trees "use death to remain alive" by nourishing their roots in permafrost, the speaker realizes that "To take away the dignity / of any living thing [...] is to make life itself trivial" (Purdy 1986, 85). The self-criticism develops from being suddenly able to "see I've been carried away / in my scorn of the dwarf trees / most foolish in my judgments" and it is the initial mistaken way of seeing that becomes the central theme of the poem:

> I have been stupid in a
> poem I will not alter the
> poem
> but let the stupidity remain
> permanent as the trees are
> in a poem
>
> the dwarf trees of Baffin Island (Purdy 1986, 85)

Such a beautiful moment when the outsider confronts his own lack of knowledge and experience! "Trees at the Arctic Circle" celebrates the autonomy of living things that persist "unbothered by any human opinion" (Purdy 1986, 85), offering an environmental view of the Arctic that counters the mundane utilitarian image of the Arctic as resource: to appreciate its natural world independent of humans and the Arctic for its own sake.

Like Purdy, Paulette Jiles was deeply moved by her experiences travelling and working as a journalist for the indigenous communities in which she lived, as she recounts in the memoir *North Spirit: Travels Among the Cree and Ojibway and their Star Maps*. In the long poem *Song to the Rising Sun*, Jiles cautions against the environmental destruction wreaked by a Modernity that is failing the indigenous peoples of the North and all humanity with them. The speaker awaiting the return of the sun remembers the flight north:

> When I first came up here
> I came up in a Twin Otter, a cargo plane
> carrying four barrels of aviation fuel.
> I like living on the edge.
>
> We flew over the Black Angel Lead and Zinc
> Mine at 8,000 feet,
> and the stories the pilot told me were always full of money.
> Junk is lethal.
> We flew through a thickening, dirty pollution
> haze that comes up from the megacities.
> Junk is lethal. (Jiles 1989, 16)

While the pilot, who tells the speaker that he thinks "Big Stuff will save us", voices the blind faith in development and technology that his occupation serves, the poem envisions Arctic air not just as space to be travelled or height to be reached. It is also a substance polluted with "100 million tons of sulphur dioxide / staining the polar air / into a haze that pilots cannot fly through" (Jiles 1989, 17). It is also "dirty with signals": "Brand names, long wave, short wave, the morse / of supply ships coming up the Davis Strait with their cargoes / or summer fuel / and dreaming sailors" (Jiles 1989, 17). The consumer society that produces the brand names and "Big Stuff", the radio signals and even poems have created the "black hole in the ionosphere" (Jiles 1989, 17), and it is this message that the speaker implores the listener to hear:

> Wake up and start again.
> Wake up and free yourself. […]
> Big Stuff will not save us. Wake up. Look. Keep watch.
> (Jiles 1989, 18)

In *Song to the Rising Sun*, the “bituminous fires” of the “Black Angel Lead and Zinc Mine” glow in the background as the speaker cautions “you are sick with power / you are sick with the deep acids of power / heavy metals washing ashore” (Jiles 1989, 18). Throughout the poem, the “air” is not just a backdrop for human ascendancy through modern technology, the invisible medium of flight that buoys the air traveller who gives no thought for what it is; rather, the atmosphere is a substance affected by the contamination of pollutants arising from the mine. Only recently has it been more widely recognized that air acts as a kind of sink when these gases are dispersed, as humans flush gaseous waste, or that air can also be “mined” for gases that humans find useful (Everard 2015, 42). The pollution generated by mining endangers the air on which aviation is “utterly dependent” to feed internal combustion and to provide the lift that sustains flight (Everard 2015, 41).[6] Ironically, it is flight that gives the speaker a perspective on what is endangering the atmosphere, the environmental destruction imposed by Modernity, and the politics of dispossession.

One legacy of F. R. Scott’s poetry may have been the elevated language of northern expansion and development, but another part of it was the greater consciousness of northern ecology manifested in these later poems by Purdy and Jiles. Scott’s “ambiguous vision” (Querengesser 1992, 102) of Canada’s colonies also records the solidifying effects of resource extraction and industrial development and gives rise to an image of a fragile environment that would shape future environmental policy. [7] It was Scott’s travelling companion, the future Prime Minister Pierre Trudeau who would proclaim Canada the “trustee for all mankind” of the Arctic archipelago in 1969 and would institute Northern Land Use Regulations designed to protect its delicate ecosystem (Bocking 2007, 878). As Lackenbauer and Farish contend, the “*delicate* Arctic” evoked in the 1969 Speech from the Throne would also serve as a reason to strengthen Canadian sovereignty in the region (Lackenbauer and Farish 2007, 933).[8] The federal government’s commitment was concretized with Bill C-187 *An Act respecting inland water resources in the Yukon Territory and Northwest Territories* and Bill C-202 *An Act to prevent pollution of the areas of the arctic waters adjacent to the mainland and the islands of the Canadian Arctic* as well as an amendment to the Territorial Lands Act rezoning and asserting land use regulations in the North.

The movement of international capital across borders was already a reality when Scott and Trudeau travelled the Mackenzie district in 1956. The power of the nation-state was already succumbing to the global flow of capital, as Grant observed, even as the literary nationalism of works like

"Mackenzie River" were attempting to draw the cultural and ideological border between Canada and the United States. Paradoxically, internationalism could create a stronger national identity as the Trudeau government tried to translate it into a Canadian policy based on care for the Arctic. In a globalizing world, this environmental vision could grant Canada a distinct role. By 1972, a year before the complete suite of poems in "Letters from the Mackenzie River" appeared in Scott's collection *The Dance is One*, environmentalists were already referring to the "Trudeau Doctrine" as a potential corrective to the "violated vision" of northern development (Woodford 1972, 99–105). The Trudeau doctrine, which was based on the idea of environmental stewardship, was international in outlook, presaging an increasingly liquid flow of political action. The survival of human beings rested on the ecological fate of the Arctic, a common cause that transcended national borders. As if anticipating Sheila Watt-Cloutier's recent argument for *The Right to Be Cold* (2014), Trudeau linked ecology to human rights, calling on the international community to "ensure to human beings the right to live in a wholesome natural environment" (quoted in Woodford 1973, 105). Trudeau himself remained an advocate of environmental protection throughout his life, for example, publishing a collection of essays entitled *Energy for a Habitable World* (1991). Many of the commitments made in the 1970s have changed as the Government of Canada (GOC) led by the Conservative party pursued a policy of unilateral action determined by global economic and security interests early in the new millennium; however, even if the newly elected Liberal government changes policy, it is still likely that material and utilitarian impulses will animate Canadian involvement in the Arctic as they always have. From many perspectives, the Arctic is imagined to be in need of protection, either to safeguard Canada's economic interests or to protect its fragile environment. In northern communities, the need for economic development cannot outweigh the pressing challenges posed by technological and environmental change on their homeland. By opening Arctic regions to the twentieth century, aviation played an important part in generating these competing interpretations of Arctic Modernity, and the "poetry of the aeroplane" played a role in increasing awareness of the northern environment even as it recorded and extended the conquest and colonization of northern lands, leaving an ambivalent legacy of celebration, critique, and complicity.

Acknowledgments

I am grateful to Janine Rogers for alerting me to the phrase that provides the title for this essay. The development of this essay was supported by travel funding from the Norwegian Research Council through a grant to the Arctic Modernities Research project under the direction of Anka Ryall. Many thanks to Anka and to the project members for the stimulating discussions that guided the writing of this essay and also to Marionne Cronin and Heidi Hansson for their thoughtful and helpful suggestions on the manuscript.

References

Achebe, Chinua. 1977. "An Image of Africa." In *Chant of Saints: A Gathering of Afro-American Literature, Art, and Scholarship*, edited by Michael S. Harper and Robert B. Stepto, 313–325. Chicago, IL: University of Illinois Press.

Adey, Peter. 2010. *Aerial Life: Spaces, Mobilities, Affects*. London: Wiley-Blackwell.

—. *Air*. 2014. London: Reaktion.

Adey, Peter, Mark Whitehead and Alison J. Williams, eds. 2013. *From Above: War, Violence and Verticality*. Oxford: Oxford University Press.

Alt, Christina. 2004. "'Recreating What is Generally Forgotten': F. R. Scott and Native History." *Essays on Canadian Writing* 81: 1–16.

Anderson, Ben. 2013. "Targeting Affective Life from Above: Morale and Airpower." In *From Above: War, Violence and Verticality*, edited by Peter Adey, Mark Whitehead and Alison J. Williams, 247–261. Oxford: Oxford University Press.

Armitage, John. 2013. "Flying Into the Unknown: Cinematic Cultures of War and the Aesthetics of Disappearance." In *From Above: War, Violence and Verticality*, edited by Peter Adey, Mark Whitehead and Alison J. Williams, 164–184. Oxford: Oxford University Press.

Bauman, Zygmunt. 2000. *Liquid Modernity*. Cambridge: Polity Press.

Bentley, D. M. R. 1997. *The Gay[Grey Moose: Essays on the Ecologies and Mythologies of Canadian Poetry, 1690-1990*. Ottawa: University of Ottawa Press.

—. 1980. "A New Dimension: Notes on the Ecology of Canadian Poetry." *Canadian Poetry* 7: 1–20.

Bocking, Stephen. 2009. "A Disciplined Geography: Aviation, Science, and the Cold War in Northern Canada, 1945-1960." *Technology and Culture* 50 (2): 265–90.

—. 2007. "Science and Spaces in the Northern Environment." *Environmental History* 12 (4): 867–894.

Campbell, Wanda. 1990. "The Ambiguous Vision of F. R. Scott." *Canadian Poetry* 27: 1–14.

Coates, Kenneth. 1985. *Canada's Colonies: A History of the Yukon and Northwest Territories*. Toronto: Lorimer.

Cronin, Marionne. 2006. "Flying the Northern Frontier: The Mackenzie River District and the Emergence of the Canadian Bush Plane." PhD diss., University of Toronto.

—. 2007. "Northern Visions: Aerial Surveying and the Canadian Mining Industry, 1919–1928." *Technology and Culture* 48 (2): 303–330.

—. 2010. "Polar Horizons: Images of the Arctic in Accounts of Amundsen's Polar Aviation Expeditions." *Scientia Canadensis: Canadian Journal of the History of Science*. 33 (2): 99–120.

—. 2013. "Technological Heroes: Images of the Arctic in the Age of Polar Aviation." In *Northscapes: History, Technology, and the Making of Northern Environments*, edited by Dolly Jørgensen and Sverker Sörlin, 57–81. Vancouver: University of British Columbia Press.

Djwa, Sandra. 1987. *The Politics of the Imagination: A Life of F. R. Scott.* Toronto: McClelland and Stewart.

Dunbar, Moira and Keith R. Greenaway. 1956. *Arctic Canada From the Air*. Ottawa: Canada Defence Research Board.

English, John. 1992. *Shadow of Heaven: The Life of Lester Pearson. Volume I: 1897-1948*. Toronto: Vintage.

Everard, Mark. 2015. *Breathing Space: The Natural and Unnatural History of Air*. London: Zed.

Geller, Peter. 2004. *Northern Exposures: Photographing and Filming the Canadian North, 1920-1945*. Vancouver: University of British Columbia Press.

Goodyear, Anne Collins. 2003. "The Effect of Flight on Art in the Twentieth Century." *Reconsidering a Century of Flight*, edited by Roger D. Launius and Janet R. Daly Bednarek, 223–241. Chapel Hill, NC: University of North Carolina Press.

Grant, George. 1969. *Technology and Empire.* Toronto: House of Anansi.

Grant, Shelagh. 1988. *Sovereignty or Security: Government Policy in the Canadian North, 1936–1950*. Vancouver: University of British Columbia Press.

Grant, Shelagh. 2010. *Polar Imperative: A History of Arctic Sovereignty in North America*. Toronto: Douglas & McIntyre.

Hill, Jen. 2008. *White Horizon: The Arctic in the Nineteenth-Century British Imagination*. Albany, NY: State University of New York Press.

Jiles, Paulette. 1989. *Song to the Rising Sun: A Collection*. Vancouver: Polestar.

—. 1995. *North Spirit: Travels Among the Cree and Ojibway and Their Star Maps*. Toronto: Doubleday.

Jones, D. G. 1983. "Private Space and Public Space." In *On F. R. Scott: Essays on his Contributions to Law, Literature, and Politics*, edited by Sandra Djwa and R. St. J. MacDonald, 44–54. Montreal: McGill-Queens University Press.

Lackenbauer, P. Whitney and Matthew Farish. 2007. "The Cold War on Canadian Soil: Militarizing a Northern Environment." *Environmental History*. 12 (4): 920–950.

McCannon, John. 2010. "Winged Prometheans: Arctic Aviation as Socialist Construction in Stalinist Russia, 1928–1939." *Scientia Canadensis* 33 (2): 75–97.

Pearson, Lester B. 1946. "Canada Looks 'Down North'." *Foreign Affairs* 24 (4): 638–647.

Potter, Russell A. 2007. *Arctic Spectacles: The Frozen North in Visual Culture, 1818-1875*. Montreal: McGill-Queens University Press.

Purdy, Al. Purdy, Al. 1986. *The Collected Works of Al Purdy*, ed. Russell Brown. Toronto: McClelland.

—. 1993. *Reaching for the Beaufort Sea: An Autobiography*. Madeira Park, BC: Harbour Publishing.

Querengesser, Neil. 1992. "Canada's Own Dark Heart: F. R. Scott's 'Letters from the Mackenzie River." *Essays on Canadian Writing* 47: 90-104.

Rogers, Janine. 2014. "A Note on Science and Canadian Literature." *Canadian Literature* 221: 6-16.

Scott, Duncan Campbell. 2000. "Poetry and Progress" (1922). In *Duncan Campbell Scott: Addresses, Essays, and Reviews*, ed. Leslie Ritchie, 299-319. London, ON: Canadian Poetry Press, 2000.

Scott, F. R. 1954. *Events and Signals*. Toronto: Ryerson Press.

—. 1966. *F.R. Scott: Selected Poems*. Toronto: Oxford University Press.

—. 1945. *Overture*. Toronto: Ryerson Press.

—. 1973. *The Dance is One*. Toronto: McClelland and Stewart.

Trudeau, Pierre Elliott, ed. 1991. *Energy for a Habitable World: A Call for Action.* New York: Crane Russack.

Vance, Jonathan F. 2002. *High Flight: Aviation and the Canadian Imagination*. Toronto: Penguin.

Wadden, Marie. 1991. *Nitassinan: The Innu Struggle to Reclaim their Homeland*. Vancouver: Douglas and McIntyre.

Wohl, Robert. 1994. *A Passion for Wings: Aviation and the Western Imagination, 1908–1918*. New Haven: Yale University Press.

—. 2005. *The Spectacle of Flight: Aviation and the Western Imagination*. New Haven, CT: Yale University Press.

Woodford, James. 1972. *The Violated Vision: The Rape of Canada's North*. Toronto: McClelland and Stewart.

Zaslow, Morris. 1975. *Reading the Rocks: The Story of the Geological Survey of Canada, 1842–1972*. Toronto: Macmillan.

Notes

[1] An excellent historical account of these events is Marie Wadden's *Nitassinan*.

[2] Jen Hill's *White Horizon: The Arctic in the Nineteenth-Century British Imagination* (2008) provides another comprehensive look at the representation of Arctic exploration in the period.

[3] It bears noting here the masculinist language and principles the Modernists critique: Scott personifies the clapped-out Victorian poetry as a failed "poetess," and Klein envisions the poet as the "first man" naming creation.

[4] In "Flying into the Unknown: Cinematic Cultures of War and the Aesthetics of Disappearance," John Armitage takes different view of filmic representations of flight by applying Virilio's concept of disappearance. Rather than creating and controlling through the image, Armitage argues, film represents the ephemeral nature of the image: "Modern-day images thus apparently move across but actually and repeatedly vanish from the fundamentally immaterial support of the screen as part of a cinematic sequence. In the case of film, such images disappear at twenty-four frames per second or, in the case of special effects, at sixty frames per second and above" (169).

[5] Christina Alt argues that Scott's "Letters from the Mackenzie River" and "The Founding of Montreal" both "demonstrate his engagement with the subject of Native history" (6), but looks to his found poems for specific examples.

[6] The study of the cultural relevance of "air" is an exciting and growing field. In addition to Everard's *Breathing Space*, see Peter Adey's *Aerial Life* (2010) and *Air* (2014).

[7] The Yukon and Northwest Territories were first identified as such in Kenneth Coates's *Canada's Colonies: A History of the Yukon and Northwest Territories* (1985).

[8] For an authoritative history of Arctic sovereignty in Canada, see Shelagh Grant's *Sovereignty or Security: Government Policy in the Canadian North, 1936-1950* (1988) and *Polar Imperative: A History of Arctic Sovereignty in North America* (2010).

The Arctic Pastoral

Henning Howlid Wærp

This chapter focuses on Nordic polar literature published at the end of the nineteenth and beginning of the twentieth century – primarily by Fridtjof Nansen, Helge Ingstad and Knud Rasmussen, but also by other explorers and hunters. The aim is to investigate whether a term such as pastoral may be appropriate for illuminating aspects of both modernity and counter-modernity in this non-fiction genre. The term is initially counter-intuitive, since the concept of the Arctic is normally associated with words like cold, icy waters, shipwrecks, frostbite, hunger, suffering and death, bringing up associations with the numerous Arctic expedition failures and the long list of tragedies connected with hunting and over-wintering.

Against this background, how can it make sense to talk about an "Arctic pastoral"? It seems that the forbidding Arctic is not the whole picture. Thus, when the polar explorer Vilhjalmur Stefansson wrote about his five-year journey (1913–1918) across northern Canada and Alaska, he entitled his book *The Friendly Arctic* (1921). Stefansson's Arctic is not an inhospitable desert, but a fruitful and welcoming place if you manage to discard your prejudices and embrace what the Arctic has to offer (Gaupseth 2011, 203). Similarly, when Helge Ingstad's book *Pelsjegerliv* (1931) was published in English in 1933, it was under the title *The Land of Feast and Famine*. Although "famine" is often associated with the polar regions, the "feast" is a much less familiar image. Polar narratives as a whole are more characterized by wilderness discourses than pastoral ones. However, some of the written accounts contain passages in a pastoral mode, or describing a pastoral moment, in which the traveller, hunter or scientist seems to reflect upon their ties to nature. In these moments, it is not only the traditional heroic discourse that is challenged, but also the ideology of progress embedded in modernity. In the following, I focus on this pastoral side of polar literature, arguing that certain aspects of Arctic literature constitute a type of modern pastoral displaying affinities with the genre of nature writing, ecocriticism and vitalism.

In Fridtjof Nansen's travel account *Through Siberia* (1914), he reports enthusiastically about a trip he has undertaken to develop trade and expand trade routes between Siberia and Western Europe. He travels by the new Trans-Siberian railway, which was still under construction, and enjoys the speed and comfort. At the same time, he wishes to see the traditional forms of rural living remain unspoiled, and he worries about the damage to the landscape. On the train, when the group of merchants and developers of which he is a part discusses progress in the area, Nansen at one point closes his eyes and listens to the wind in the woods outside the compartment: "I can hear the murmur of the forest out there in the darkness – a familiar sound. It is like the voice of the forest at home. Here, too, the forest whispers eternally over the endless plain" (Nansen 2014, 163).[1] There is an everlasting conflict between the primitive and the civilized, the instinctive and the analytic in Nansen's work. This is a rift – produced by modernity – that he shares with many others.

Traditional polar literature is sometimes regarded as nostalgic or even reactionary in its critique of civilization and its celebration of outdoor life. However, with our new awareness of climate and ecology, it might be time to re-evaluate this literature, as a corrective or alternative to the consumerism and struggle for economic progress in Western societies today. The traveller or the dweller in the north often advocates a kind of "simple living" that in itself is an aspect of modernity. In this respect, the pastoral side of polar literature has a critical potential.

The pastoral

In Greg Garrard's book *Ecocriticism* (2004), the pastoral and the apocalypse are viewed as the two most powerful genres when it comes to rousing people to action for environmental protection. The pastoral generates images of a better life, a rural existence in which humans are in harmony with their natural surroundings, while the apocalypse depicts an urban hell after the collapse of the ecosystem and civilization. Apocalyptic fiction emerges in the late 1800s. Since then, the genre has taken on motifs such as pandemics, environmental disasters, alien invasions, nuclear wars, data collapses, the ascendancy of machines (the cyborg) and so on. In contradistinction to the complex and unpredictable world of these depictions of doom, the pastoral offers apparent simplicity and transparency. In this context, the pastoral is much more than a historical artefact or ossified convention of Greek and Roman Antiquity, the Renaissance, or the Romantic Age. As Garrard emphasizes, the pastoral may be "any literature that describes the country with an implicit or explicit contrast to the urban" (Garrard 2004, 33).

The terms pastoral and idyll are often used interchangeably. *Pastoral* derives from the Latin word for shepherd, *pastor*, while *idyll* refers to the short depictions, or *eidyllion*, written by the originator of the genre, the Greek poet Theocritus, during the first half of the second century BCE. His idylls were short, poetic texts about shepherds resting in the shade, talking, singing and playing the flute. Historically, the idyll has nothing to do with being idyllic, although there is a perception that the genre is sentimental and nostalgic for a lost golden age, often located as Arcadia, an idealized mountain region in Greece.

In *What Is Pastoral* (1996), Paul Alpers argues that we should not limit the concept to such a Golden Age or to a particular landscape. For Alpers, the pastoral is not a particular genre but rather a mode in the sense of the attitude or tone of the text. This attitude is not by definition sentimental or nostalgic. The pastoral, with the shepherd and his life at its centre, is a way of treating loss and decline at the same time as maintaining a human community. The shepherd is no hero, "he is able to live with and sing out his dilemmas and pain, but he is unable to act so as to resolve or overcome them or see them through to their end" (Alpers 1996, 69). He is a representative of human vulnerability.

Alpers might not, however, recognize the texts I treat here as pastoral. The shepherd in the classical pastoral lives in a *locus amoenus*, a garden-like landscape featuring trees, grass and water, an idealized place of safety and comfort. In the Arctic, the landscape itself may constitute a danger. Nevertheless, there are several similarities in terms of their sensibility towards nature.

"Thus I find the birth of spring as I fare through the forest [...]" Helge Ingstad writes from northern Canada in the late 1920s (Ingstad 1992, 96).[2] What he is celebrating here is his participation in the processes of nature, the passing of the day and the changing of the seasons and the weather, a type of transformation that displays an underlying unity by being cyclical rather than progressive. The wilderness is the polar opposite of the city and the progress of civilization, but with the simultaneous recognition that the development of society cannot be stopped.

The pastoral contrasts with the dominant masculine discourses of polar literature, in which a man proves himself in a series of tests that, if passed, will confirm his superiority over the North (rivers, cold, emptiness), its wild animals and other human beings (Grace 2007, 185). The pastoral has more to do with being at peace in a place, transforming the wilderness into a dwelling place to care for as home, in the Heideggerian sense – experiencing immersion rather than confrontation, recognition rather than challenge (cf. Ryall, this volume). In this way, the pastoral is a challenge to the heroic discourse of the Arctic.

However, the ideal of rural tranquillity in the pastoral has been criticized for being a vehicle for sublimated bourgeois ideology, a nostalgic retreat into the past, or an imperial/colonial desire manifesting fantasies of domination, more than a catalyst for the active transformation of established social structures. Some critics state that the pastoral is heavily dependent on the very class system it claims temporarily to suspend, silencing less privileged groups. Huggan and Tiffin, however, point to the fact that the pastoral might have an imaginative potential for the assertion of a new and better world, and that former settler colonies, like the Caribbean, seem to adapt the pastoral to their own interests, re-imagining identity as conditioned by a dynamic interaction between place and displacement (Huggan and Tiffin 2010, 112). Similarly, Terry Gifford in *Pastoral* (1999) sees an alternative pastoral emerging, based on the idea of land as a collective resource, as opposed to an individual luxury.

Fridtjof Nansen's healthy Arctic

In 1892, Fridtjof Nansen was interviewed by the British newspaper the *Pall Mall Gazette*. Nansen was then in the process of planning the *Fram* expedition to the North Pole, and had behind him, in 1888–1889, a crossing of the Greenland ice sheet from east to west, as well as a winter stay in Nuuk/Godthåb. Moreover, in 1882 he had travelled with a sealer to the hunting grounds in the West Ice near Jan Mayen. In answer to the reporter's question about what he associates with the Arctic, Nansen says:

> I think of the Arctic summer rain. I think of the sunshine, reflected from mountains of snow and ice, shining upon little lakes of clear, rippling water, where hundreds of seals playfully splash the water into glistening sprays of rainbow hues. What is the charm of the Arctic? Health, glorious health! Your muscles twitch with a desire for action. You eat like a horse, and sleep twelve or fourteen hours without a dream. Before you is the vast unknown: all around you is silence and solitude […] In winter the scene is almost as beautiful as in summer. The nights are clear, the moon and stars shine brightly upon the sea of soft white snow. (Nansen 1892, 117)

The north is not depicted here as dark and menacing, but as bright, beautiful and healthy. Nansen emphasizes the positive side of the divided image of the Arctic.

Peter Davidson acknowledges this positive image in *The Idea of North* (2005), in which he shows how the concept of North evokes two opposing associations: light and darkness, which is understandably based on whether you choose to focus on the summer or the winter in the North, infinite day or endless night. The light may be associated with virginity, purity and innocence, while the dark points to something dangerous, sinister and diabolical. This has led to a split perception of the people in the North. On the one hand, they can be seen from a perspective critical of civilization as bearers of purity, authenticity and naturalness, on the other they may be viewed as representatives – or victims – of dangerous, uncontrollable dark forces.

In traditional images of the North, however, the negative aspects seem to dominate. Maybe Nansen has this in mind when, in the book *Farthest North* (*Fram over Polhavet*, 1897), he discusses earlier representations of the Arctic, both in conversations he has had and expeditionary books he has read. He quotes from the American explorer Elisha Kane's book *Arctic Explorations* published in 1856: "I feel that we are fighting the battle of life at disadvantage, and that an Arctic night and Arctic day age a man more rapidly and harshly than a year anywhere else in this weary world" (1897, 281). Nansen's answer to this is that Kane has only himself to blame for the expedition's problems: "his preparations were miserably inadequate, it seems to me to have been a reckless, unjustifiable proceeding to set out with such equipment" (Nansen 1999, 143).[3] Unfortunately, Kane is not alone in his view, Nansen continues: "I remember I once spoke with an English polar explorer on this point, he also had a grim outlook on life in the Arctic regions" (Nansen 1897, 281).[4] Nansen's experience – and point of view – is diametrically opposite: not only is the summer in the Arctic energizing, but so is the winter. On board the *Fram*, Nansen writes in his diary at the end of December 1893:

> For my own part I can only say that I have not noticed any ageing or weakening influence of the polar night in any direction – on the contrary I become younger. This quiet and orderly life suits me exceedingly well, and I cannot recall that I have ever been so physically sound and in full equilibrium as right now. I would rather recommend these regions as a perfect sanatorium for nervous and weakened people. (Nansen 1897, 281)[5]

One of the attractions of the Arctic is the simple way of life: "Bah, this life of research", he writes in his diary during the summer of 1900, after a scientific expedition, adding: "I long indescribably for the simple life" (Nansen in Jølle 1900, 384).[6] Some years later, as a Norwegian diplomat in London, he asks in his diary whether luxury really is the goal of civilization, answering in the negative.

For Nansen, simple living means life in nature in general, but the Arctic has an extra dimension, because of both its isolation and its beauty.

In the Arctic, Nansen is confronted with something that challenges habitual ways and standards of life. While living with the Inuit in West Greenland during the winter of 1888–89, he makes the following statement: "Civilization, steam, great thoughts and great misery – everything is distant – to travel freely and experience the joy of life – that is all" (Nansen 2003, 477).[7] In a eulogy to Nansen given in 1930, the Danish polar explorer Knud Rasmussen declares, with particular reference to *Farthest North* (1897):

> For the first time arctic nature is brought into literature; before it was just an image of terror and suffering. Now it shines forth in the richness of unimagined colours. The first white man has glimpsed the heart of the polar lands and understood their beauty. (Rasmussen 1930, quoted in Hastrup 2010, 692)[8]

In Rasmussen's view at least, the arctic pastoral begins with Nansen.

Helge Ingstad's Arctic simplicity

In the book *East of the Great Glacier* (1935), Helge Ingstad describes his winterings in East Greenland in 1932–33 as taking place in "a lonely landscape, in all its sober strength [...] in the fairy light of stars and the shimmering aurora" (Ingstad 1937, 158).[9] For Ingstad, this landscape inspires another way of living, "a life of conscious simplicity", to use Alpers' term. "For three days we strode on, slept in a tent, and boiled the water for our tea over a fire of fragrant heather [...] the silent joy of vagabond life in the mountains of Greenland" (Ingstad 1937, 67).[10] The simplicity of life is reflected in the style of writing: "Here is a river, and I am tired!"[11]

The cabin they build, to have a place to return to, is described as a natural object:

> The result was surely no palace, but a solid, authentic cabin, with an air of friendliness about it. The usual bunkers of earth and sod were piled high against the outside walls. The shed was built largely of driftwood logs [...] The location was [...] incomparably beautiful. (Ingstad 1937, 36)[12]

"Heimen" – "Home" – as they significantly name the cabin, is made of what is available on site, peat and driftwood, and becomes part of its natural surroundings. The work, the daily routine, is not depicted as alienating, but as part of the cycles of the day and the seasons. It is the joy of work that is emphasized again and again: "each has his special place and his particular duties to perform. [...] Stockings to be darned, breeches to be patched, boots to be stitched, harness to be repaired" (Ingstad 1937, 147, 156).[13]

A habit of orderliness has to be learned, not as a convention, but as a necessity, because the work and preparations have an immediate effect on their lives. If, for example, they are careless about where they pitch their tent, they might awaken completely snowed under and have to embark on a wild dig for skis, sleds, dogs and equipment. The choice of camp site depends on many considerations; the snow must be evenly distributed and not too loosely packed, the camp must not lie wholly in the lee of a hillcrest, for there it would be caught in the drift and snowed under, but at the same time the location must be such that the tent can ride out a storm. To master all this brings a certain joy.

> Our pleasures arose from small happenings: a drive up the valley at a dizzy trot with empty sleds beneath a starry sky; the excitement we felt upon approaching each trap we had set for fox; the constant shifts in snow formations and the quality of the air; the sight of small brooks which, despite the biting cold, in summer guise had broken through the snow. (Ingstad 1937, 159)[14]

The landscape is depicted as both simple and majestic, the daily tasks are portrayed one by one without conflicting considerations, and the silence stands in contrast to the urban cacophony.

According to Greg Garrard, the pastoral "celebrates a bountiful present" (Garrard 2009, 37). Numerous examples of this are to be found in another book by Ingstad, *The Land of Feast and Famine* (*Pelsjegerliv*, 1931), set in northern Canada, where he lived as a hunter for four years:

> Here is my canoe, here my nets, my guns, and my dogs. Forest and lake are at my disposal. I sit in my shirt-sleeves in front of my tent and feel like a millionaire. [...] On the table in our cabin stands a pot of steaming venison stew, a bowl containing two thick beaver tails saved for this special occasion, and a cup of cranberry sauce. We begrudge ourselves nothing... (Ingstad 1992, 145, 89)[15]

Whereas civilization is characterized by words such as corrupt, impersonal and hectic, nature is categorized as quiet and plentiful. "These people are not so much interested in hunting for its own sake as for the sake of the free and rugged life in the wilds it entails", Ingstad writes about his companions (Ingstad 1992, 124).[16] The same is true of his own life in the Arctic:

> I rambled about wherever I chose, and nothing required me to hurry. If my fancy prevailed upon me to sit down on some hillside for a rest and a quiet

> smoke, I did so; if it crossed my mind to climb a ridge far off in the distance for the sake of the vista which lay beyond, I did so. (Ingstad 1992, 178)[17]

In the classical pastoral, the shepherds meet, have conversations and sing, recite verses and talk about love. If polar literature and the life it describes is characterized by "silence and solitude", as indicated by Nansen's account, this aspect of the pastoral would seem to lack an Arctic counterpart. As part of expedition life and hunting life there will, however, be conversations and gatherings in the evenings, among the immediate group of men themselves and when they meet with other men or expeditions. In Helge Ingstad's books about the Arctic there are countless meetings and conversations, songs and dance, both from meetings among the trappers themselves, and from meetings between trappers and the indigenous population. Here from *The Land of Feast and Famine*:

> It was a hard and fast custom which took me, each evening, over to my neighbours to pay them a visit. [...] The hunters would assemble, and far into the night we would lie there, drawing on our pipes and chatting about hunting and distant lands. [...] As the hours pass by, the mood of the party changes. Mac walks to the centre of the floor and, in spite of his bow-legs, begins dancing a jig which would do credit to a youth of twenty. Then Bablet bursts into song. He knows hundreds of ballads. (Ingstad 1992, 172, 133)[18]

Although the trappers live separately, alone, and the First Nations people live in their extended families and tribes, they experience a commonality based on living in the Arctic. As Ingstad writes: "In spite of the fact that we differ in many respects, the country seems to bind us together" (Ingstad 1992, 124).[19] In a similar way, the shepherds in the classical pastoral are united by their life within the landscape.

Knud Rasmussen's Arctic vitalism

Knud Rasmussen (b. 1879) grew up in Jakobshavn (Ilulissat) in West Greenland and learned as a boy how to handle hunting weapons and dog sleds, in the tradition of the Inuit. From the age of twelve, however, he was schooled in Copenhagen. There are indications of a tragic discord within him between his white and his Inuit identity, but it could just as easily be said that there is a competency in both cultures of which he makes use as a polar scientist (Thisted 2010, 59). The union of instinct and intellect that he experienced among the Inuit guided him towards the life force with which the vitalism of the time was preoccupied.

Vitalism, with its worship of the life force, was an intellectual current at the end of the nineteenth century and beginning of the twentieth that was influenced by modern thinkers like Nietzsche and Bergson and may be detected in much polar literature, both fiction and poetry. This vitalism can be connected with the pastoral mode, through the desire to attain a lifestyle more in keeping with the natural surroundings. An example of this strain of thought appears in Knud Rasmussen's *Min Rejsedagbog: Første Thule Ekspedition 1912* (My Travel Diary: The First Thule Expedition 1912), which describes the crossing of northern Greenland in dogsleds. The expedition members had to sustain themselves by hunting and Rasmussen feels as though he has been transported to the Stone Age, and that it is the life force – which can be defined as a natural energy, an active principle forming part of any living thing – that connects him to his surroundings. Rasmussen's descriptions of "the lungs of the storm" and "the great winter's surge of forces in its powerful chest!" (Rasmussen 1912, 22, 63), are more than rhetorical flourishes; the imagery expresses the idea that everything and everybody is part of the stream of life.[20]

Rasmussen's book begins:

> 6 April 1912
> Today the journey begins, the great route north around Greenland, in the springtime sun and the joy of departing. Hello, comrades, happy men on the threshold of cheerful revelations! The mornings will lift the veil of the great unknown, and with the sun we run to meet our yearnings!
> Now we're off, full of appetite for the dawning day! With muscles tensed, greedy like roving wild animals we greet the beginning of our journey towards the north!
> Clear of mind, ready to run into new worlds! [...]
> Could anyone be richer? (Rasmussen 2005, 11)[21]

The exclamation marks, the salutations, the choice of adjectives and the rhythm combine into a celebration of life and power – an aesthetic vitalism, condensed into language.

Vitalism builds on the view that organic life not only has physical and chemical causes, but is also underwritten by an unknown life force that is contained in the material of life. "Life" becomes a key word for a critique of civilization and culture around the turn of the twentieth century. Vitalism manifested itself in different discourses – in the natural sciences as well as in philosophy – but also in concrete actions, such as hiking in the mountains, swimming nude, doing gymnastics, keeping communal gardens, etc. The following statement by Knud Rassmussen can be situated in this kind of discourse about health: "This roving life makes us lords over our bodies, and each day we feel the joy that accompanies the powers of health" (Rasmussen 2005, 95).[22]

Henri Bergson describes in *L'évolution créatrice* [Creative Evolution, 1907] the sympathy among all living things as one of the characteristics of the life force and notes that approaching the life force cannot occur through intellect alone, but requires intuition, which is a combination of intellect and instinct. Western rationalism has, however, suppressed the instinctual side. Rasmussen finds the latter among the Inuit; for example, his description of the expedition's encounter with a seal on the ice in the Danmarksfjord: "At the sight of it, both Eskimos were gripped by the mystical feeling of joy with which only the born hunter is familiar, in a way that can perhaps be compared to the war stallion's rearing up at familiar decampment signals" (Rasmussen 2005, 153).[23] This observation does not mean that Rasmussen saw only the element of instinct in the Inuit: he was quite preoccupied with their intellectual culture and assembled a large collection of their legends, myths and stories.

According to Rasmussen, a polar scientist who travelled and lived as the Inuit would be a happy and complete person: "Perhaps it is because his risk is more obvious, more present, his effort is more tangible; and all of this simplifies existence into lines that go either with or against the grain" (Rasmussen 2005, 153).[24] And, he concludes: "He is constantly on his way towards the marvellous" (Rasmussen 2005, 153).[25]

Just as the genre of the pastoral has been criticized for being reactionary, so the word vitalism seems to trigger associations with reactionary ideology, irrationalism and even fascism, as well as positive ones such as health, natural presence and what is now referred to as deep ecology. Both concepts can, however, be seen as challenging the notion of progress as embedded in modernity.

It is probably true to say that if Nansen is the great Nordic polar scientist, Knud Rasmussen, with his many books about Greenland and Canada, is the great Nordic polar writer. *Den 5. Thule-ekspedition* [The Fifth Thule Expedition], a book about Knud Rasmussen's journey in 1921–1924 from Baffin Bay to the Bering Strait on dog sleds, is his major work. The Danish writer and Nobel Prize winner Johannes V. Jensen, in a speech on Knud Rasmussen's fiftieth birthday, observes with regard to this book:

> Let me tell you now, that when I read your great work about the sled journey from Greenland to the Pacific Ocean, I felt obliterated, like a little student faced with an acclaimed and eminent literary arbiter; it was much worse, for it was certainly not literature that was speaking, but a quarter of all the earth, existence itself and the human, from a primordial state up

> until the appearance of culture, direct *contact.* And what can one write that can stand comparison with this? (Jensen 1941, 47)[26]

As Kirsten Hastrup points out, Johannes V. Jensen's verbal imagery was influenced by Knud Rasmussen's early books. Jensen's *Bræen. Myter om Istiden og det første Menneske* (The Glacier: Myths About the Ice Age and the First Human) came out in 1908, which is to say after Rasmussen's three books *Nye Mennesker* (1905; New People), *Under Nordenvindens Svøbe* (1906; Under the Whip of the North Wind) and *Lapland* (1907) – books with a significant, arctic and primitive aesthetic (Hastrup 2010, 683).

Trappers' memoirs: Arctic continuity

"At the root of pastoral is the idea of nature as a stable, enduring counterpoint to the disruptive energy and change of human societies," Greg Garrard writes (2004, 56). This feeling of continuity is an aspect of many travelogues from the Arctic. In *Eskimo Life*, for instance, Nansen depicts Greenland thus: "It is strong and wild, this Nature, like a saga of antiquity carved in ice and stone" (Nansen 2003, 2).[27] During a stay with a First Nations tribe in northern Canada in 1929, Helge Ingstad experiences not only the landscape, but also the people as part of an unbroken tradition from the beginning of time: "Here was the hard cold struggle for existence as it has developed amongst these Indians for thousands of years" (Ingstad 1992, 218).[28] The time that clocks are designed to measure does not count here, the way of living in the northern wilderness follows the seasons, the daylight and the weather.

In the pastoral, the hubris of man and civilization is scaled down to the shepherd, a figure of humility, who lives "a life of conscious simplicity", in Alpers' terms (1996, 188). There are several parallels with life in the Arctic as depicted in polar literature, for instance in how the Norwegian trapper Wanny Wollstad describes her residence in Hornsund, Svalbard, in 1932, in her book *The First Woman as a Trapper in Svalbard* (Første kvinne som fangstmann på Svalbard, 1956):

> Up here we build neither storehouse, brewery nor boathouse. The chalet consists only of a low house, about eight metres long and three metres wide [...] On the side facing the endless ocean is a large window – three square windowpanes wide and two high [...] In stormy weather we cut and chop firewood indoors, as well as skinning the fox and thawing the polar bear and preparing meat and fur, everything is done in our living room. (Wollstad 2005, 7)[29]

The importance of tending to one's daily tasks with diligence is well described in an autobiography by another Norwegian trapper, Henry Rudi, from one of his reflections about trapper life in Svalbard:

> Freedom does not mean to relax for days in a warm and cosy cabin [...] freedom means taking care of one's tools and gear every single day, so you could say to yourself that now I have done what should be done, I have prevailed over the external forces, which are bad weather and darkness, I have prevailed over the inner problems, indifference and melancholy. I have won over powerful forces, therefore I am free. [...] Up in the north the days stand out in their stark reality: Keep on working or surrender. (Rudi 1958, 160)[30]

Life in the Arctic is portrayed in terms of simplicity and unequivocalness, as a contrast to the complexity of modernity. The popularity of these polar accounts is probably due to the fact that they present the combination of explorer and hunter as suggesting a true, authentic way of life that modern civilization only knows as utopia. In the Arctic it is possible to enjoy "a delightful sense of freedom," because there is no other jurisdiction than nature, and no other laws than those agreed upon between men, the Danish polar traveller Jørn Riel writes (Riel 1975, 45).[31]

While modernity postulates heterogeneity and change, the Arctic landscape is perceived as a continuous tradition from time immemorial. The Norwegian polar explorer John Giæver writes in his book *Animal Trails and Bird Migration at 74° North* (*Dyretråkk og fugletrekk på 74° nord*, 1955): "What made me come back to these lands beyond sea and ice, year after year, was first of all the wildlife. There the Creator's elementary laws lay naked and open on snow and rock, and they were easy to read and understand" (Giæver 1967, 10).

Conclusion

On the one hand, polar expeditions during the late 1800s and early 1900s are expressions of modern science and rationality, with polar voyagers as exponents of civilization's very latest technology. On the other hand, yearnings towards the desolate Arctic are based upon a desire to escape modernity. This paradox is discussed by Michael F. Robinson in *The Coldest Crucible: Arctic Exploration and American Culture* (2006). He shows that the Arctic expeditions from 1890 onwards no longer necessarily argue for the expansion of civilization, with science as companion, but just as often take a critical stand. According to this line of thinking, explorers sailed north not to extend but to escape the reach of civilization, to find a route that returned them, in a symbolic sense, to the original state of nature (Robinson 2006). This frame could contain Nansen, Rasmussen and Ingstad.

The type of investigation of and negotiation with modernity present in the writings of Nansen, Rasmussen and Ingstad are reminiscent of ideas that a number of modernist authors such as Knut Hamsun (for instance in *Growth of the Soil*, 1917), were engaging with at the same time. In a pointed remark in *Eskimo Life*, Nansen criticizes western civilization: "the Europeans [are] a corrupt and dishonourable race, who ought to come to Greenland in order to learn morals" (Nansen 2003, 345).[32] At the same time, there is in Nansen and later also in Rasmussen and Ingstad a recognition of indigenous people's vulnerability in the face of modernity, and an awareness that their traditional way of life is about to be lost.

As "a way of connecting the self to the environment" (Bate 1991, 115), Jonathan Bate observes, with reference to William Wordsworth, the pastoral is not a myth, but a psychological necessity. The reflections taking place in the pastoral can be seen as an attempt to develop a human economy that is responsive in its relationship with the economy of nature. The pastoral is defined by the problem of man's relation to nature, and in that respect it should not be dismissed as nostalgic. Or, if the pastoral indeed looks back, it might be as a necessary corrective to contemporary life and thought.

References

Alpers, Paul. 1996. *What Is Pastoral*. Chicago, IL: University of Chicago Press.

Bate, Jonathan. 1991. *Romantic Ecology: Wordsworth and the Environmental Tradition*. London: Routledge.

Buell, Lawrence. 1995. *The Environmental Imagination: Thoreau, Nature Writing and the Formation of American Culture*. Cambridge, MA: Harvard University Press.

Davidson, Peter. 2005. *The Idea of North*, London: Reaktion Books.

Finch, Robert and John Elder. 2002. *Nature Writing: The Tradition in English*. New York: Norton.

Garrard, Greg. 2004. *Ecocriticism*. London: Routledge.

Gaupseth, Silje. 2011. "Naive naturbarn eller ren klokskap? Inuittene i Vilhjalmur Stefanssons vennlige arktis" [Naivity or wisdom: The Inuits in Vilhjalmur Stefansson's friendly Arctic]. In *Reiser og ekspedisjoner i det litterære Arktis* [Travels and expeditions in the

Literary Arctic], edited by Johan Schimanski, Cathrine Theodorsen and Henning Howlid Wærp, 201–219, Trondheim: Tapir.
Giæver, John. (1955) 1967. *Dyretråkk og fugletrekk på 74° nord* [Animal trails and bird migration at 74° North]. Oslo: Den norske bokklubben.
Hastrup, Kirsten. 2010. *Vinterens hjerte. Knud Rasmussen og hans tid* [The heart of the winter: Knud Rasmussen and his time]. Copenhagen: Gad.
Huggan, Graham and Helen Tiffin. 2010. *Postcolonial Ecocriticism: Literature, Animals, Environment*. London: Routledge.
Ingstad, Helge. (1931) 2010. *Pelsjegerliv. Blant Nord-Canadas indianere* [A fur-hunter's life among North Canada's Indians]. Oslo: Gyldendal.
—. 1992. *The Land of Feast and Famine*. Translated by Eugene Gay-Tifft. Montreal: McGill-Queens's University Press.
—. (1935) 2006. *Øst for den store bre* [East of the Great Glacier]. Oslo: Gyldendal.
—. 1937. *East of the Great Glacier*. Translated by Eugene Gay-Tifft. New York: Alfred A. Knopf.
Jensen, Johannes V. 1941. *Mindets-Tavle* [Memories]. Copenhagen: Gyldendal.
Jølle, Harald Dag. 2011. *Nansen. Oppdageren* [Nansen: The explorer]. Oslo: Gyldendal.
Knutsen, Nils Magne. 1991. *Fra Ottar til Oluf. Artikler om nordnorsk litteratur* [From Ottar to Oluf: Articles on northern Norwegian literature]. Tromsø: Cassiopeia.
Nansen, Fridtjof. 1891. *Eskimoliv* [Eskimo Life]. Kristiania: Aschehoug.
—. 2003. *Eskimo Life*. Translated by William Archer. Honolulu, Hawaii: University Press of the Pacific.
—. 1914. *Gjennem Sibirien* [Through Siberia]. Kristiania: Jacob Dybwads forlag.
—. (1890) 2003. *På ski over Grønland. En skildring av den norske grønlands-ekspedisjon 1888–89* [Skiing across Greenland: An account of the Norwegian Greenland Expedition 1888–89]. Oslo: Kagge.
—. 2002. *The First Crossing of Greenland*. Translated by Hubert Majendie Gepp. Edinburgh: Birlinn.
—. 1897. *Fram over polhavet. Den norske polarfærd 1893–1896.* [*Fram* across the Arctic Sea. The Norwegian Polar expedition 1893–1896] Kristiania: Aschehoug.
—. 1999. *Farthest North*. New York: Random House.
—. 1914. *Through Siberia – The Land of the Future*, London: William Heinemann.

Rasmussen, Knud. (1915) 2005. *Min Rejsedagbog. Første Thule Ekspedition 1912* [My Travel Diary: First Thule Expedition 1912]. Copenhagen: Gyldendal.

—. 1933. "Radioforedrag for børn, d.26. marts 1933" [Radio lecture for children, Mar. 26 1933].

Riel, Jørn. 1975. *En arktisk safari og andre skrøner* [An Arctic safari and other tall tales]. Copenhagen: Lademann.

Riffenburgh, Beau. 1993. *The Myth of the Explorer: The Press, Sensationalism, and Geographical Discovery*. London: Belhaven Press.

Robinson, Michael F. 2006. *The Coldest Crucible: Arctic Exploration and American Culture*. Chicago, IL: University of Chicago Press.

Rudi, Henry. 1958. *Isbjørnkongen* [The Polar Bear King]. Told to Lars Normann Sørensen. Oslo: Gyldendal.

Schimanski, Johan, Cathrine Theodorsen and Henning Howlid Wærp. 2011. *Reiser og ekspedisjoner i det litterære Arktis* [Travels and Expeditions in the Literary Arctic]. Trondheim: Tapir.

Thisted, Kirsten. 2010. "Voicing the Arctic: Knud Rasmussen and the Ambivalence of Cultural Translation." In *Arctic Discourses*, edited by Anka Ryall, Johan Schimanski and Henning Howlid Wærp, 59–81. Newcastle: Cambridge Scholars.

Vassenden, Eirik. 2012. *Norsk vitalisme. Litteratur, ideologi og livsdyrking 1890–1940* [Norwegian Vitalism: Literature, Ideology and the Worshipping of Life, 1890–1940]. Oslo: Scandinavian Academic Press.

Wollstad, Wanny. 1956. *Første kvinne som fangstmann på Svalbard* [First woman has trapper in Svalbard]. Oslo: Tanum.

Notes

[1] "Jeg hører susen av skogen der ute i mørket – gammel kjenning. Det er som storskogen der hjemme. – Skogens evige sus også her over de endeløse flater" (Nansen 1914, 133).

[2] "Slik er jeg med på våren fra dag til dag" (Ingstad 2010, 69).

[3] "det synes mig uforsvarligt og letsindigt at dra afsted med slige greier" (Nansen 1897, 281).

[4] "Jeg mindes en gang jeg talte med en engelsk polarfarer om dette punkt; han hadde også et mistrøstig syn på livet i de arktiske egne" (Nansen 1897, 281).

[5] "For min egen del kan jeg bare si at jeg ikke har merket noget spor af ældende eller svækkende indflydelse af polarnatten i nogen retning – tvertimod blir jeg yngre. Dette rolige regelmæssige liv passer mig overordentlig godt, og jeg mindes ikke at jeg har været så legemlig sund og i fuld ligevegt som netop nu. Jeg vilde

snarere anbefale disse egne som et fortrinlig sanatorium for nervesvage og nedbrudte mennesker"(Nansen 1897, 281).

[6] "Uf dette forskende liv"; "jeg lenges usigelig tilbake til det enkle" (Nansen in Jølle 1900, 384).

[7] "Samfunn, damp, store tanker og stor elendighet – alt ligger fjernt – å ferdes fritt og gledes ved livet – det er alt" (Nansen 2003, 477).

[8] "For første gang drages her den arktiske natur ind i litteraturen; den var før kun et billede på rædsler og lidelser. Nu stråler den i uanede farvers rigdom. Den første hvide mand har set ind i Polarlandenes hjerte og forstået deres skønhed" (Rasmussen 1930, in Hastrup 2010, 692).

[9] "det enkle, sterke landet [...] under et flimmer av nordlys og stjerner" (Ingstad 2006, 78).

[10] "Vi sov i telt og kokte vår te med duftende lyng. [...] det fri landstrykerliv inne mellem Grønlands fjell" (Ingstad 2006, 36).

[11] "Her er en elv, og jeg er sliten" (Ingstad 2006, 89). The translation by Eugene Gay-Tifft reads "Tired and worn out, I sit down on a stone by the river" (Ingstad 1937, 180), but this formulation does not do justice to Ingstad's style.

[12] "Noe praktverk blev det ikke, men en helstøpt og ordentlig hytte som det var noe vennlig ved. Torvklædningen gikk langt opefter veggene. Sjåen var for en stor del bygd av rekvedtømmer [...] Beliggenheten var [...] makeløs vakker [...]" (Ingstad 2006, 21).

[13]"hver ting har sin plass og hver av oss sitt bestemte arbeide. [...] Strømper skal stoppes, bukser lappes, støvler syes igjen, seletøi repareres"(Ingstad 2006,73, 77).

[14] "Det var de små hendelsene som skapte gledene for oss: kjøringen med tomsleden innefter dalen i susende trav under stjernehimmelen, spenningen ved hvert redskap vi hadde satt for reven, det stadig skiftende i sneformasjoner og luftens tone, småbekker, som tross den bitende kulde brøt sommerlig op over sneen" (Ingstad 2006, 78).

[15] "Her er min kano, mine garn, min børse og mine hunder. Skogene og vannene står til min rådighet. Jeg sitter i skjorteermene foran teltet og føler meg som en millionær" . "På hyttebordet står en dampende gryte reinkjøtt, en skål med to digre beverhaler gjemt for anledningen, og en tallerken med tyttebær. Vi misunner ingen" (Ingstad 2010, 101 & 64).

[16] "Det som da opptar disse folkene, er ikke så meget pelsjakten som selve det sterke, fri liv i villmarken" (Ingstad 2010, 88).

[17] "Jeg streifet omkring som jeg ville, og det hastet ikke med noen ting. Fikk jeg lyst til å ta en pust i bakken og kveike en pipe, gjorde jeg det, kom det over meg at jeg ville klatre over en fjellrygg langt borte for å få greie på hvordan det så ut bakom, gjorde jeg det" (Ingstad 2010, 122).

[18] "Det var fast skikk at jeg utpå kvelden tok meg en tur over til naboene. [...] Jegerne kom da sammen, og langt utover natten lå vi der og dampet på pipene, pratet om jakt og fjerne land". "Efter som timene går, blir tonen en annen. Mac stiger frem på gulvet og gir seg til å steppe som en ungdom på tyve. Bablet synger viser. Han har hundre av dem på lager [...]") (Ingstad 2010, 118, 94).

[19] "Til tross for at vi er forskjellige på mange måter, binder landet oss sammen" (Ingstad 2010, 87).
[20] "stormens lunger," "den store vinters sus af kræfter i dens mægtige bringe" (Rasmussen, *Min Rejsedagbog* 22, 63).
[21] "6. april 1912. I dag begynder rejsen, den store fart norden om Grønland, i forårssol og herlig udfartsglæde. Hej, kammerater, lykkelige mænd på tærskelen til muntre åbenbaringer! Morgenerne skal løfte fligen for det store ukendte, og med solen løber vi vore længsler i møde! Nu i vejret, fulde af appetit på de gryende dage! Muskelspændte, grådige som farende rovdyr hilser vi starten, farten mod nord! Klare i sindet, for alle klude rede til at lænse ind i nye verdener! [....] Kan nogen være rigere? (Rasmussen 2005, 11)
[22] "Dette farende liv gør os til herrer over vore legemer, og vi føler hver dag den glæde, som følger med sunde kræfter" (Rasmussen 2005, 95).
[23] "Begge eskimoer blev ved synet af den grebet af den mystiske glædesfølelse, som kun den fødte jæger kender til, på en måde, der måske kan sammenlignes med krigshingstens stejlen ved kendte opbrudssignaler" (Rasmussen 2005, 86).
[24] "Det ligger måske i, at hans risiko er mer åbenbar, mer nærværende, hans indsats håndgribeligere; og alt dette forenkler tilværelsen i linjer, der direkte går med eller imod" (Rasmussen 2005, 153).
[25] "Han er bestandig på vej mod det vidunderlige" (Rasmussen 2005, 153).
[26] "La mig sige dig nu, at da jeg havde læst dit store Værk om Slæderejsen fra Grønland til Stillehavet, følte jeg mig tilintetgjort, som en lille Student overfor en Navnkundighed og literær Sværvægts-Dommer; det var meget værre, for det var slet ikke Literaturen, der talte, men en Fjerdedel af Jorden, Tilværelsen selv og Mennesket, fra Urstadiet og op til Kulturen, direkte *Berøring* – og hvad har man af skrevne Ting at sætte op mod sligt?" (Jensen 1941, 47).
[27] "Den er sterk og vild, denne natur, som en oldtidssaga, skaaret i is og sten" (Nansen 1891, 1).
[28] "Her var den rendyrket, kampen for tilværelsen slik som den har formet seg for disse indianerne gjennom tusener av år" (Ingstad 2010, 147).
[29] "Her oppe bygger vi verken stabbur, bryggerhus eller naust. Eneboligen består bare av et lavt hus, ca. åtte meter langt og tre meter bredt [...] På forsiden ut mot det endeløse hav er et stort vindu – tre ruter i bredden og to i høyden [...] I styggevær sages og hogges brenneveden inne, rev flåes, bjørn tines, flåes og spekkes også i vår lille salong" (Wollstad 1956, 14).
[30] "Friheta består ikke i å legge seg over døgnvis i ei varm og god hytte [...] Friheta består i å røkte sine redskaper hver bidige dag, så en kan si til seg sjøl at nå har jeg gjort det som gjøres skulle, jeg har vunnet over de ytre motstandskreftene, som er styggvær og mørke, jeg har vunnet over de indre som er oppgitthet, likesælhet eller noe slikt. Jeg har vunnet over sterke krefter, derfor er jeg fri. [...] der oppe trer de så klart i dagen, de står fram i sine enkle fakta: Klemme på eller gi seg over" (Rudi 1958, 160).
[31] "en herlig fornemmelse af frihed" (Riel 1975, 45).
[32] "europæerne er en uhæderlig, fordærvet rase, som burde komme til Grønland for at lære moral" (Nansen 1891, 288).

EXOTIC ARCTIC

EARLY MASS TOURISM AT THE NORTH CAPE: INFRASTRUCTURE, ENVIRONMENT AND SOCIAL PRACTICES

ULRIKE SPRING

Today, the North Cape is the leading tourist attraction in Northern Norway (Barstad and Nilsen 2010, 3), providing tourists with an extensive tourist infrastructure, including a large parking lot, a huge globe that has adorned the Cape since the 1970s and a so-called North Cape Experience Centre (Nordkapphallen). The Cape is at the northernmost edge of the European continent and consists of a steep cliff emerging from the polar sea. Its development from a seldom-visited place to the most popular tourist destination in Northern Norway started during the final decades of the nineteenth century, as a result of rapidly spreading modernization processes across Europe. As I will show in this chapter, during this short period it became part of modern tourism discourse through being included in a complex network of narratives, perspectives and approaches, as well as becoming part of sightseeing rituals that placed it in various relations to its surroundings (for a similar argument on tourism in general see Hannam, Sheller, and Urry 2006, 13). This discourse evolved around, and in dialogue with, a rapid expansion of modern infrastructure in the North and on the Cape itself.

The extensive development of infrastructure was a major driving force of the modernization processes that took place during the nineteenth century and a necessary precondition for the development of tourism (Zuelow 2016, 44, 59). Infrastructure influences how tourist destinations are perceived and, hence, knowledge of its development in the North and on the Cape itself is vital for understanding tourist evaluations of the Cape. My focus will be on the period spanning the late 1870s to the early twentieth century. This was a time when infrastructure in Norway was developing rapidly and, as a consequence, the accessibility of the North changed radically. Examples of tourism infrastructure in the North include: new roads, electricity, telegraph stations and hotels, but also steamships which made it possible to transport large numbers of passengers. These developments affected the way in which the North was perceived. I have shown earlier that in the tourist imagination it gradually became part of Europe and European culture, although it was at the same time located in the Arctic or, as one would say at the time, Northern nature and culture (Spring 2016).

Over the past few years, inspired by the growth of information and environmental studies, scholars have focused on the relevance of infrastructure to perceptions of the environment (Edwards 2002; Star and Bowker 2006; Edwards 2009; Schipper and Schot 2011; Kaijser, van der Vleuten, and Högselius 2015). In addition, and informed by Science and Technology Studies (STS) and Mobilities Studies, there has been a shift towards the concept of *infrastructuring*, in an effort to move away from the idea of infrastructure as something static and stable and instead to emphasize its processual character and social and cultural embeddedness (Pipek and Wulf 2009; Blok, Nakazora, and Winthereik 2016). As Blok, Nakazora, and Winthereik point out, infrastructuring "involves attending to how 'the environment' is managed and known, through what material and conceptual means, and to what effects" (2016, 3). In other words, it looks at the material, social, cultural and political practices involved in how society approaches and arranges the environment, and the effects this has on the use and imagining of the environment. An advantage of applying this verb rather than the noun "infrastructure" is that it conceptualizes infrastructure as a hybrid and dynamic process. It also emphasizes that infrastructure is always relational (Blok, Nakazora, and Winthereik 2016, 11): infrastructure develops in relation to something or someone. Hence, the concept of infrastructuring can include a consideration of the wider context in which the infrastructure in question is placed. Inspired by Actor-Network Theory (ANT), researchers envisage environments as themselves active in infrastructuring processes (Näser-Lather and Neubert 2015, 5–6; Blok, Nakazora, and Winthereik 2016, 5–6). In the case of the North, this means that the Arctic or Northern environment responds to the man-made infrastructure, both affecting, challenging and hindering it. The third component in this relational triangle is the human intervention, in my case the visitors: they use the infrastructure available to them, they walk through and interpret the landscape surrounding them and they apply concepts of what constitutes the North based on these experiences.

What exactly is the interaction between the new infrastructure leading to the Cape, the Northern landscape and the travellers? Whereas I wish to find out more about the man-made physical structures at the North Cape and how they interacted with the travellers' experiences, I am less interested in the agents behind the infrastructuring processes or in the political and economic strategies behind building this infrastructure. The questions I wish to address are: How did travellers use infrastructure, what social and cultural practices did they perform using them, and how did this influence and define their understanding and experience of the Cape? How did infrastructuring contribute to making the Cape part of modern tourist discourse? As I will argue, in the years around 1900, a specific North Cape tourist identity emerged as a result of the interplay between novel infrastructuring processes, specific tourist practices and the Northern landscape.

Research on travel to Norway in the late nineteenth and early twentieth centuries has primarily focused on the history of travel in general terms and on travellers' perceptions of the North (Birkeland 2000; Fjågesund and Symes 2003; Povlsen (ed.) 2007; Büchten 2009; Spode 2010; Gaupseth, Federhofer, and Aspaas (eds.) 2013; Walchester 2014), while infrastructure has commonly functioned as a background explanation.[1] By focusing explicitly on infrastructure and its relation to travellers and the Northern environment, I thus foreground a previously neglected topic in research on early mass tourism to the North. My starting point is the argument that material changes go hand in hand with conceptual changes, and that alterations in infrastructure will affect people's experience of the North. In this way, I propose to contribute to the discussion on how modernization processes in the nineteenth century shaped tourist perceptions of the North.

At the same time, however, I do not want to suggest that the results of this particular case study are representative of a Northern or Arctic discourse that was dominant around 1900. Such an approach would mean drawing a single line from the specific to the general and to deny the fact that discourses of the North are played out in different ways at different times and as different responses to the available infrastructure. For example, similar studies of towns in Northern Norway, such as Tromsø or Hammerfest, with a longer and more complex history of infrastructuring, may reveal different tourist experiences of the North than those I find at the North Cape. My approach is in line with studies on infrastructuring that, rather than encouraging global applications and generalized theses, focus on the complexity of the relations between specific environments, infrastructures and humans (Blok, Nakazora and Winthereik 2016).

Travelling to the North

My main sources are reports on journeys to the North Cape that were published in German by travellers and journalists in newspapers, journals and books between the 1870s and early 1910s, a period during which Northern Norway was gradually becoming accessible to tourists. I have included guidebooks and accounts by travellers from Germany, Switzerland and the Austro-Hungarian Empire; they belong to what might be called a corpus of North Cape literature in German that is often characterized by cross-references. There is a wide range of publications on individual tourists travelling to the North, and important research has been done on tourism from the German-speaking countries to the Norwegian North during the nineteenth century (e.g. Göttsch-Elten 2001; Büchten 2009; Kinzler and Tillmann (eds.) 2010; Spode 2010, 20–7; Fjågesund 2014, see in particular his chapter "The Case of Germany"). However, while many of them mention travels to the Cape, there is little systematic research on early tourist visits from these countries to the North Cape.

In addition to guidebooks such as Bædeker, which was reprinted in several revised editions, as well as travelogues published in book form, I have found many articles on tours to the North Cape in local, regional and national newspapers and journals. My focus has been on reports by or about tourists who visited the North Cape as part of an organized trip; in other words, not individual travellers or scientists on expeditions. This is because I am interested in the experience of the North Cape as a tourist destination and a collective experience, and because that experience was arguably different from those of individual tours or journeys with the aim of scientific research. However, I do occasionally include accounts of individual tours when this serves to showcase contrasts or to explain historical developments.

While my focus is on the North Cape, I take inspiration from Actor-Network Theory and consider the Cape as a node in an extended network that is situated in relation to other places such as Hammerfest (the last town before reaching the Cape), the Arctic Circle (crossing into the Arctic) and the Arctic Sea (the unknown), as well as objects such as the midnight sun (the exotic), the steamship (representing modernity), the bodies of the travellers and the travellers' homes in Europe, to name only those that were mentioned most often in the material. It is not only the relationships between these nodes, but also the alternative forms of mobility enacted by each of them that influenced narratives of the Cape. Hence, it makes a difference to the traveller's perspective whether the Cape is approached from the sea or from land and whether one uses boats or cruise ships or, later, bikes or cars. For Jacobsen (1997, 347), some of what he calls the "sacralization" of the Cape was diminished when a new road was built in 1956 leading directly to the top, making the North Cape easily accessible and the walk up from Hornvik Bay superfluous. While my sources only sporadically refer to a sacred experience, many of them emphasize the exceptionality and sublimity of the cliff when first glimpsed from the sea. Likewise, only looking at the Cape and the midnight sun from a ship will create alternative perspectives than walking uphill to the main plateau and gazing at the sun from the top.

Travel to the North Cape became popular during the final decades of the nineteenth century. Soon hundreds of tourists would embark on these trips, although it was still a long way before one could speak of "mass tourism". The trip to the Cape in 1873 by King Oscar II of Sweden and Norway is often cited as an important spur for the increase in tourism to the cliff (e.g. "Am Nordkap" 1908, 147; Richter Hanssen 1990, 239). Certainly, the widely publicized trips to Norway and the Cape by the German Emperor Wilhelm II after 1889 (Büchten 2009, 115; Gammelien 2010), and also by other prominent royalty such as Austro-Hungarian Archduke Karl Ludwig in 1890, contributed to its attractiveness. For many German speakers, theories stressing a shared origin between Germans and Scandinavians, Richard Wagner's music, Viking-inspired handicrafts, the German and Austro-Hungarian polar expeditions of the 1860s and 1870s, followed by the International Polar Year 1882–3 and the spectacular polar expeditions of Fridtjof Nansen during the 1880s and 1890s may also have contributed to a growing interest in the North (see Spode 2010, 27; Henningsen 2010, 37, 39; Halén 2010; Fjågesund 2014, 471–6; Büchten 2009, 114–6). Moreover, during the final decades of the nineteenth century, the Arctic was among the few places close to Europe that was still largely unexplored. Travelling to the North, even if one did not reach the unknown Arctic itself, carried an aura of the mysterious, sublime and pristine. As many travel reports imply, it was considered a break from modernity, urbanism and the sense of a general acceleration of life.

German-language newspapers and journals repeatedly referred to the growing attraction that the North and the Cape held for tourists from German-speaking countries. As the Austrian provincial newspaper *Vorarlberger Landes-Zeitung* affirmed in 1890: the Cape had become fashionable ("Kurze Nachrichten," 3). In the same year, German tourists would constitute the third largest number of foreign travellers to Norway ("Opgave over Antallet" 1890, 189–90), although they were few in number compared to visitors to more Southern destinations (Spode 2010, 24).[2] Still, the sensationalist Viennese newspaper *Neuigkeits-Welt-Blatt* stated in 1895 that the North had become the new South in terms of attracting tourists ("Das nördliche Ende Europas," 10; for later examples see Büchten 2009, 116 and Spode 2010, 24, 27).

In order to understand the way in which infrastructure interacted with tourist experiences, it is necessary to look at the development of infrastructure for travel to Northern Norway during the second half of the nineteenth century. A key characteristic of modern tourism, as of modern society in general, is technological knowledge and mobility. The influx of tourists to the North has to be seen in the context of the invention of new technologies that made travelling to faraway places possible – and affordable. Changes in modes of transport are one of the key features that explain mass tourism. This process began in the late 1830s when the first steamship began transporting mail to the North (Gulowsen 2011, 394). In 1853, regular summer routes were introduced between Hammerfest and Vadsø (Johannessen 1997, 223), and in 1865, the two Norwegian steamship companies Det Bergenske Dampskipsselskab and Det Nordenfjeldske Dampskipsselskab initiated a collaboration, arranging regular routes throughout the year between Trondheim and Hammerfest (Johannessen 1997, 225). While the original aim was to transport goods and local people, the companies soon realized the potential of such tours for the growing tourist industry in the North. Whereas steamships had only sporadically passed the North Cape on their way to the north-east of Norway, preferring to take a safer and quicker route closer to the coast, in 1879 Bergenske and Nordenfjeldske Dampskipsselskab collaborated and decided to send their steamers past the Cape during summer (Bakka 1997, 22). In 1888, a tourist route between Bergen and the Cape was established (Richter Hanssen 1990, 250). A new development took place when the North Norwegian company Vesteraalens Dampskipsselskab started also using steamships consistently at night, introducing in 1893 what is today known as *Hurtigruten* or the Coastal Express (Gulowsen 2011, 405) and drastically reducing the time it took to travel to the North. The journey between Trondheim and Tromsø, for instance, now took 52 rather than 84 hours (Gulowsen 2011, 405). Improving economic connections with the wider European continent also prepared the path for tourism. In 1870, for example, Bergenske and Nordenfjeldske Dampskipsselskab started regular services from Hamburg to Vadsø; in fact, as Richter Hanssen (1990, 307) points out, for decades Germany was the only country that had regular direct traffic to the region of Finnmark in Northern Norway.

As a result of this development, European travellers and tourism companies gradually discovered the European North as a tourist destination. In 1879, Bædeker published the first German edition of its guidebook on Sweden and Norway, an illustration of the growing interest in these countries by German-speaking visitors. The British travel company Thomas Cook organized tours to the North Cape as early as 1875, symbolizing the start of modern tourism to the landmark (Birkeland 2000, 215), the first German cruise ship set off from North Germany to the North Cape in 1890 (Kludas 1988, 185), and the Norddeutsche Lloyd offered regular tours to the North Cape from 1908 onwards (Tillmann 2010, 158). During the early 1910s, a number of different agencies organized journeys to the Cape during the summer months (Bædeker 1911, 429). By the late 1800s, tourists could reach the Cape via luxurious cruise ships, the so-called swimming hotels, the modern and well-outfitted postal steamers and steamers combining tourism and goods transport. All these ships had modern and comfortable facilities, and thus formed a contrast to the wild and sparsely populated North-Norwegian coastline.

The Cape was both a natural and symbolic destination for the journey to the North. It was perceived as a border between the Arctic and Europe or Norway and, with the exception of tourists travelling on to Spitsbergen, it was the northernmost point of their trip, the highlight and the turning point of the journey to the Northern lands, the "Nordlandsreise". The first sight of the Cape was from the sea, and its shape not only made it stand out from the coastline, but also turned it into a sublime experience. The Austrian cleric P. Aemilian Mayr (or Mayer) ("Correspondenzen" 1896) found his gaze arrested by the grey-black cliff, Haffter (1912, 50) experienced its sight as "uncanny", and the Bædeker guide (1879, 358) declared the first sight of the Cape "a majestic view!"[3] According to the Austrian lawyer Julius Reisch (1909, 3), the word "North Cape" sufficed to evoke "a certain sensation" among his fellow travellers: "One was soon to enter a place that is said to be the most Northern one of our part of the world and where certainly not all too many stray." On glimpsing the Cape, he described it as "bleak yet impressive" (Reisch 1909, 3). In 1909, when Reisch travelled, journeys to the North Cape were no longer a unique experience; yet, by insisting on its uniqueness and perceived inaccessibility, he retained an image of the Cape and the North as the end of the known world. Moreover, considering that he approached the cliff on a moving ship that was characterized by the latest technology and up-to-date modern infrastructure, the view of the bleak and lonely cliff served to emphasize the contrast between the tourist's embeddedness in modernity and the surrounding sublime Northern landscape. Right from its first edition, Bædeker (1879, 359) quoted the words of the early nineteenth century Italian traveller Giuseppe Acerbi about the sublime character of the ocean under the midnight sun, thus setting the tone for a Romantic discourse of

Northern nature and the sublime, which is still crucial for an understanding of tourist experiences of the Cape (see Birkeland 2005, 40–2).

Before tourist infrastructure had been developed there, the Cape was already a site that derived its meaning from travel discourse, with the exception of fishermen who had long used it as a landmark for navigation (Richter Hanssen 1990, 170, 176). This becomes apparent when it is recognized that it is in fact *not* the northernmost point of mainland Europe, this being Kinnarodden on the peninsula Nordkyn (today Nordkinn), whereas the Cape lies on an island. Moreover, the low promontory Knivskjellodden next to it actually reaches further to the north than the North Cape itself. This was already mentioned in Bædeker's first edition of 1879 (358–9), but the Cape was considered the most spectacular point (e.g. "Am Nordkap" 1908, 147). Because of its height and prominent position, it best served the purpose of symbolizing the border of the known world, while the geographical fact of being "northernmost" or belonging to the European mainland became less important. In 1895, the Norwegian geologist Hans Reusch even called Knivskjellodden prosaic in contrast to the poetical Cape (Birkeland 2005, 39). The "spatial dualism" of these two Northern edges was hence "also hierarchically organized" (Birkeland 2002, 72), and one might add that the aesthetic appreciation of the North Cape depended also on the construction of Nordkyn as being of lesser interest (see Birkeland, 2002, 72).

Walking uphill – Northern environment and human control

Materiality, meaning and practice are interlinked and together produce place (Cresswell 2009, 3). Accordingly, an analysis of the role of infrastructure and the making of meaning at the North Cape must also take into account the social and cultural practices performed by tourists on their journey.

Before the 1880s, reaching the Cape was in general a strenuous, expensive and lengthy endeavour which only the most dedicated travellers would undertake. Cruise ship tours such as those organized by Thomas Cook were infrequent and, accordingly, the Cape as a travel destination played only a minor role in travel books. In Frisch's travel guidebook of 1869, the trip to the North Cape only received a few lines of description, and the only way to approach it was by hiring a boat (1869, 109). For most travellers on steamships, like the writer and translator Louis Passarge in 1877, walking uphill was not a part of the journey, as the ships on their way to Vadsø would not stop there and only, if at all, allow a view from the sea (Passarge 1881, 176–8). In its 1879 edition, Bædeker suggested two possible approaches, both starting at Magerøy, the island on which the Cape is placed. One could either hire a rowing boat at Gjesvær (a village west of the Cape), crossing the sea for many hours before arriving at the foot of the Cape at Hornvik Bay and from there walking up to the top for circa 45 minutes (Bædeker 1879, 358). Alternatively, one could walk from the village of Kjelvik (south-east of the Cape) across the island, or travel on foot and by boat from the village of Honningsvaag (359–60). In its next edition, of 1882, Bædeker informs its readers that steamships would make a stop of several hours at Hornvik, thus greatly facilitating the approach and ascent of the Cape (341). Soon, the Cape was to become a tourist spot, although it was mainly accessible through arranged tours on steamships: individual access via Gjesvær or Kjelvik, although still available, was hardly mentioned any more, probably because of the large amount of time and effort it took.

While modern steamships thus seemed to have conquered the wild Northern sea around the Cape, they were often at its mercy as well, especially when bad or windy weather prevailed, as one can read in many reports (e.g. Schwimmer 1895, 3; Beyer 1902a, 27).

As a result of the growth in organized tourism, reaching the Cape was no longer an individual, solitary and strenuous endeavour but part of a modern collective experience. One can trace this shift from an individual to a collective experience in the reports published in newspapers and journals. In the early 1880s, newspapers still reported of individuals who had reached the Cape, for instance the Hungarian Eugen Stoll in 1883 ("Aus Hammerfest…" 1883, 8) or Dr. Tunner from Graz in 1884 ("Kühne Bergsteiger," 4). Neither of these men was very well-known in their time, and their main achievement for the reporters consisted of having climbed the North Cape. This changes in later reports on the Cape, after tours became more common. If individuals were mentioned at all, it was because travel clubs sent telegrams about their journeys to newspapers and referred to some of the travellers by name (e.g. "Die Nordkapfahrt des Touristenklubs" 1887, 4). They might also be part of scientific expeditions or well-known people in society (e.g. "High-life. General A. v. Kodolitsch..." 1887, 9; "Der Direktor des Hofburgtheaters..." 1892, 14). Examples of the latter include trips made by royalty, which were extensively discussed, such as that by the crown prince and crown princess of Sweden and Norway ("Der Kronprinz und die Kronprinzessin..." 1887, 3), by Emperor Wilhelm II, who travelled to the Cape several times (e.g. "Prag, 19. Juli" 1889, 2), the Austro-Hungarian Archduke Karl Ludwig and his son Ferdinand ("Erzherzog Karl Ludwig..." 1890, 3), or the widow of the

Austro-Hungarian crown prince, Stephanie (e.g. "Die Kronprinzessin-Witwe auf Reisen" 1893, 5).

Despite the increasing touristification of the Cape experience, the actual ascent remained a challenging endeavour. Infrastructuring also means ordering the environment according to one's needs and as a result gaining some control over it; the history of the accessibility of the Cape illustrates this dialogue between human intervention and the wild and unruly Northern landscape. After landing at the foot of the cliff by boat, one had to climb for about 45 minutes up a steep zig-zag path to the top (Figure 1). This path was gradually improved over the decades. However, as my material shows, it retained its character as a borderscape, a strenuous and challenging Northern space that had to be traversed on the way from the modern and luxurious ship to the plateau at the top where once again bourgeois culture would be enacted. On this path, the power balance in the triangle of material infrastructure, people and environment tipped towards the environment. This became particularly conspicuous when bad weather did not allow travellers to walk uphill ("Correspondenzen" 1896).

Figure 1. Anders Beer Wilse: "Hornviken – Nordkap," 1909. Photo: Anders Beer Wilse/Norwegian Museum of Cultural History

Throughout the decades covered by my material, the reports described the path as difficult to traverse, although the infrastructure was improving to allow easier ascent. In 1873, a decade before the onset of early mass tourism to the Cape, ropes were secured along the path on the occasion of King Oscar's visit (Richter Hanssen 1990, 240). In 1882, Bædeker described the path as leading across stones, with the last third in particular being difficult to walk on (375). Its character as a border was apparent in the description of the ascent by the Danzig/Gdansk-born journalist and later historian Robert Davidsohn. During the summer of 1884, together with 26 other people from various countries, he had climbed to the top of the Cape. In his description of the walk uphill, the path appears to be devoid of infrastructure and using it was experienced as a frightening activity. He even stated that wild reindeer, which had climbed up here before, would probably have had difficulties. He described the walk up literally as a climb, with the 27 climbers "hanging on the cliff, […] half floating in the air, seeing each other almost vertically above each other." His exclamation: "For Heaven's sake, no looking down!" reinforced this dramatic experience. At the same time, he wrote that some participants experienced the path in a contrasting, more relaxed way, like a Cambridge student who helped him downhill and a Norwegian crew member who carried a basket of champagne bottles to the top (Davidsohn 1884, 514–5). In his text, the absence of any mention of infrastructure made the uphill trek appear to be a significant achievement. When, in 1896, tourists complained to their captain about the poor condition of the path, he allegedly defended it by claiming it to be more poetic that way (Ritz 1897b, 1). Here, the sublime character of the path and its location in the wild Northern landscape is apparent, in spite of the captain's obvious sarcasm.

Most reports, however, refer to infrastructure that would help on the way up, making the actual ascent less strenuous and hence less of a wild Northern experience. Norwegian steam companies had gradually improved the path by placing stones, iron poles and strong ropes and had installed wires as orientation to guide tourists on the often foggy plateau at the top (Ritz 1897a, 1; Beyer 1902a, 27; Bædeker 1903, 454). Even though Bædeker pointed out in 1903 (454) that it was still a demanding journey to the top and that good shoes were necessary, Davidsohn's exclamations would probably have been perceived as exaggerated by most later travellers.

In 1888, the scientist Ferdinand Krauß recognized several difficult sections on the "small, stony path", which had however been made "harmless" by the wired ropes (Krauß 1888, 854). When dermatologist

Ernst Schwimmer travelled to the Cape as part of a group of 62 people in the summer of 1895, he also described the path as “exceedingly bad, since it leads across swampy ground that is covered with slate and gravel”, but at the same time he reassured his readers that the ropes facilitated the ascent of the steepest parts (Schwimmer 1895, 4). Likewise, the philologist and writer Beyer, who travelled in the summer of 1902, described the path as “a stepped pathway that was easy to use and that consisted of loose boulders and strong ropes at the side” (Beyer 1902a, 27). These examples suggest that the walk uphill, though strenuous and considered to be a part of a sublime Northern tourist experience, was by no means impossible to overcome. This interpretation is supported by the repeated mention in some travel reports that the existing infrastructure on the path also made the trip safe for female tourists.

We know from passenger lists and travel reports that there were many women among the steamship passengers who embarked on the journey to the North Cape (e.g. “Nordlandsreise des Wiener Lehrerhausvereines” 1908, 3). Organized leisure travel is a phenomenon of the mid-nineteenth century, and together with the development of available accommodation and transport, it contributed to female independence by making it easier for women to travel and to experience adventures (see Walchester 2014, 85–6). At the same time, expeditions and adventurous travel were still a male domain, and the sublime Norwegian landscape, and in particular the Arctic, were regarded as essentially masculine, in contrast to the Southern Mediterranean’s “feminine” landscape (Fjågesund and Symes 2003, 292–6). Birkeland (2000, 227–31) on the other hand describes images of Northern nature as a female virgin, waiting to be conquered by the tourist-explorer. Both forms of representation assume a gender hierarchy that gives priority to male agency. The statements about the path and its accessibility to women, however, indicate that the Northern environment had now also become available to women. This may be an illustration of an emerging discourse on the North that began during the second half of the nineteenth century and which opened up space for female ascriptions of the formerly exclusively masculine landscape (Fjågesund 2014, 466–8; Hansson 2007), a discourse which no doubt owed a lot to the increase in modern infrastructure.

In the travel reports, women were often mentioned as passengers who required special treatment. Beyer (1902a, 27) remembered that female passengers received particular attention during the windy journey to the Cape, but he did not mention any specific challenges when describing the visit to the cliff. Prefect Johann Huber, who walked uphill with circa 50 passengers plus several crew members in 1895, recalled that ropes had been installed in the steepest places "to make the ascent of the mountain also possible for ladies" (1895, 1). Doctor Elias Haffter, who travelled to the Cape in 1899, reported that of 360 passengers barely 100 reached the top (1912, 53); some of the ladies he attended to turned back halfway, something which he attributed not least to the bad securing of the path, with the rope in some particularly dangerous places lying on the ground (1912, 51). Here, we may still discern the ascent of the Cape as a challenging, adventurous and primarily male experience.

However, we also learn of the German Princess von Sachsen-Meiningen, who travelled incognito to the Cape in 1886 and, according to the newspapers, "was the first one [of the passengers] to climb up the North Cape" ("Eine interessante Reisende" 1886, 16). The point here is that it was considered normal that she was the first of her group to scale the cliff. Likewise, in July 1896, the Austrian church employee Franz Ritz (1897a, 1) allowed a female traveller to pass him on the way uphill, as he realized that she wished to be the first to the top. Julius Reisch, who had visited the Cape in the summer of 1909, described the ascent as difficult, but did not differentiate between the ladies, the gentlemen and the crew, who according to him all walked uphill to the plateau (Reisch 1909, 3).

Notwithstanding the increasing number of tourists making the climb, the fact that one had to embark on a ship, then take a boat and finally walk uphill for about an hour meant that reaching the top was perceived as an achievement regardless of gender, and something that helped to establish the Cape as an exceptional attraction (see Birkeland 2002, 83–4).

A North Cape tourist identity

The travel reports demonstrate that travellers positioned the infrastructure built at the top of the Cape in relation to their experience of the path and the viewing of the midnight sun. In addition, the physical constructions erected at the top of the cliff and the socio-cultural practices enacted there created imaginaries that linked the tourists' present experiences to previous and future ones and evoked associations and identifications across space and time. These imaginaries contributed to the performance of what I call a specific North-Cape tourist identity. In addition, they are often divided into hierarchically structured dichotomies (see Salazar 2012, 864), such as Northern nature versus European culture, modern versus sublime experiences, and male versus female. We have already seen that the infrastructure installed on the path could evoke imaginaries that differentiated between men and women; at the same time, its bad condition and steepness enabled it to represent the wild Arctic landscape, and it formed a contrast to the luxury and modernity of the steamer. The top of the Cape, on the other hand, functioned as a symbol of both Arctic sublimity and exoticism and of European modernity; as the various texts imply, the relationship between these two became stronger the more infrastructure was built and used to enact tourist practices at the top. In other words, the type and size of the infrastructure available on the Cape was related to the meanings ascribed to it and to the practices the tourists enacted there.

The material infrastructure on the plateau was characterized by two types: commemoration and commercialization (see Figure 2). In the early years of organized travel to the Cape, the first category dominated: a granite column commemorating the visit by the Swedish-Norwegian King Oscar II in 1873,[4] a memorial in honour of the German Emperor Wilhelm II's visit in July 1891, and in 1908, the Vienna Teachers' Association (Wiener Lehrerhaus-Verein) unveiled a commemorative plaque to celebrate 60 years in office of the Austro-Hungarian emperor and king, Franz Joseph I ("Der Lehrerhausverein am Nordkap" 1908, 3; Reisch 1909, 3).[5]

The second category, commercialization, became fully developed in 1891–92 when an octagonal wooden building, the later so-called Stoppenbrinck's (or Stoppenbrink's) champagne pavilion, was erected at the top of the North Cape cliff (Richter-Hanssen 2011, 140; Huber 1895, 1). Benches and tables were placed along the walls inside the pavilion, around "a kind of buffet" where champagne and biscuits and probably other supplies were stored (see Huber 1895, 1). There was also a musical clock which functioned as a kind of orchestra (Ritz 1897a, 1; Bædeker 1903, 454). Louis Passarge, who had travelled to the Cape in 1877, when he did not come ashore, and again in 1893, wrote that shortly before his arrival in 1893, Stephanie, the widow of the Austro-Hungarian Crown Prince Rudolph, had danced in the pavilion to the musical clock and that champagne was consumed (c. 1900, 270).

Both types of infrastructure served several purposes: on the one hand practical ones, as they functioned as resting places;[6] on the other, they created a direct link to the outside world in time and space. The monuments reminded the visitor that he or she was part of a royal tradition of travelling to the North, and for the Swedish, German and Austro-Hungarian tourists, they could also provide a direct connection to home. Moreover, the pavilion became a space of European bourgeois and early modern tourist culture. The fact alone that monuments had been erected on top of a hard-to-reach cliff signalled the triumph of modern civilization.

Figure 2. "Oscarsstötten, Nordkap [Oscar Column, North Cape]," 1898–1914. Photo: North Cape Museum/Honningsvåg
The postcard shows the two main types of infrastructure on top of the Cape, monuments and the champagne pavilion. The Oscar column to the left was used as a gathering place before the pavilion was erected.

The pavilion also functioned as a protective space against the weather (Ritz 1897a, 1). By doing so, it changed the experience of the Cape, since tourists no longer had to be exposed to the unreliable weather of the Northern environment. Hence, it turned the Cape into a tourist spot where the same practices were enacted as at other tourist destinations around the world: the writing of postcards, the drinking of champagne to celebrate a particular achievement or experience, and communal listening to music (see Figure 3). In nineteenth-century Cape tourism, however, these practices were not linked to a potentially socially egalitarian mass tourism identity, but to a bourgeois and upper-class culture.

As we have seen, the number of trips organized to the North Cape increased steadily during the final decades of the nineteenth century. As early as 1887, the respected Viennese newspaper *Neue Freie Presse* stated that the journey to the Cape had become available "even for the more modest traveller" ("Nach dem Norden," 28) and travel agencies advertised "cheap" journeys to the North during the 1890s (e.g. "Nordlandreisen" 1897, 8). While this suggests an increased social stratification of Northern tourists, passenger lists and descriptions in the travelogues suggest that they mostly belonged to the nobility and upper-middle classes; that is, the classes that could afford the several weeks' long tour to the North in terms of money and time. In addition, the reports conveyed an image of a bourgeois and wealthy upper-class culture that was enacted at the top: for a long time, there was only expensive champagne to be had, and while some suggested that this was a sign of tourist exploitation (Reisch 1909, 3), others pragmatically explained it with the fact that "mostly wealthy and genteel travellers use to visit the North Cape" (Beyer 1902a, 27). The drinking infrastructure was continuously developed, showing the Cape to be a commercially important place. During the early 1880s, the ship's crew still provided the drinks (Davidsohn 1884, 514), while later a caretaker lived at Hornvik during the summer months and sold glasses or bottles of champagne to the tourists (Ritz 1897a, 1; Bædeker 1903, 454). According to Richter Hanssen (1990, 334), permission to sell wine at the Cape was granted in 1912, provided that food and drinks with little or no alcohol were available as well. Even before that, people from Skarsvåg had sold coffee at Hornvik (Richter Hanssen 1990, 334). There are, however, many more reports of drinking champagne than coffee, no doubt because of the celebratory nature of champagne as opposed to the more mundane status of coffee.

Many travellers also reported that they sang together with the other tourists, and Reisch (1909, 3) recounted having listened to the Austrian imperial anthem by Joseph Haydn. It may be that in his case the ship's orchestra walked up the cliff to perform, as Haffter (1912, 51) remembered from his journey in 1899. The pavilion became the centre of a network of contacts and narratives, a place where tourist rituals such as writing postcards, so-called "Cape-cards", and signing one's name in the visitor book were performed (Huber 1895, 1). A photograph from 1907 shows a "post-office" at the foot of the access path, making the Cape less exotic and positioning it within a global network (Wilse 1907). In his humorous travel report, Haffter (1912, 50) described the ordeal of the postmaster, who had to carry 4,000 postcards to the top in order to stamp them, getting blisters as a result. As Reilly (2009, 10) points out, postcards were immensely popular souvenirs at the time, and even though he is referring to Arctic tours, my examples show that this also applied to North Cape tourists. Moreover, postcards refer to a type of infrastructure that I have not so far mentioned as such, but which was crucial for the tourist experience: the postcards and other equipment such as food, drinks and cameras had to be transported to the top as well. In other words, it was not only the fixed infrastructure installed on the Cape itself, but also the infrastructure taken along to the Cape, that influenced the travellers' experiences and all of these have to be taken into consideration in an appraisal of the North Cape tours of around 1900.

Figure 3. Tourists drinking champagne (?) in Stoppenbrinks Champagnerpaviljong, 1890s–1914. Photo: North Cape Museum/Honningsvåg
Protected from wind and weather, here the travellers could gather to enact specific tourist practices. The photograph shows that besides many nationalities, as reported in the travelogues, also different genders and age groups were represented on board.

The social and cultural practices performed as a joint activity by the tourists demonstrate that the experience of the Cape had become a collective and increasingly common tourist experience and that infrastructuring reinforced this experience and in fact premised it. If there had been no pavilion at the top, some leisure activities would still have been possible, as the case of the tea party described by English steamship traveller Henrietta Kent in 1876 proves (Hansson 2007, 91), but in many cases they would have been impossible due to the Northern weather.

Tourism, and particularly mass tourism, is by nature a cross-cultural and cosmopolitan activity. Many travel reports provide evidence of an idealized version of cross-cultural understanding being performed at the top of the Cape. The communal social and cultural practices, but also the fact of actually having reached the Cape, helped to reinforce this. The diversity of nationalities aboard the steamships (which could transport several hundred people) was an oft-repeated topic in the travel reports. In 1884, Davidsohn travelled together with passengers from England, Scotland, Russia, Germany, Austria and Norway (514); in 1895, Schwimmer's fellow travellers came from France, England, America, Sweden, Norway, Russia, Peru, Spain, Brazil and Austria-Hungary (3); in

1902, Beyer experienced the Cape with tourists from the Netherlands, Germany, America and Hungary (1902b, 29, 31); while in 1909 Reisch travelled with men and women from Austria-Hungary, Germany, Russia, England, Italy, Portugal, Romania, Switzerland and Japan (1909, 1). Many reports mention the easy familiarity that would soon develop among the travellers on the ships and during the collective experience of standing on the Cape, drinking and making toasts together. Davidsohn (1884, 515) toasted with his fellow travellers to Germany, to the nations the travellers came from, to Norway and Old England. Beyer (1902a, 27) gathered with his group in the wooden pavilion, toasting at midnight to the god of light, that is, to the sun, to the tour's participants and to loved ones at home. In 1909, some passengers even posed with a large signboard at the top of the cliff announcing them to be from the steamer *Oceana* (Figure 4). A common identity was reinforced through this activity and made its way into the visual memory of the tour via souvenir photographs.

Figure 4. Anders Beer Wilse: "Nordkap Oceaneas skilt på Toppen 23/7 1909 [Signboard Oceana at the Top of the North Cape]," 18 July 1909. Photo: Anders Beer Wilse/Norwegian Museum of Cultural History
In the background one can see the pavilion.

The most important social practice of their visit to the Cape for nineteenth-century tourists, however, would have been gazing at the midnight sun from the top (Birkeland 2005, 36, 39). The combination of the summit and the midnight sun was considered to be the climax of the tour: it was something, so Huber writes, that he "would certainly not be able to forget during my lifetime" (1895, 2), and Schwimmer pronounced it a "sublime spectacle" (1895, 4).

The travel reports also imply that seeing the sun as part of a collective experience helped to foster the North-Cape tourist identity. Schwimmer described how he and his fellow passengers waited for the sun to appear and that afterwards they all celebrated together on board (1895, 4). Beyer toasted with his fellow travellers to the "God of Light", indicating a shared sacred experience (1902a, 27).

To some extent, due to the improving infrastructure, the Cape developed features of what Marc Augé (1995) has called "non-place"; that is, a particularly modern place without history, where a similar infrastructure is installed and a specific identity common to all users is enacted. Augé applies his concept to late twentieth-century places that can be found all across the globe and are similarly executed, such as airports, shopping malls and hospitals. From this perspective, the Cape lost its particularity and became a symbol similar to other nodes in the tourist network characterized by their extremity, such as the North and South Poles, Cape Horn or the most westerly Cape of Europe in Portugal. In all these cases, the primary quality of the destination was its geographically extreme character, while the relevance of its specific topographic location was minimized. I do not, however, wish to draw this analogy too far: the travel reports illustrate that, across the non-place North Cape, social networks were formed and performed and they indicate that the memories of the trip, through souvenirs such as stones and photographs, would create the specific identity of the North-Cape tourist that would endure long afterwards.

Infrastructuring the North Cape

Tourist infrastructure in the late nineteenth century could come in many different forms: in the shape of invisible matter such as electricity or, the focus of this chapter, in material form such as roads, steamships, tourist buildings and postcards. As the examples of the tours to the North Cape illustrate, they were not just immobile material assemblies, but changed in meaning and usefulness in relation to their users and their environment. The relationship between the components of the triangle of human actor –

environment (or non-human actor) – infrastructure changed continuously and defined the tourists' experiences of their tour. However, as the examples show, certain patterns emerged, with the path uphill being a space where the Northern environment largely retained control, and the top of the cliff became a space where the travellers essentially defined their experience.

Tourist destinations are thus dynamic places. They are performed continuously: through materiality, the tourists' bodies and their social practices, but also through what is physically absent; that is, narratives of the past or imaginaries of Western travel culture (Bærenholdt 2015). The examples above show that the increasing infrastructuring of the North Cape led to the development of a specific modern tourist identity that cut across cultures, assisted by the performance of tourist practices that were similar throughout the emerging European mass tourism industry. The texts also illustrate that previous visits to the Cape by famous people like King Oscar II, the widowed Crown Princess Stephanie and, in particular, Emperor Wilhelm II played a role in tourists' experience of the Cape.

Modern ships functioned as containers in which cross-cultural understandings could take place and where a specific collective tourist identity, not necessarily just a German one, could be enacted and reinforced. At the same time, they were representations of modernity and its increasing mobility. The oscillation of these tourist experiences between the dichotomies: Northern–European, known–unknown, modern–sublime, mobile–immobile, helped to reinforce the identity of a Cape tourist. One could even say that the tourists created a specific "Arctic discourse" (Ryall, Schimanski, and Waerp, 2010) that vacillated between these dichotomies and which is still very much alive (Birkeland 2002; Jacobsen 2015). It probably also helped that the lonely and insular Cape was at once difficult to reach (because of its distance from human settlement) and easily accessible (because of the modern infrastructure) and that it was situated on the route between thriving towns: Hammerfest, Vardø and Vadsø. It was both a solitary and a sublime place, simultaneously on the border with the unknown Arctic and firmly located within the Norwegian landscape, seafaring and discourses of civilization. To Richter Hanssen (1990, 170), this ambivalence explains some of the Cape's appeal; according to him, if the Cape had been a cliff somewhere in the Polar Sea, its attraction would probably not have been the same.

Wild Northern nature would still play a crucial role in the experience of the Cape – with descriptions of seasickness, stormy weather, the loneliness of the North and the still largely uncharted regions of the polar North – but developing infrastructure and the increase and repetitious performance of social and cultural tourist practices would eventually lead to it becoming an organized tourist spot and would prepare the way for the mass tourist destination it is today.

References

"Am Nordkap" [At the North Cape]. 1908. *Dillinger's Reiseführer* Jul. 1.

Augé, Marc. 1995. *Non-Places: Introduction to an Anthropology of Supermodernity*. Translated by John Howe. London: Verso.

"Aus Hammerfest…" [From Hammerfest...]. 1883. *Prager Tagblatt*, Jul. 27.

Bakka Jr., Dag. 1997. *Hurtigruten: sjøveien mot nord* [The Coastal Express: The sea route to the North]. Bergen: Seagull.

Bædeker, K. 1879. *Schweden und Norwegen nebst den wichtigsten Reiserouten durch Dänemark* [Sweden and Norway, and the most important travel routes through Denmark]. Leipzig: Verlag von Karl Bædeker.

—. 1882. *Schweden und Norwegen nebst den wichtigsten Reiserouten durch Dänemark* [Sweden and Norway, and the most important travel routes through Denmark]. Leipzig: Karl Bædeker.

—. 1903. *Schweden und Norwegen nebst den wichtigsten Reiserouten durch Dänemark* [Sweden and Norway, and the most important travel routes through Denmark]. Leipzig: Karl Bædeker.

Bædeker, Karl. 1911. *Schweden, Norwegen nebst den Reiserouten durch Dänemark und Ausflügen nach Island und Spitzbergen* [Sweden, Norway, and travel routes through Denmark and excursions to Iceland and Spitsbergen]. Leipzig: Karl Bædeker.

Bærenholdt, Jørgen Ole. 2015. "Designede turiststeder: Hva får dem til å virke?" [Designed tourist destinations: What makes them effective?]. In *Med Sans for Sted. Nyere Teorier* [With a sense of place: Newer theories], edited by Marit Aure, Nina Gunnerud Berg, Jørn Cruickshank, and Britt Dale, 65–81. Bergen: Fagbokforlaget.

Barstad, Åse Lill, and Eli Nilsen. 2010. "Nordkapp: et fyrtårn for reiselivet i Nord-Norge?" [North Cape: A beacon for tourism in Northern Norway?]. MBA thesis, University of Tromsø. http://munin.uit.no/handle/10037/2988

Beyer, C. 1902a. "Eine Nordlandfahrt auf dem norwegischen Salondampfer 'Kong Harald'" [A Journey to the North on the Norwegian saloon steamship 'King Harald']. *Sport und Salon. Illustrirte Zeitschrift für die vornehme Welt*, Aug. 30.

—. 1902b. "Eine Nordlandfahrt auf dem norwegischen Salondampfer 'Kong Harald'." [A journey to the North on the Norwegian Saloon Steamship, "King Harald"]. *Sport und Salon. Illustrirte Zeitschrift für die vornehme Welt*, Sep. 6.

Birkeland, Inger J. 2000. "Representasjoner av Nordkapp og norsk natur 1860–1900" [Representations of the North Cape and Norwegian nature 1860–1900]. In *Menneskers rom*, edited by Inger J. Birkeland and Bjørg Lien Hanssen, 213–233. Oslo: Unipub.

—. 2002. "Stories from the North: Travel as Place-Making in the Context of Modern Holiday Travel to North Cape, Norway." PhD diss., University of Oslo.

Birkeland, Inger. 2005. *Making Place, Making Self: Travel, Subjectivity and Sexual Difference*. Aldershot: Ashgate.

Blok, Anders, Moe Nakazora, and Brit Ross Winthereik. 2016. "Infrastructuring Environments." *Science as Culture* 25 (1): 1–22. doi:10.1080/09505431.2015.1081500

Brevig, Hans Olaf. 2010. "Wege ins Reiseland Norwegen. Die Infrastrukturentwicklung um 1900" [Paths to the tourist country Norway: The development of infrastructure around 1900]. In *Nordlandreise. Die Geschichte einer touristischen Entdeckung. Historien om oppdagelsen av turistmålet Norge*, edited by Sonja Kinzler and Doris Tillmann, 98–105. Hamburg: Mare.

Büchten, Daniela. 2009. "Midnattssolens land. Turister på vei mot nord 1880–1940" [The Land of the Midnight Sun: Tourists on their way North 1880–1949]. In *Kompassrosen: Orientering mot nord*, edited by Benedicte Gamborg Briså and Bente Lavold, 113–123. Oslo: Nasjonalbiblioteket.

"Correspondenzen. Wels, 5. Nov." 1896. *Salzburger Volksblatt* Nov. 6.

Cresswell, T. 2009. *Place*. n.p.: Elsevier. http://booksite.elsevier.com/brochures/hugy/SampleContent/Place.pdf

"Das nördliche Ende Europas" [The northern end of Europe]. 1895. *Neuigkeits-Welt-Blatt*, Sep. 7.

Davidsohn, R. 1884. "Am Nordkap" [At the North Cape]. *Militär-Zeitung*, Aug. 15.

"Der Direktor des Hofburgtheaters..." [The Director of the Hofburg Theatre…]. 1892. *Neuigkeits-Welt-Blatt*, Jul. 19.

"Der Kronprinz und die Kronprinzessin…" [The Crown Prince and the Crown Princess…]. 1887. *Neuigkeits-Welt-Blatt*, Jul. 24.

"Der Lehrerhausverein am Nordkap" [The Teacher's House Association at the North Cape]. 1908. *Neue Schlesische Zeitung*, Jul. 30.

"Die Kronprinzessin-Witwe auf Reisen" [The widowed crown princess travelling]. 1893. *Pester Lloyd*, Jul. 16.

"Die Nordkapfahrt des Touristenklubs" [The North Cape Tour of the Tourist Club]. 1887. *Neuigkeits-Welt-Blatt*, Jul. 30.

Edwards, Paul N. 2002. "Infrastructure and Modernity: Scales of Force, Time, and Social Organization in the History of Sociotechnical Systems." In *Modernity and Technology*, edited by T. J. Misa, P. Brey, and A. Feenberg, 185–225. Cambridge, MA: MIT Press.

—. 2009. "Introduction: An Agenda for Infrastructure Studies." *JAIS – Journal of the Association for Information Systems* 10 (5): 364–374. http://www.pne.people.si.umich.edu/PDF/EdwardsEtAl2009AgendaForInfrastructureStudies.pdf

"Eine interessante Reisende" [An Interesting Female Traveller]. 1886. *Neuigkeits-Welt-Blatt*, Aug. 5.

"Erzherzog Karl Ludwig…" [Arch Duke Karl Ludwig…]. 1890. *Neuigkeits-Welt-Blatt*, May 30.

Fjågesund, Peter. 2014. *The Dream of the North: A Cultural History to 1920*. Amsterdam: Editions Rodopi.

Fjågesund, Peter, and Ruth A. Symes. 2003. *The Northern Utopia: British Perceptions of Norway in the Nineteenth Century*. Amsterdam: Editions Rodopi.

Frisch, C. F. 1869. *Norwegen. Praktisches Handbuch für Reisende* [Norway: A Practical Handbook for Travellers]. Grieben's Reise-Bibliothek 13. Berlin: Albert Goldschmidt.

Gammelien, Stefan. 2010. "Kaiserliche Nordlandfahrten. Die Reisen Wilhelms II. nach Skandinavien" [Imperial Journeys to the North: Wilhelm II's Journeys to Scandinavia]. In *Nordlandreise. Die Geschichte einer touristischen Entdeckung. Historien om oppdagelsen av turistmålet Norge*, edited by Sonja Kinzler and Doris Tillmann, 68–83. Hamburg: Mare.

Gaupseth, Silje, Marie-Theres Federhofer, and Per Pippin Aspaas, eds. 2013. *Travels in the North: A Multidisciplinary Approach to the Long History of Northern Travel Writing*. Hannover: Wehrhahn.

Göttsch-Elten, Silke. 2001. "Populäre Bilder vom Norden im 19. und 20. Jahrhundert" [Popular Images of the North in the Nineteenth and Twentieth Centuries]. In *Ultima Thule: Bilder des Nordens von der Antike bis zur Gegenwart* (=Imaginatio borealis 1), edited by Annelore Engel-Braunschmidt, Gerhard Fouquet, Wiebke von Hinden, and Inken Schmidt, 123–143. Frankfurt am Main: Peter Lang.

Gulowsen, Jon. 2011. "Kysten som skapte hurtigruten" [The coast that created the Coastal Express]. *Historisk tidsskrift* 90 (3): 391–412.

Haffter, Elias. 1912. *Briefe aus dem hohen Norden. Eine Fahrt nach Spitzbergen mit dem HAPAG-Dampfer "Auguste Viktoria" im Juli 1899* [Letters from the High North: A journey to Spitsbergen with the HAPAG steamship "Auguste Viktoria" in July 1899]. 3rd edition. Frauenfeld: Huber & Co.

Halén, Widar. 2010. "Der Drachenstil. Ein deutsch-norwegischer Stil um 1900" [The Dragon Style: A German-Norwegian Style around 1900]. In *Nordlandreise. Die Geschichte einer touristischen Entdeckung. Historien om oppdagelsen av turistmålet Norge*, edited by Sonja Kinzler and Doris Tillmann, 106–115. Hamburg: Mare.

Hannam, Kevin, Mimi Sheller, and John Urry. 2006. "Editorial: Mobilities, Immobilities and Moorings." *Mobilities* 1 (1): 1–22. doi:10.1080/17450100500489189.

Hansson, Heidi. 2007. "Henrietta Kent and the Feminised North." *Nordlit* 22: 71–96. http://septentrio.uit.no/index.php/nordlit/article/view/1572/1477, accessed 21 Jan. 2017.

Henningsen, Bernd. 2010. "'Norwegen ist unwahrscheinlich lang und weilig.' Von Bildern und Stereotypen" ["Norway is incredibly long and boring." Of images and stereotypes]. In *Nordlandreise. Die Geschichte einer touristischen Entdeckung. Historien om oppdagelsen av turistmålet Norge*, edited by Sonja Kinzler and Doris Tillmann, 34–47. Hamburg: Mare.

"High-life. General A. v. Kodolitsch..." 1887. *Wiener Salonblatt*, Jun. 11.

Huber, Johann. 1895. "Meine Nordlandsreise" [My Journey to the North]. *Salzburger Chronik*, Dec. 19.

Jacobsen, Jens Kristian Steen. 1997. "The Making of an Attraction: The Case of North Cape." *Annals of Tourism Research* 24 (2): 341–356.

—. 2015. "North Cape: In the Land of the Midnight Sun." In *The Tourism Imaginary and Pilgrimages to the Edges of the World*, edited by Nieves Herrero and Sharon R. Roseman, 120–140. Bristol: Channel View Publications.

Johannessen, Finn Erhard. 1997. *Alltid underveis. Postverkets historie gjennom 350 år* [Always on the move: The history of the post through 350 years]. Vol. 1. *1647–1920*. [Oslo:] Elanders forl.

Kaijser, Arne, Erik van der Vleuten and Per Högselius. 2015. *Europe's Infrastructure Transition: Economy, War, Nature*. London: Palgrave Macmillan.

Kinzler, Sonja, and Doris Tillmann, eds. 2010. *Nordlandreise. Die Geschichte einer touristischen Entdeckung. Historien om oppdagelsen av turistmålet Norge* [Journey to the North: The history of a touristic discovery]. Hamburg: Mare.

Kludas, Arnold. 1988. *Die Geschichte der deutschen Passagierschiffahrt. Sprunghaftes Wachstum 1900 bis 1914* [The history of German passenger seafaring: Soaring Growth 1900–1914]. Vol. III. Hamburg: Ernst Kabel.

Kolltveit, Bård. 1978. "Med turister ombord" [With tourists on board]. *Norsk sjøfartsmuseum Årsberetning*: 45–104.

—. 2005. "Dekkstol-oppdagere. 130 år med turistreiser til arktiske og antarktiske farvann" [Deckchair explorers: 130 years of tourist travels to the Arctic and Antarctic Seas]. *Årbok Norsk sjøfartsmuseum*: 221–261.

—. 2016. *Eventyret om norsk cruisefart. Fra vestlandsfjorder til verdens hav* [The adventure of Norwegian cruise travel. From the Vestland fjords to the world's seas]. [Oslo]: Orfeus.

Krauß, Ferdinand. 1888. *Von der Ostsee bis zum Nordcap* [From the Baltic Sea to the North Cape]. Neutitschein: Rainer Hosch.

"Kühne Bergsteiger" [Bold mountaineers]. 1884. *Neuigkeits-Welt-Blatt*, Jul. 15.

"Kurze Nachrichten" [Brief news]. 1890. *Vorarlberger Landes-Zeitung*, Jun. 3.

"Nach dem Norden" [To the North]. 1887. *Neue Freie Presse, Abendblatt*, Apr. 26.

Näser-Lather, Marion, and Christoph Neubert. 2015. "Traffic – Media as Infrastructures and Cultural Practices: Introduction." In *Traffic – Media as Infrastructures and Cultural Practices*, edited by Marion Näser-Lather and Christoph Neubert, 1–27. Amsterdam: Brill / Rodopi.

"Nordlandsreise des Wiener Lehrerhausvereines" [Northern journey of the Vienna Teacher's House Association]. 1908. *Znaimer Tagblatt*, Jul. 17.

"Nordlandreisen" [Journeys to the North]. 1897. *Pester Lloyd*, April 2.

"Opgave over Antallet af udenlandske Reisende, ankomne til Norge i Sommeren 1890" [List of the number of foreign travellers, arrived in Norway in the summer of 1890]. 1891. *Meddelelser fra Det statistiske Centralbureau* 8 (12): 189–190. https://www.ssb.no/a/histstat/sm/sm_1890.pdf

Passarge, L. 1881. *Drei Sommer in Norwegen. Reiseerinnerungen und Kulturstudien* [Three summers in Norway: Travel memories and cultural studies]. Leipzig: Bernhard Schlicke.

—. [c. 1900]. *Sommerfahrten in Norwegen. Reiseerinnerungen, Natur- und Kulturstudien* [Summer journeys in Norway: Travel memories, nature and culture studies]. Vol. 1, 3rd revised edition. Leipzig: B. Elischer Nachfolger.

Pipek, Volkmar, and Volker Wulf. 2009. "Infrastructuring: Toward an Integrated Perspective on the Design and Use of Information Technology." *JAIS – Journal of the Association for Information Systems* 10 (May): 447–473. http://aisel.aisnet.org/cgi/viewcontent.cgi?article=1503&context=jais

Povlsen, Karen Klitgaard, ed. 2007. *Northbound: Travels, Encounters, and Constructions 1700–1830*. Aarhus: Aarhus University Press.

"Prag, 19. Juli" [Prague, 19 July]. 1889. *Prager Abendblatt,* Jul. 19.

Reilly, John T. 2009. *Greetings from Spitsbergen: Tourists at the Eternal Ice 1827–1914*. Trondheim: Tapir Academic Press.

Reisch, Julius. 1909. "Bis zum ewigen Eise. (Nordlandsreise August 1909 auf der 'Thalia')" [To the eternal ice. (Journey to the North August 1909 on the 'Thalia')]. *Czernowitzer Allgemeine Zeitung* Sep. 25.

Richter Hanssen, Einar. 1990. *Nordkapp: en fiskerikommune. Fra de eldste tider til i dag* [North Cape: A fishing county. From ancient times until today]. [Honningsvåg:] Nordkapp kommune.

Richter-Hanssen, Einar. 2011. *Nordkapp – ytterst i verden* [North Cape: Most distant place in the world]. Honningsvåg: Arctic Suvenir AS.

Ritz, Franz. 1897a. "Von Drontheim zum Nordcap. (7. Fortsetzung.)" [From Drontheim to the North Cape. (7th sequel.)]. *Linzer Volksblatt*, Apr. 2.

—. 1897b. „Von Drontheim zum Nordcap. (Schluss.)" [From Drontheim to the North Cape. (End.)]. *Linzer Volksblatt*, Apr. 3.

Rogan, Bjarne. 1998. *Mellom tradisjon og modernisering: kapitler av 1800-tallets samferdselshistorie* [Between tradition and modernization: Chapters of nineteenth-century transport history]. Oslo: Novus.

Ryall, Anka, Johan Schimanski, and Henning Howlid Wærp, eds. 2010. *Arctic Discourses*. Newcastle upon Tyne: Cambridge Scholars Publishing.

Salazar, Noel B. 2012. "Tourism Imaginaries: A Conceptual Approach." *Annals of Tourism Research* 39 (2): 863–882. doi:10.1016/j.annals.2011.10.004

Schipper, Frank, and Johan Schot, eds. 2011. "Infrastructural Europeanism, or the Project of Building Europe on Infrastructures." *History and Technology, Special Issue* 27 (3).

Schwimmer, Ernst. 1895. "Bis zum Endpunkte Europas. II." [To Europe's endpoint. II.]. *Pester Lloyd*, Aug. 30.

Spode, Hasso. 2010. "Nordlandfahrten. Norwegen als Idee und als touristische Destination" [Journeys to the North: Norway as idea and tourist destination]. In *Nordlandreise. Die Geschichte einer touristischen Entdeckung. Historien om oppdagelsen av turistmålet Norge*, edited by Sonja Kinzler and Doris Tillmann, 16–33. Hamburg: Mare.

Spring, Ulrike. 2016. "Arctic and European In-Betweens: The Production of Tourist Spaces in Late Nineteenth-Century Northern Norway." In *Britain and the Narration of Travel in the Nineteenth Century: Texts, Images, Objects*, edited by Kate Hill, 13–36. Farnham: Ashgate.

Star, Susan Leigh, and Geoffrey Bowker. 2006. "How to Infrastructure." In *Handbook of New Media: Social Shaping and Social Consequences of ICTs*, edited by Leah A. Lievrouw and Sonia Livingstone, 230–244. Updated student edition. London: SAGE.

Tillmann, Doris. 2010. "Mit der Kamera in die Fjorde. Die Polarfahrten des Fotografen Richard Fleischhut" [With a Camera into the fjords: The Polar journeys of the photographer Richard Fleischhut]. In *Nordlandreise. Die Geschichte einer touristischen Entdeckung. Historien om oppdagelsen av turistmålet Norge*, edited by Doris Tillmann and Sonja Kinzler, 158–167. Hamburg: Mare.

Walchester, Kathryn. 2014. *Gamle Norge and Nineteenth-Century British Women Travellers in Norway*. London: Anthem Press.

Wilse, Anders Beer. 1907. "Nordkap postkontor" [North Cape post office]. http://digitaltmuseum.no/021015777183?page=4&type_filter=Photograph&query=nordkapp&from_year=1870&to_year=1910&advanced_search=1&pos=85, last accessed 17 Feb. 2017.

Zuelow, Eric G. E. 2016. *A History of Modern Tourism*. London: Palgrave.

Notes

[1] Bjarne Rogan has written extensively on travel infrastructure in Norway during the nineteenth century, but without a special focus on the North. See e.g. Rogan 1998. Bård Kolltveit (2016, 2005, 1978) has published on tourism and steamship traffic to the North. See also Brevig (2010) and Richter Hanssen (1990) on the development of travel infrastructure in Norway and the North Norwegian coast around 1900.

[2] Even today, the Cape holds an attraction for German-speaking travellers. According to a recent study, 40% of all tourists visiting the North Cape in 2010 came from Germany and approximately 2.5% were from Austria (Barstad and Nilsen 2010, 4).

[3] All translations are mine.
[4] According to Richter Hanssen (1990, 240), 300 people from the surrounding villages attended the unveiling of the column by the king, who designated it in his speech as the border of the Norwegian kingdom.
[5] The stone with the plaque no longer exists, and it is not known when it disappeared (Richter-Hanssen 2011, 148).
[6] "On the steps of the plain column we sit down and look out onto the open polar sea, radiating away in shades of red" (Davidsohn 1884, 514).

"LAVISH LAPLAND": GENDER, PRIMITIVISM AND THE PICTORIAL IN OLIVE MURRAY CHAPMAN'S TRAVELOGUE *ACROSS LAPLAND*

JAN BORM

The British painter and traveller Olive Murray Chapman (1892–1977), author of several travelogues, including books about Iceland, Cyprus and Madagascar, travelled through Lapland during the early 1930s. Taking photographs and filming the Sami while travelling the region in winter, she cuts a striking figure as the first female British author to have done so and to have produced both a documentary film and an account of her adventure in writing, *Across Lapland* (1932), the book this chapter is dedicated to. In what follows, I will argue that Chapman's journey and her narrative are both an illustration of modernity increasingly leaving its stamp on a region perceived by outsiders as remote at the time, if not coming straight out of a fairy tale, and the desire to find and document the traditional life of the Sami, Chapman's primitivism being her own principal driving force.

In the introduction to his anthology *No Particular Hurry: British Travellers in Finland 1830–1917*, Tony Lurcock suggests that many British travellers to Finland in the early nineteenth century were still adventurers, while political and social interest became the main reason to travel to Finland in later years (Lurcock 2013, 1–2). But what was the case for Finnmark and Lapland? A similar trend can be noted here, at least as far as the interior parts of Lapland are concerned, while the coasts of Finnmark had become a destination spectacularly more accessible with the introduction of the steamboat in 1837, which reduced the journey from Trondheim to Hammerfest from at least a month to eight days, as Xavier Marmier notes in his account of the "La Recherche" expedition.[1] Nonetheless, almost a hundred years later, Olive Murray Chapman notes in her travelogue *Across Lapland* that "Finnmark is visited by very few, and in the winter time, so I was told, by none" (Chapman 1939, 15;

hereafter cited parenthetically). The region was thus clearly perceived as remote as far as tourism destinations go and therefore potentially as an exciting venue for adventure of the kind armchair travellers liked to read about.

The railway to Kiruna, completed in the late nineteenth century and extended to Narvik in the early twentieth, does not seem to have brought many visitors to Lapland in wintertime. Chapman mentions an official of a "well-known Swedish travel bureau" telling her in Stockholm in 1930 that her project to cross northern Lapland in winter was "altogether beyond the bounds of reason" (18), implying to her readers that the author was planning a real adventure trip, an idea not likely to be encouraged by many at the time. The official certainly does not seem to have done so, insisting, it appears, on the severe cold in Lapland in winter and the fact that the only means to cover great distances during March was to travel by sledge and reindeer with the threat of frequent "terrible blizzards" during that month. He is described as "working himself up into a state of great agitation" who ends up replying: "'No, no! What you propose to do is an adventure, not an ordinary tourist's journey. We cannot help you here at all nor be responsible'" (8). The words of the official stand her in good stead, announcing, as they do, the dangers of the journey and enhancing the author's image as a daredevil, a well-known trope in travel writing that consists in promising an exciting tale of adventure right from the start.

It is also to be remembered that the discourse on Lapland was very largely male-dominated, as Heidi Hansson has pointed out in an article on Alexander Hutchinson's *Try Lapland*: "The male-coded adventure discourse of the first half of the nineteenth century disqualifies the North – Canadian or Scandinavian – as a tourist destination and defines it as inaccessible to women" (Hansson 2012, 148). No wonder then, that Chapman presented a very unexpected figure as "the lady [...] who was travelling alone" in the words of a Norwegian forester encountered at Jotka fjellstue in Finnmark *(*67). Her adventure was remarkable in a number of respects. The author of the blurb on the back cover of the 1939 Penguin edition of her travelogue certainly seems to have tried his best to purvey this idea in a number of ways:

> Olive Murray Chapman, artist, traveller and explorer in many remote parts of the world, started her career as a student at Heatherley's Art School. In recent years she has become known chiefly through her Arctic journeys in Iceland and Lapland.
>
> In the winter and spring of 1931, Mrs. Murray Chapman made a journey of some hundreds of miles across the frozen wastes of Norwegian, Finnish, and Swedish Lapland. She travelled alone with reindeer sledges

> and a Lapp guide, risking the dangerous blizzards that sweep the country in March. The object of her journey was to study the life of the primitive nomad Lapps, and to obtain a pictorial record of the people and their customs. Her venture was completely successful, and her remarkable film, the first of its kind to be taken by a traveller in Lapland, from which are taken the photographs which illustrate this volume, was shown before the Royal Geographic Society.

Part tourist, part explorer, Olive Murray Chapman is here presented as an artist, a daredevil and a pioneer of visual ethnography, and thus cuts an eminently modern figure. The idea of cinematography reaching the Sami presented here as "the primitive nomad Lapps" suggests one of Chapman's central themes, if not predicaments, that is the supposedly archaic European Far North where one could study "the life of the Lapps, that picturesque race of nomads who inhabit the far north of Norway, Sweden, Finland and Russia" (15) being confronted with modernity.

At the same time, her primitivism seems to be well in line with what Paul Fussell has identified as a chief trait of literary writing between the wars in his seminal study *Abroad:* "the rejection of industrialism and everything implied by the concept of 'modern northern Europe'" (Fussell 1980, 210). To what extent Olive Murray Chapman's *primitivism* may appear to be at odds with her own progressive stance as a woman cinematographer breaking new ground in these faraway places is the key issue I purport to discuss in what follows, arguing that her text is poised between a primitivism of the picturesque and pictorial kind (it is to be remembered that Chapman was also a painter), a repeatedly pictorial attempt to dream Lapland in fairy-tale, at times transcendental terms, while never losing sight of the time frame of the traveller writing home to an audience adhering to modern ideals of comfort that she is quite relieved herself to rediscover towards the end of the journey.

Her account is an adventure tale with all the difficulties that crossing Lapland in winter in the early 1930s entailed related in some detail. At the same time, her narrative is a "second journey" in some respects, to borrow Maria Lindgren Leavenworth's phrase (2010), drawing on other descriptions of Lapland as intertexts while exploring what Karen Lawrence called the "tension between the thrilling possibilities of the unknown and the weight of the familiar, between a desire for escape and a sense that one can never be outside a binding cultural network" (1994, 19).

Laplandscapes

It is hardly surprising that Chapman's verbal representation of Lapland can be described as eminently pictorial. An artist herself, engaged in drawing and filming the Sami as much as she determined to portray them in writing, her vision of Lapland and its people is very much that of a painter. Indeed, her view is expressly artistic and ethnographic, two approaches that are obviously not mutually exclusive in one person, as the work of Emilie Demant Hatt, one of her closest predecessors in the North, also shows. In Chapman's brief foreword, she announces her goal:

> My object in making the journey was in order to study the life of the Lapps under winter conditions and to obtain a pictorial record, by means of sketches, photographs and a cinematograph film of the people and their customs, and especially of the famous Lapp Market at Bossekop at the beginning of March, and later the Easter Festival at Karasjok, near the border of Northern Finland. (Chapman 1939, vii)

The film mentioned here is entitled "Across Lapland: The Story of a Woman's Adventurous Journey in the Arctic (1933), photographed, edited and described by O.M.C.". Barbara Sjoholm has suggested that Murray was perhaps the first foreigner to film the Sami" (Sjoholm 2007, 83). Whatever the case may be, this film, which we have not been able to visualize unfortunately, is certainly to be considered one of the earliest non-Fennoscandian cinematographic documents representing Lapland and the Sami. As to the black and white photographs published in Chapman's travelogue, they are no doubt of ethnographic interest, though their facture is rather static and not experimental in any "modernist" sense of the term, i.e. lacking in spontaneity and not expressing the idea of movement unlike some of her pictorial passages that appear more creative in this respect, a point we will return to below.

As to Chapman's winterscapes, then, and her portraits of indigenous people who inhabit them, they are clearly designed to reflect the fairy-tale trope applied time and again to the North of Norway and Lapland which she very evidently expects her readers to share. For instance, Tromsø is described as looking "most fairy-like, with its twinkling lights and frosted roofs" (24). Hammerfest in the snow appears to have been "an eerie and awe-inspiring sight" (25), the author introducing a shade of Burkean sublime to enhance the impression of a faraway, if not out-of-this world place.[2] As to Finnmark's Altafjord in moonlight, it occurred to her as "ethereal and unreal" (27). In other terms, Lapland comes across as somewhat out of this world, an image enhanced by the description of her first encounter with the Sami:

> Then, to my surprise, I saw a number of what appeared to be gnomes running swiftly down the steep slope to the jetty. I was beginning to wonder if I had indeed sailed into Fairyland; but, on closer inspection, these quaint little people proved to be Lapps on their way to the Bossekop market. […] Many were under five feet in height, and with their quaint pointed blue and scarlet caps, and gaily embroidered tunics, might have stepped straight out of a Hans Andersen fairy-tale. (28)

At first sight and from a twenty-first-century point-of-view, this description of the Sami seen as "gnomes" appears condescendingly ethnocentric, just as her use of the term "Lapps" would be considered problematic today, though in common use at the time.[3] As Sami scholar Vuokko Hirvonen points out in *Voices from Sápmi*: "the Sámi have felt that the term *Lapp* is associated with a variety of negative attributes that have been associated with the Sámi throughout history, and, therefore, they have demanded that the term be replaced by the Sámi term *sápmelaš* and terms coined from it in the dominant languages (*en same* in Norwegian and Swedish)" (Hirvonen 2008, 31).

However, it needs to be specified that one can hardly accuse the author of not appreciating the Sami and their culture. Her apparent "belittling" of them needs to be put into perspective by bearing in mind these further remarks she makes in the foreword: "If I am able to interest the readers in the wonderfully brave life of the nomad Lapps, and the cheerfulness, fortitude and endurance shown by their women, often under the most trying and rigorous Arctic conditions, I shall feel that my journey was not made in vain" (vii). There is no doubt, therefore, about the author's admiration for the people she set out to portray, announcing her primitivism, a notion we will return to below. It is important to stress that this introductory appeal to her readership includes a gendered view of the Sami, praising the courage of the men while insisting on the endurance of women, a dual focus that seems rare even by contemporary standards according to Hirvonen, who considers that Sami women have been marginalized thrice, as Sami, women and Sami women, at least from an academic point of view (Hirvonen 2008, 36).

Though Chapman's fairy-tale image of Lapland and the Sami may have little if anything to do with the way in which the Sami consider themselves, her use of the image is mainly a literary technique aimed at helping the reader shift from familiar terrain, that of the Lapland fairy tale, to the largely unknown regions of Sami ethnography no matter how strong the "Euro- and ethnocentric under-pinnings" of Sami studies in those days may have been (Hirvonen 2008, 30). As to readers' horizon of expectations, it needs to be added that an Anglophone readership in the 1930s would have been likely to assume that the author would respect the maxim of *pleasure* and *instruction* that appears to underscore Anglophone travel writing in an almost generic way. Hence the number of pictorial scenes designed to reflect not only the author's mood but also to satisfy the reader's interest in exotic destinations and people.

That Lapland resembled a country out of a fairy tale is notably suggested by Graham Greene's few comments on Chapman's book in one of his reviews of travelogues for the *Spectator*: "She writes attractively of a land which it is difficult to realize exists outside the pages of Hans Andersen" (Greene 1932, 21). Here is an excerpt of Chapman's description of Bossekop market, "a more picturesque sight it would be hard to find" (35):

> By midday the sun was shining brightly from a clear sky. In the background the dazzling ranges of snow-covered mountains, sloping down to the deep-blue waters of the bay, were in striking contrast to the riot of gorgeous colour in the market. The Lapps revel in bright warm shades as a relief, no doubt, to the cold austerity of their surroundings; and their dress is very gay, with splashes of scarlet, yellow, green and bright blue, the colours of which look wonderful against the pure, unbroken, whiteness of the snow. (35–6)

Though the aim consists in painting the scene with words, drawing on pictorial elements such as composition (*the background*), colour and even effects of the brush (*splashes*), the result can hardly be called "picturesque" in the received way, the notion of a "riot of gorgeous colour" corresponding more to a Modernist, asymmetrical and expressionist vision of experience, not to say an unexpected variant of Bakhtinian *carnivalesque* (we are at a fair, after all…). Be that as it may, no matter how striking the contrast between Sami dress and snow – represented here hardly surprisingly as an image of purity – may have seemed to the author, her description once again draws on the fairy-tale trope as the term "wonderful" suggests.

Primitivism and Lapland

Chapman's rhetorical disposition to share her admiration with the reader is to be born in mind also when discussing her *primitivism*, a notion introduced by the author almost immediately when accounting for her motivation:

> What a chance, thought I, of studying the manners and customs of these primitive folk, and, like them, of journeying across country with sledge and reindeer, the only mode of travel possible in Lapland during the long winter, for there is no spring as we understand the word beyond the Arctic Circle.

Once again, her use of the expression "primitive folk" appears hardly unusual for her period, but it does introduce the theme of the primitive that she returns to repeatedly in her travelogue, partly engaged, as she was, in what one might term a proto-ethnographic enterprise in the sense of non-institutional ethnography (study the manners and customs). At the same time, the dichotomy of the domestic and the foreign (spring as we understand it versus beyond the Arctic Circle) enhances the impact of the notion of *otherness* that the word "primitive" is likely to have raised at the time.

Having seen the Sami at Bossekop market, Karasjok and Kautokeino, Chapman refers to a personal notion of varying degrees of the primitive, considering the Kautokeino Sami as "a more primitive type than most of those I had seen in Karasjok" (117), while calling the Swedish Sami near Kiruna "far more sophisticated than those in Norwegian Lapland" (Chapman 1939, 166). It might be tempting to read these passages as voicings of an ethnocentric stance, othering the other into a primitive, as it were. Such an interpretation would appear excessive, however, in view of Chapman's manifest admiration of certain Sami qualities in the foreword. Similarly, her expression "these children of nature" (55) can be seen to work in a twofold way: on the one hand, the author thus manages to introduce the myth of the noble savage, one of the standard tropes in the repertoire of travel literature, admittedly idealizing the Sami but trying to render them even more praiseworthy in the minds of those readers who might wish to share her primitivist feelings, dreaming of the age of natural man that the Enlightenment was so fond of reflecting on (Borm, Cottret and Münzel 2010).

This reading could be reinforced by drawing a parallel with her *nomadism*, as expressed in this passage on Sami life, described as "their hard, courageous, but on the whole happy and free nomad life" (47). On the other hand, the expression "children of nature" is used in the context of Sami shamanism and Christian reactions against it, but also in reference to what she calls "the strange manifestations that sometimes occur among the Laestadians" (55), an allusion to visually and audibly exalted forms of expressing belief during a service, one of the traits of Laestadian spirituality considered as a movement of pietistic enthusiasm.[4] Taken together with her qualifying the Sami mind as "simple" (59), this view appears clearly paternalistic reflecting a philosophy of history based on the notion of progress:

> In ancient times the Lapps believed these lights formed the bridge by means of which the gods descended to earth, and who knows what strange fancies they conjure up even now in the simple minds of these people who live so close to the great heart of Nature, and to whom the supernatural has ever remained so vivid a reality? (59)

One may easily be reminded of the image of the conjurer (or wizard) that Christian missionaries and travellers used so frequently to condemn shamanism. The idea of a "simple mind" is however clearly limited to Sami shamanism in the context of the passage and to her observations of the Sami during a Laestadian service.

It is to be noted that Chapman introduces a time frame when discussing shamanism from an ethno-historical point-of-view: "Some of the Lapps appear to have been so skilled in what was then regarded as magical arts that they were able to work quite independently of the drum. In the light of present knowledge many of these people would have been called 'mediums', for such they undoubtedly were" (54). No matter how paternalistic some rare comments of hers may seem today, this personal remark suggests a proto-scientific approach to the field when it comes to observing shamanism that one is unlikely to notice frequently in older sources. Some of the best known of these are referred to by Chapman herself, such as Scheffer's seminal *Lapponia* (1673).[5] She also refers to an early French visitor, the playwright Jean-François Regnard who reached Swedish Lapland in the 1680s and described his experience in a text published posthumously. Having arrived in Kiruna, Chapman decides to visit Jukkasjärvi church, noting the famous inscription that Regnard left there and which still can be admired today, on standing with his fellow travellers de Fercourt and de Corberon "'where our world ends'" (172).

Swedish Lapland is obviously far from being the *ultima Thule* Regnard made it out to be, but this was a convenient point to leave a sign of the French in the European Arctic before the Duke of Orléans, future King Louis-Philippe, was to reach the North Cape and have his bust sent up there, a replica of which is still to be found in the visitors' centre. In the days of Chapman, the idea of having reached the end of the world in Lapland could hardly have convinced her readership. She does appear, if not as a tourist, as a literary traveller, following in the footsteps of previous adventurers, but also those of Johan Turi in the Jukkasjärvi region, her principle authoritative source on Sami ethnography.[6]

It needs to be pointed out, however, that Chapman's primitivist stance is far less interactive – if at all – than that of Turi's principal interlocutor and editor/translator of his famous account of the Sami, the Danish artist Emilie Demant Hatt (1873–1958). Demant Hatt's approach has been described by Hanna Eglinger as "mediating between civilized (or even colonial) and indigenous perspectives, and migrating between insider and outsider as well as male and female positions" (Eglinger 2016, 193). Eglinger considers primitivism as one of the driving forces of Demant Hatt's work which she qualifies as "artistic ethnography" (Eglinger 2016, 197). It is not my purpose here to discuss the nature of the relations between art and ethnography, but there is no doubt that Demant Hatt's art is more avant-garde in the modernist sense of the term than Chapman's. However, the latter's paintings participate in the effort of mediation between "civilized and indigenous perspectives" identified by Eglinger in Demant Hatt's work.

Back in Arctic modernity

Crossing Lapland in winter accompanied by two Sami men, Chapman takes her time to narrate how "the going now became very hard" (Chapman 1939, 66). Despite all the hardships of her journey in a pulka, a sledge for one person drawn by a reindeer and not always easy to manage, the crossing also offers opportunities for *reveries* about peace and quiet by somebody temporarily on the run from the noises of motorized modernity: "vast deserted snow fields stretched white and untrodden as far as the eye could reach, until they finally merged into the blue distance of the mountain peaks that fringed the horizon. I lingered a short while longer held by the peace and beauty of my surroundings" (114).[7] The author visibly embraces Lappish space with the eye of the painter, appropriating it momentarily as an aesthetic and therapeutic time-space. One of those moments of great personal joy she accounts for in the following terms in another passage: "the cold, exhilarating air and bright sunshine gave me a wonderful sense of well-being and happiness" (74). Travelling to Lapland becomes a way of clearing or emptying the mind, getting closer to the heart of Nature. At this point, her account seems to create that "permeable membrane between home and the foreign, domestic confinement and freedom on the road" as Karen Lawrence puts it (1994, 19).

Paradoxically, however, the very fact that Chapman traverses Lapland in winter announces another step in modernity impacting the region, which is the effects of summer tourism. She accounts for the relatively greater degree in sophistication she attributes the Sami in Swedish Lapland in the following way:

> The reason, no doubt, lies in the fact that many of the Swedish Lapps come far more into contact with the civilized world, Kiruna being easily reached by rail from Stockholm and drawing crowds of visitors during the short summer, […] The result of this invasion is that many of the Swedish Lapps, particularly those around Kiruna or others who have migrated to the Norwegian coast near Tromsö, become quite spoilt and often refuse to be photographed unless they are paid, and even sometimes tell their children to run away, unless money is offered them. (166–7)

That's it, then, for the existential *angst* of indigenous "children of nature" in front of a camera. Cultural tourism, and even anthropology in the field up to a certain extent, is a business like any other business. The passage is revealing in a number of respects. First of all, the dichotomy between the civilized usually opposed to the barbarian, and here more to those she initially refers to as "the primitive Lapps", is a standard trope in travel literature.

Of more immediate interest in the present context is the idea that modernity is clearly seen as the driving-force of this process, for better or worse; in fact, rather the latter, it seems, since the arrival of (summer) tourism is depicted as a form of territorial conquest, be it still temporary and relatively short at that time. The idea that civilization spoils indigenous people is also a well-known trope, frequently taking the form of *ethnographic mourning* where the narrator claims to be one of the last to have the sad privilege of seeing some forms of cultural practice disappear. Here, the matter appears somewhat differently though, since the "spoiling" of the Swedish Sami has rendered them also more sophisticated, a quality that does not necessarily need to be read in a negative way even if it does imply a particularly developed degree of cunning (asking to be paid to have one's photograph taken), associated in the immediate context with the notion of spoiling (i.e. a practice to be regretted).[8] Admittedly, by this time, the Sami were experiencing a social and cultural crisis as Asbjørn Nesheim notes in his preface to the French edition of Johan Turi's narrative: "Modern times had definitely reached Lapland and were there to stay. Better connections definitely put an end to the isolation of the Sámi" (1974, 8).

Do we have to conclude that Chapman rejects the impact of modernity, that her primitivism makes her mourn an almost bygone age – or long to return to the more isolated places of Lapland, both for the purposes of ethnography and her personal well-being? Whatever the case may be, she goes on to point out that one needs to travel to the parts of

Lapland “unvisited as yet by the tourist and preferably to live among the people during the winter season” in order to “get into touch with primitive Lapp life” (167), an idea frequently voiced, after all, in ethnographic writings and travel accounts, that-is-to-say that the most archaic forms of a culture are to be observed in the remotest parts of the terrain under consideration. She evidently admired the Sami encountered at Karasjok, “those people, in their gay clothes and picturesque pointed hats and scarlet bonnets” (78), since they seemed to perfectly fit the colourful type of Sami in traditional dress she had hoped to encounter.

Then again, Chapman can hardly be seen as condemning the idea of civilization and modernity outright. Both notions are key aspects of her “cultural network” to use Karen Lawrence’s expression once more, and even of her own field work, since he has chosen to not only photograph, but also film Sami life, witnessing varying degrees of the impact of modernity on the Sami as in this scene describing her filming a group of them at Bossekop: “some of the bolder spirits, taking it for an ordinary camera with which many of them were familiar, would solemnly pose for me when I tried to explain that they could move about” (39). The passage witnesses the arrival of modernity in stages, some of the Sami being manifestly used to posing in order to have their photograph taken (an allusion, incidentally, to the static mode of representing the Sami in photographs at the time, leaving little room for spontaneity), but none of them expecting to be a part of moving pictures.

The contrast is striking: on the one hand, Chapman wishes to represent her primitivist view of the Sami by drawing on modern technology; on the other, the “primitivized” Sami have evidently become used to such modern devices as photography, while not being accustomed to film. Chapman’s stay in Kiruna also includes a visit of the mines in the company of a Mr. Holmquist, presented as “a Swedish railway official belonging to the town” (170), speaking “excellent English” *(*170). The story of Kiruna’s industrial development up until 1930 is called “dramatic” by Chapman, not qualifying her adjective any further but summing up the “blasting operations”, an activity that clearly also serves her as a metaphor for the arrival of modern times, Mr. Holmquist being quoted as quietly shuffling her away from a particular spot in the mines: “‘Should we remain where we are standing now,’” said Mr. Holmquist calmy, “‘we should be blown to pieces’”. Modernity makes a lot of noise in this part of the Arctic.

To round off her experience, Chapman stops over at Lake Torneträsk in Sweden “for a week or two’s rest and painting at this lovely lake”, enjoying this “wonderful Northland under winter conditions” (176) before the summer crowd would return in mid-June: "From then onward, until about the end of August, it would be crowded with visitors of many nationalities, who would travel up here in order to see the glories of the Midnight Sun, and also to gain an impression of Swedish Lapland" (175-6). By this time, Lapland evidently has become an attraction for mass-tourism, the construction of the railway facilitating not only the transport of natural resources but also providing access to the region for holiday-makers, the train being one of the most powerful vectors of the arrival of modernity up North. Perhaps not surprisingly, her rendering of the seasonal invasion of modern tourism is also focused on noise: "Motor-boats would steam to and fro across the lake; sirens would blow; mosquitoes would buzz; and the wonderful beauty of the mysterious ice-bound solitude in which the place now lay would be no more. The spell would be broken […]" (176).

Admittedly, mosquitoes have been buzzing there for the longest time, but still, modernity is clearly an audible phenomenon too, introducing mechanical or industrial noises into a space imagined by primitivist travellers like Chapman as noiselees, the noisiest intrusion being no doubt represented by the train she took to reach "these mountain solitudes" (175) on her way to Narvik. Though she likes to see herself as different from the summer crowd and even privileged in witnessing the scene – "I congratulated myself that I should no longer be there" – she is of course using modern means of transport herself at this stage, enjoying a holiday of sorts after her remarkable winter crossing. "The little electric ore-train" (179) then takes her into Norwegian landscapes, the "luxuriant green grass" of which (180) strikes her as a "miracle" (179): "The world had become transformed as if by magic into a green and lovely flower garden and already the cold white solitude of the Arctic Winter was a dream of the Past" (180). This is the end of her account. The effect of magic is of course caused by sudden change, the speed – be it modest from our contemporary point-of-view – of the train being one of the chief traits of modernity.

The capital letters used for the "Arctic Winter" make the expression stand out as a concept or an experience in its own right and a phenomenon somehow associated with the Past. Not only is the author's own winter crossing a thing of the past, but modernity has already made its way into the Arctic summer and winter sports are on the verge of being developed across the border, in Finnish Lapland, by the time Olive Murray Chapman leaves.[9] Seen in this light, her account is a lament of the modern traveller witnessing the invigorating solitude of the "frozen North" (179) just a few years before the Pallas hotel open in winter was built in the Finnish fjells

(Tuulentie and Lamkila 2014, 210). Having crossed Lapland herself, she had pointed the way, perhaps despite herself.

References

Borm, Jan, Bernard Cottret and Mark Münzel, eds. 2010. *Christentum und der Natürliche Mensch, Beiträge zur französisch-deutschen Tagung "Christianisme et l'homme naturel" (Journée d'études, Marburg, Januar 2007)*. Marburg: Förderverein "Völkerkunde in Marburg" e.V., "Curupira Workshop" series, vol. 13.

Burke, Edmund. (1757) 1998. *A Philosophical Enquiry into the Sublime and the Beautiful.* London: Penguin.

Chapman, Olive Murray. (1932) 1939. *Across Lapland.* Harmondsworth: Penguin Books.

Eglinger, Hanna. 2016. "Nomadic, ecstatic, magic: Arctic primitivism in Scandinavia around 1900." *Acta Borealia* 33 (2):189–214.

Fussell, Paul. 1980. *Abroad: British literary traveling between the Wars.* New York: Oxford University Press.

Greene, Graham. 1932. "The Unsentimental Journey". *The Spectator*, 11 June.

Hansson, Heidi. 2012. "An Arctic Eden: Alexander Hutchinson's *Try Lapland* and the Hospitable North." *Northern Review* 35 (Spring): 147–166.

Hirvonen, Vuokko. 2008. *Voices from the Sápmi: Sámi Women's Path to Authorship*. Guovdageaidnu/Kautokeino: DAT.

Kulonen, Ulla-Maija, Irja Seurujärvi-Kari and Risto Pulkkinen, eds. 2005. *The Saami* Vammala: SKS.

Lawrence, Karen R. 1994. *Penelope Voyages: Women and Travel in the British Literary Tradition*. Ithaca and London: Cornell University Press.

Leavenworth, Maria Lindgren. 2010. *The Second Journey: Travelling in Literary Footsteps*. Umeå: Umeå University.

Lurcock, Tony. 2013. *No Particular Hurry: British Travellers in Finland 1830-1917*. London: CB editions.

Marmier, Xavier. 1857. *Lettres sur le Nord* (5th edition). Paris: Hachette.

Nesheim, Asbjørn. 1974. "Préface". In Johan Turi, *Récit de la vie des Lapons*. Paris: François Maspero.

Regnard, Jean-François. 1731. *Les Œuvres de Mr. Regnard. Nouvelle édition revuë, corrigée & augmentée. Tome I. Contenant ses Voyages de Flandres, d'Hollande, Suède, Dannemark, la Laponie, la Pologne & l'Allemagne*. Paris: Pierre-Jacques Ribou.

Sjoholm, Barbara. 2007. *The Palace of the Snow Queen: Winter Travels in Lapland.* Berkeley, CA: Counterpoint.

—. 2013. "Translator's Notes." In Emilie Demant Hatt, *With the Lapps in the High Mountains: A Woman Among the Sami, 1907-1908*, tr. and ed. B. Sjoholm. London: University of Wisconsin Press.

Tuulentie, Seija and Jenni Lamkila. 2016. "A Hotel Waiting for Renovation: Pallas as a Challenging Case for Tourism Development in Finnish Lapland." In *Tourism Destination Development: Turns and Tactics*, edited by Arvid Viken and Brynhild Granås, 209–26. London: Routledge.

Notes

[1] "Il fallait un mois au moins pour aller de Hammerfest à Drontheim [...] Maintenant le bateau à vapeur va de Drontheim à Hammerfest en huit jours." (Marmier 1857, 283).

[2] Edmund Burke links the feeling of awe to our sense of divinity. See his *A Philosophical Enquiry into the Sublime and the Beautiful* [1757]. London: Penguin, 1998, p. 112 (Part II, section V, "Power"): "It were endless to enumerate all the passages both in the sacred and profane writers, which establish the general sentiment of mankind, concerning the inseparable union of a sacred and reverential awe, with our ideas of the divinity."

[3] Interestingly enough, the American author Barbara Sjoholm has decided to use the term "Lapp" in her translation of Emilie Demant Hatt's narrative about her stay among the Sami and the word "Sámi" in her own comments on that work, a choice she accounts for in the following way: "It would have been possible to change 'Lapp' throughout my translation to 'Sami,' but to do so would change the reality that *With the Lapps in the High Mountains* is a historical document from the early twentieth century and valuable for that reason. We wouldn't republish or retranslate the books of the early Arctic explorers and substitute 'Inuit' for 'Eskimo.' [...] When the voice is mine, I use 'Sami.' Otherwise I faithfully translate Demant Hatt's *lapper* as 'Lapps'" (Sjoholm 2013, xxxix-xl).

[4] Olive Murray Chapman seems to have had mixed feelings about Lars Levi Laestadius and Laestadians whom she refers to as "a curious Christian sect" (*AC* 40). Her descriptions of Karesuando do not mention Laestadius' hut nor the church he was in charge of. She sums up her experience of a Laestadian service as having consisted of witnessing "extraordinary and rather terrible scenes of wild hysteria" (*AC* 104). For an English introduction to Lars Levi Laestadius see Juha Pentikäinen's entry on the eminent botanist and Lutheran minister in the encyclopedia *The Saami* (Ulla-Maija Kulonen, Irja Seurujärvi-Kari and Risto Pulkkinen eds.), Vammala: SKS, 2005, pp. 170-172.

[5] Risto Pulkkinen notes: "*Lapponia* was fundamentally important in disseminating information about the Saami, but it also engendered a romantic view of them as extreme 'pacifists' who complaisantly gave way to their conquerors. [...] Contrary

to its purpose, the book in fact reinforced the reputation of the Saami as a people of sorcerers [...]", (Ulla-Maija Kulonen, Irja Seurujärvi-Kari and Risto Pulkikinen 2005, 192). As to the history of disseminating information about the Sami, Vuokko Hirvonen argues that "the first texts dealing with the Sámi from a Lappological point of view were written almost two thousand years ago" (in Vuokko Hirvonen, *op. cit.*, p. 31).

[6] The English edition of his best-known text, *Turi's Book of Lappland*, was published by Jonathan Cape in 1931. Olive Murray Chapman refers to this book as an "excellent translation" (Chapman 1939, 76).

[7] Murray Chapman *does* mention (relatively) good travel conditions whenever applicable, as in this example: "the 'före', as the Norwegians call it, was good, and travelling should be easy in consequence" (*AC* 74).

[8] In a different context, the idea of the Sami having been spoilt by their contacts with Europeans is voiced in relation to the subject of alcoholism.

[9] According to Tuulentie and Lamkila, "the contemporary mode of tourism started to grow in the region in the 1930s as winter sports made their breakthrough. In particular, Pallas fells came to be known as the most suitable place for cross-country and alpine skiing in Finland" (Tuulentie and Lamkila, 2014, 210).

THE ROMANCE OF THE FAR FUR COUNTRY: INDIGENOUS LIFE BETWEEN TRADITION AND MODERNITY

ROSWITHA SKARE

The Romance of the Far Fur Country (1920) and Nanook of the North (1922) are two films about the same topic and from the same period but with surprisingly different legacies. Robert Flaherty's *Nanook of the North* is considered a classic and a milestone of silent cinema and ethnographic film. Enormously popular when released in 1922, the film is still "screened throughout the world, by culturally diverse audiences" (Berger 1995, 177), and is thus "familiar to every school child as well as every film buff" and "widely studied and written about" (Shepard 1974, 60). In contrast, *The Romance of the Far Fur Country* went into the archives after its theatrical release, was forgotten for many years and has only recently been re-discovered and restored. The reason for these diverging histories is the impetus behind this chapter.

The making of *The Romance* was initiated by the Hudson Bay Company's London office to celebrate the company's 250th anniversary. As described by Peter Geller, who was the first to research the film and who includes a chapter about it in his study *Northern Exposures*, "[a] variety of materials relating the story of the company's growth and contributions to Canada and the Empire was produced in honour of the occasion" (Geller 2004, 85). The film portrays the HBC's fur trade industry, showing different parts of Canada and the effort required to cross great distances to visit those remote outposts where trade with indigenous people[1] is taking place. Educational Films Corporation, a New York City-based company, undertook the production of the film with Harold Wyckoff as cameraman, joined by Bill Derr as fellow operator.

The Romance premiered in 1920, two years before Robert Flaherty's *Nanook of the North* was screened for the first time. As Peter Geller argues, "the publicity and interest surrounding the HBC's moving picture did influence Revillon Frères, [...]. Thinking ahead to its 200th anniversary in 1923, the Paris-based company backed the making of *Nanook of the*

North" (Geller 2004, 178). The two films are therefore related to each other, not only because of the many similarities in topic, period and setting, but also in being sponsored by fur companies. However, while Flaherty's intention in making *Nanook* was to capture traditional Eskimo life before it disappeared – "[w]hat I want to show is the former majesty and character of these people, while it is still possible – before the white man has destroyed not only their character, but the people as well" (Flaherty cited in Ruby 1989, 450) – *The Romance* is a travelogue or a road movie, in which the progress across the country is the main objective.

Because *The Romance* was made with a focus on the HBC and its northern involvement, contact between indigenous peoples and white people and their civilization is not omitted, as in *Nanook*. On the contrary, the fur trade and Christian missionaries share the screen with the Inuit and other indigenous people. The film presents images from remote areas, but also from large cities with department stores, restaurants and cars. There are hunters with guns and indigenous people in western clothes attending church services, as well as posing in front of the camera in their traditional costumes. Indigenous people are shown engaging in their daily activities, mainly in relation to the fur trade, the trading posts and the supply boats of the HBC. Even though the film's title suggests a romantic picture of faraway areas and the people living in it, it can be argued that *The Romance* presents life in the Arctic much more realistically than Flaherty does in *Nanook*.

Geller's chapter about *The Romance* is based on his extensive study of archival material. Focusing on two narrative sequences within the mainly non-fictional film, Geller mentions Flaherty's *Nanook* and the relationship between the two competing companies in the area, the HBC and Revillon Frères. He also discusses the screening of the film in connection with the other anniversary activities of the HBC in Western Canada and stresses the film's place "within the history of the north's appeal" (Geller 2004, 106). Its attraction was not only in its images of the still wild and untamed North, but especially in the "film's portrayal of the seldom seen Inuit" (Geller 2004, 102). Thus, the non-modern lifestyle of the Inuit and other indigenous people of the Canadian North had become the subject of a modern technology that could bring new knowledge to the public. The costumes and habits of indigenous people became important elements in a network of happenings that made the HBC's celebration of their 250th anniversary a modern media event. As part of the "publicity 'stunt…' a block of one hundred seats [at the Allen Theatre] nearest the orchestra were roped off for accommodation of the Red Men" (Geller 2004, 100), all dressed up in their traditional costumes. Photographs were taken, and we can assume that the press reported about this and other events during the anniversary celebration. Thus the "primitive" people of the North became part of the event.

The question of how *The Romance* presents the area and indigenous life between tradition and modernity is the main focus of this chapter. I start with some general remarks about the film, before focusing on the representation of indigenous people.

Continuity and progress

The focus on the HBC is immediately established in the film's opening sequence, with the logo of the company and an intertitle informing the viewer about the occasion of the film's production and the name of the film company (Figure 1 and 2).

Figure 1 and 2. Screen shots from *The Romance of the Far Fur Country*

Two longer intertitles describe King Charles II signing "the charter incorporating the Governor and Company of Adventurers of England" in 1670 and the company's development into "a vast trading concern". Pictures of "Prince Rupert, the first Governor" and the ship that "sailed from England on the first Hudson's Bay expedition" are shown before the fourth intertitle, "Two hundred and fifty years later", takes the viewer to the present time of 1919/20.

Much has changed over 250 years: a steamer has replaced the sailing ship, and the cargo taken into the Hudson Bay area is a "vastly different" one. But the Hudson Bay area is still far away, beyond civilization, and travelling into the area is time-consuming and dangerous: "fog and no observations" and "heavy ice" are natural forces the *Nascopie* has to negotiate before reaching the first Hudson's Bay post. Even the relationship between the HBC and the Inuit is still as it was, in which the Inuit are loyal to the HBC and dependent upon it, faithfully following the company's rules.

In the restored DVD version, a map illustrates the ship's route from Montreal to Lake Harbour. As described by Kevin Nickel in the DVD booklet, "[n]o maps of the journey appear in *The Romance of the Far Fur Country* film reels, but they were created and inserted in the UK version" (Five Door Films 2015, 23). The absence of maps in the Canadian version of the film can be related to the film's title. Although one might assume that the area was more familiar to Canadians than to the British at that time, it is more likely that the film-makers did not want to present a concrete geographical area so much as a country far away from civilization, exotic and out of reach for most of the audience, as the title a "far fur country" suggests. A similar emphasis can be observed in the promotion of the film. Although *The Romance* does not deny the presence of modern inventions, both in the large cities and in remote areas, the film "was conceived of and viewed as a 'northern film'" (Geller 2004, 90), presenting images of the "Last Wildness" and of "the Eskimos and their doings, their customs, habits and their general mode of living" never before seen in motion pictures (Figure 3).[2]

The northern setting of the pictures and the dangers involved in achieving the footage are highlighted in the promotion of the film: "Faced Death to Obtain Pictures/Wonderful Reels Taken in Frozen North to Be Shown Here" (*Vancouver Sun*, Apr. 27, 1920).

While the short introduction to the company's history at the beginning of the film establishes a relationship between past and present, thus highlighting the company's traditions, one important purpose of the film is to show innovation and modernity in the large cities of the West, such as Calgary or Vancouver. Multi-story buildings, cars and, most importantly, the department stores of the HBC are images chosen from these cities.[3] The audience is taken inside the department store and introduced to the products available in the fur department and promised "[a]n excellent repast" when visiting the restaurant. These more or less hidden advertisements, promoting the products of the HBC to potential buyers, can be considered early forms of product placements in movies. The screenings of the finished film itself can be considered part of an advertisement campaign for the HBC: tickets were given away free, but the audience had to come to the department store in order to pick them up, thus the film targeted potential customers and brought them to "the modern department store, the palace of consumption that united public relations and commerce" (Geller 2004, 98).

Figure 3. Advertisement in the *Manitoba Free Press*, Saturday, May 1st, 1920

While the content of such promotions was not new – the demand for fur products was already declining by that time – the department store was obviously a rather new innovation in western Canadian cities. Department stores were important elements of a modern city, not only selling necessary supplies, but also offering service and luxury to their customers, thereby transforming shopping into a leisure activity. By presenting fur products as desirable objects necessary for achieving a certain lifestyle, these consumption advertisements were directed towards white middle-class consumers who had the time and money to spend on cinema, newspapers and the products available in the department stores of the HBC. This is even more obvious in the UK version of the film, *The Heritage of Adventure*.[4] This version contains many more scenes in which white women are shown wearing different fur products, posing on a theatre stage while the curtain is opened and closed, allowing the viewer glimpses of both women and fur products. The presentation of these fur products nevertheless targets a mainly female audience as consumers,[5] a status that gave "an intimate familiarity with the rapidly changing fashions and lifestyles that constituted an important part of the felt experience of being modern" (Felski 1995, 62).

As presented in the film, consumerism belongs to white people's cities and department stores, while people in remote areas make their own clothes, and trading is based on the exchange of goods. Furs are payment for necessary supplies, and the trapper has to trust the trader, as explained in one intertitle: "The Chief Trader behind the counter must be a shrewd and honest man well versed in values." While in the cities fur is fashion for the well-to-do, the Inuit use fur garments for practical reasons in the cold areas of the country. But even in those outposts of civilization, the viewer can find signs that indicate tourism and consumerism, such as drug stores and a hotel. *The Romance* not only promotes the products of the HBC, however, but it also provides the audience with background information about the fur business by taking them to the areas of hunting, cleaning, packing and shipping.

Remote areas and indigenous people

As the film takes the viewer on a journey across Canada, the audience meets different indigenous people. A few intertitles suggest conflicts between white and indigenous people, but these conflicts obviously belong to the past. The present of 1919/20 seems to be characterized by a friendly, peaceful coexistence between the different groups. The Inuit perform duties for the HBC, while the HBC delivers necessary supplies to the Inuit. It seems that a mixture of western and indigenous lifestyles, including traditional rituals and Christianity, exist side by side without troubling each other: one intertitle informs the audience that Indians are "regular churchgoers" while images of young boys wearing clothing used in the Potlatch ceremony are being shown. The otherness of indigenous customs and practices is only suggested in a few rather neutral intertitles: "The Thunder Bird watches over this grave", "Totem pole beside the grave of a young chief", and "Finery donned for the Potlatch ceremony". By duplicating the information given in the moving images, the intertitles anchor the meaning of the images for the audience.[6]

The film's "'view' aesthetic" (Gunning 1997, 9) implies the claim, described for early nonfiction film by Tom Gunning, "that the subject filmed either pre-existed the act of filming (a landscape, a social custom, a method of work) or would have taken place even if the camera had not been there (a sporting event, a funeral, a coronation)" (Gunning1997, 14). *The Romance* consists of a wide range of scenes that actually might have taken place without a camera being present. Some of these scenes, however, suggest a collective, unfavourable evaluation of the Inuit's otherness, created by the film's editing and the way in which some of the intertitles are worded.

One of the most obvious examples of this unfavourable representation can be found about fifteen minutes into the film. The "natives" – as stated in the intertitle – are living in remote areas without "news of the outside world for nine months", thus seemingly unaffected by what is happening in the modern world outside, such as the First World War: "The natives know only the H.B.C. and believe that the recent European conflict was a war between the H.B.C. and some rival company. They are now delighted to learn that the H.B.C. won the war." Even though this intertitle seems to state a fact, and the ignorance of the "natives" could be explained by their isolated existence, it is also possible to read it as a representation of the Inuit as naïve and childlike, in line with Flaherty's "happy-go-lucky Eskimo".

The following scene intensifies this impression. To celebrate the end of the war, the HBC arranges a "candy scramble" in which women, men and children fight over the candy thrown out to them, not unlike the dogs that are fighting for meat in the following feeding sequence. The candy scramble scene and the cut to the feeding scene combine into a derogatory comparison of the Inuit with dogs. In addition to being uninformed about world politics, the Inuit are portrayed as primitives, displaying animal-like behaviour, belonging to the nature in which they live. Although this is probably the most negative scene in the whole film, there are several other intertitles describing the Inuit as huskies.[7] Here the film's editing, together with the intertitles, creates an image of the Inuit as close to animals and nature. In the course of the film, only a few "natives" are presented as individuals by name; for example, Battice Plakoti, "the best trapper in the region", or Ingmilayuk, who functions as the narrator in a sequence about an Eskimo's life story.

The way the Inuit are presented in *The Romance* in many ways suggests a state of hybridity. They live somewhere between traditional life and the modern western world. This is most obvious in the way they dress, but also in their style of living, which has obviously changed from nomadic to sedentary, performing duties for the HBC and waiting for the annual supply boat to arrive. Nevertheless, it seems that they do not live in houses built in a western style, but in more primitive accommodation like huts or tents. In addition, the audience can see words written in their own alphabet, and there is information about the Rev. Edmund Peck, "a missionary who makes the trip each year, compiling an Eskimo grammar" (intertitle). So, on the one hand, the Eskimo language is no longer only oral, but now has its own alphabet and grammar, having taken a step into modernity with the assistance of the white missionary. On the other hand, it is exactly an oral narrative to which the viewer is introduced in one important scene in the film, a narrative sequence entitled "REMINISCENCES. Life Story of an Eskimo, as told by himself. Scenes laid in Baffin Land." Presented as a film within the film, with opening credits but lacking any distinct end, the reminiscence is about six minutes long and consists of a series of "mainly staged scenes that were meant to represent different aspects of Inuit life at a typical arctic trading post" (Five Door Films 2015, 14).

The title "Life Story of an Eskimo, as told by himself" does not mention the name of the narrator, but the opening credits name him and the other participants in both Latin and Inuit type. Ingmilayuk and a nameless white man sit together around a fireplace, both smoking pipes, creating a sense of peaceful coexistence. Ingmilayak's story is illustrated with moving images through which scenes from the past are re-enacted. The story is told in the past tense, and from what the audience can observe even in the old days the white man was present in the area. Scenes around the fireplace and the re-enacted scenes are intercut with each other. When looking at the two men sitting together, the viewer gets the impression that it is Ingmilayak who is taking the active role by telling his story, while the white man sits below him, just listening (Figure 4).

There is an ambiguity in the way the Inuit is represented to the audience in this "film inside the film". On the one hand, they are represented as individuals with names and life stories, with their original names and Inuktitut syllabics used in the intertitles.[8] On the other hand, the audience is presented with some of the same stereotypes about the Inuit as appear in other films from the same period. In “Reminiscences”, these stereotypes primarily concern how the Inuit express emotions.

Figure 4. Screen shot from *The Romance of the Far Fur Country*

Ingmilayuk’s autobiography contains a love story, but the actors show very little excitement or emotion. The intertitles draw the viewers’ attention especially to the relationship between internal and external activities, as Flaherty described it (Figure 5, 6 and 7).[9]

The faces of the two people involved are shown in close-up, first without revealing any emotions, then, after a while, putting on an artificial smile as if they were instructed to do so. This change of facial expression, together with the intertitles’ demand for attention, makes the viewer even more conscious of the alleged absence of emotions in the face of an Inuit.

As mentioned above, the film’s viewing aesthetic (Gunning 1997) belongs mainly to early nonfiction film. Another characteristic of films prior to the 1920s is the use of expository intertitles. As Kathrine Nagels explains, “[i]ntertitles have commonly been divided into two categories: dialogue and expository” (2012, 368), but the use of the latter decreased during the late 1910s in favour of more dialogue. In the case of *The Romance*, the audience only finds intertitles that could be characterized as dialogue, or rather monologue, in the narrative sequences of the film during which Ingmilayuk tells his life story.

Figure 5, 6, 7. Screen shots from *The Romance of the Far Fur Country*, representing the young Ingmilayak

In this narrative from the past, the presence of the HBC is not denied, but celebrated. White men and the HBC play an important role in the life of Ingmilayuk and the other members of the group. Ingmilayuk's narrative calls attention to the fact that this tradition continues ("I was but a youth when I learned to hunt, as my fathers did before me"), but also that the Inuit engage in competitive sports, like wrestling and kayak racing, as leisure activities.

The film inside the film was planned and scripted by Wyckoff on his journey into the area (see Five Door Films 2015, 7). By focusing on the relationship between "then" and "now", the film's presentation of time correlates with Zygmunt Bauman's description of modernity as the history of time: "The history of time began with modernity. Indeed, modernity is, apart from anything else, perhaps even more than anything else, *the history of time*: modernity is the time when time has a history" (Bauman

2000, 110). In a way, Ingmilayuk arrives in modernity by relating past and present time to each other.

Means of transportation

As described above, *The Romance* is organized as a travelogue or road movie. It journeys across Canada, from Montreal through Hudson's Bay, and across the Canadian prairies to Winnipeg, Calgary and Vancouver in the West, while the last leg takes the filmmaker north again to Fort McMurray and Fort Chipewyan in Alberta. The HBC's trading posts in the Northern Territory are visited, along with places of historical importance for Canada and the company.

Thus, in many of the scenes there is a focus on the different means of transportation used for travel across the country. While there are cars in the cities, there is obviously only traditional transportation available in remote areas: kayaks and canoes, sledges, dogs and horses. Unexpected problems caused by nature can force travellers to take longer breaks or even require them to go back in order to find alternative transportation. The travellers are forced into a kind of pre-modernity when modern means of transportation – "non-human and non-animal means of transportation" in Bauman's words (2000, 111) – are not available or it is not possible to use them.

Even though Wyckoff's diary states that the team travelled by train across the prairies and mountains (see Five Door Films 2015, 16), there are no trains or railway stations visible in the film. Only one intertitle suggests that trains are used to transport the fur: "Large freight canoes take the fur up the Abitibi River to the railroad and return with trading goods." While there are cars in the cities, and the steamboat of the HBC comes once a year, there are apparently no vehicles in these remote areas that "could move faster than the legs of humans or horses" (Bauman 2000, 111), a lack of speed that reminds the viewer of pre-modern conditions. One explanation might be that trains could be viewed as an invasion of the landscape, destroying its romance, an experience described by Leo Marx as a "sense of the machine as a sudden, shocking intruder upon a fantasy of idyllic satisfaction" (Marx 1964, 29). Traversing these huge areas takes time and skill; space and distance are foregrounded and of great importance in the film. Travel is not for pleasure, but – as the intertitles inform the audience – hard work for which the "coolness of judgement and long experience" is needed. The men are forced to push themselves to their limits, but at the same time they experience the beauty of nature; "a magnificent view of the river", for instance, is the reward for a long walk uphill. Travelling by canoe or sledge creates a closeness to nature and a belonging "to the same space as the perceived objects" (Schivelbusch 1979, 66), a closeness that can also be experienced by the film's audience.

The Inuit and other indigenous people are not presented as nomads; they only travel over small distances to their hunting grounds. Even though the film does not make this clear, the viewer can assume that their contact with the HBC, western civilization and Christianity has been part of their development from "primitive beings, 'underdeveloped' [...] and 'behind time'" (Bauman 2000, 12) to a more modern lifestyle.

The omission of images of the railroad seems to be part of the film-makers' intention to show the hardship of travel, which requires men with skills and great persistence.[10] The audience receives an impression of huge distances where travel can last for very long periods, and where "the next point of Northern civilization" (intertitle) can be far away. Travelling by canoe or sledge is much more like an expedition than an ordinary journey following a regular route. It seems that the men are travelling through new and unknown territory where only the scattered outposts of the HBC provide pockets of civilization. Where Nanook and his family travelled through completely empty land, finding only an abandoned igloo without any traces of western civilization, the travellers in *The Romance* not only rest at these posts, they also meet other whites who show them the beauty of nature: "The daughters of the Factor at New Post, very proud of the falls near their home, take the newcomers to see them" (intertitle).

The hardship of travel, together with trapping scenes in which, for instance, Battice Plakoti "explains many of the trapper's secrets to the Factor", gives the viewer a sense of the skills and time needed to provide furs for the department stores of the HBC, thus explaining their value to the audience. That indigenous people play an important part in this economy becomes especially obvious in one of the final sequences in *The Romance*, where the audience meets Battice Plakoti. In this sequence, the film's ambiguity is observable: a few of the indigenous people, like Plakoti, are portrayed as individuals, not childish, naïve, or animal-like, but knowledgeable and skilled at what they are doing. The Factor appears as his guest; he is joining Plakoti and learning from him. As in the film within the film, the Factor – and the film's audience – learns from the native.

A message to the King

There is almost nothing in the film that indicates any negative consequences of the presence of the HBC in the area or of its 250 years of colonization. As in the picture of Ingmilayuk and the white man sitting by the fireplace, there is an impression of peaceful co-existence: "the HBC film version Inuit are celebrated as full and happy participants in the fur trade economy" (Five Door Films 2015, 15). There is only one intertitle and one scene that together hint at the problematic relationship between indigenous and white people, and the consequences of colonization. This intertitle comes early in the film and informs the viewer that the Inuit have been moved by the HBC to the best hunting grounds: "Assigning the best grounds to the best hunter. The Eskimos were first brought to Baffin Island by the H.B.C., upon its discovery of good hunting grounds there." But the fact that the Inuit were relocated by the HBC is stated in a neutral way without any comment or further explanation; no negative consequences for the Inuit or any resistance against the move are mentioned. It even presents the HBC in a positive light: not only does the HBC know about the best hunting grounds, but the company informs the Inuit and moves them to those areas.

Later in the film, there is a scene carrying a much more political message, starting with "The Chief of the Chipewyans sends a message to the King by way of the camera." This scene is a good illustration of the relationship between the moving images and the intertitles. The viewer first sees an image in which three men are in conversation with each other. The camera then pans to show the chief from the tip of his toes to the top of his head. In the last part of the scene, the camera focuses on his face in close-up. Even though the viewer cannot hear his words, he/she can see his commitment to the case presented: a treaty with the white man about the hunting season (Figure 8 and 9).

While the focus of the camera is on the chief's otherness during the pan up his body, almost measuring him inch by inch, the close-up on his face generates the possibility for a recognition of sameness, intended or not. This close-up encourages the audience – at least today's audience – to feel sympathy with the person presented, since a kind of contact is established between the audience and the person portrayed by the close-up. Nevertheless, the person investigated by the camera remains nameless and only his status as chief of the Chipewyans is communicated.

To have a conversation with the Factor, he obviously needs a translator, and this person is positioned between himself and the Factor in the frame. This could emphasize his otherness and the distance between the people living in the area and the Factor as a representative of the HBC, a royally chartered company functioning as an arm of the British Government. But the scene also highlights the contact between different cultures and thus inter-lingual communication as a facet of modernity, as did the use of Latin and Inuit lettering earlier in the film. The scene that follows does not comment on the message at all; the next intertitle just informs the audience that the HBC's Factor is introduced to the best trapper in the region. What follows is a long scene about trapping. Even though sending the message to the King "by way of the camera" has no visible consequences in the film, this scene nevertheless highlights the potential of modern film technology to overcome time and space.

Figure 8 and 9. Screen shots from *The Romance of the Far Fur Country*

Conclusion

The Arctic presented to the viewer in *The Romance* is not just a cold, empty and dangerous region, as it is in Flaherty's *Nanook*. There is also a summer season featuring the midnight sun and haymaking, and there are the trading posts of the HBC. But travelling across the country is still challenging, almost like an expedition in the old days. Flaherty wanted to present the Inuit as "real person[s]" (Rotha & Ruby 1983, 50) and not as freaks, but he was nevertheless limited by the dominant stereotypes of his time when he presented his subject, the "fearless, lovable, happy-go-lucky Eskimo". The same ambivalence can be observed in *The Romance*. While the otherness of the indigenous people is only stated in some scenes and intertitles in an informative and almost neutral way, other scenes – especially the candy scramble – are clearly racist in their presentation of the Inuit as uneducated and primitive, acting more like animals than civilized people.

As described by Jennifer Peterson, "[t]ravelogues are an acting-out of tensions between attraction and repulsion: exoticism is presented and then taken away" (Peterson 1997, 76). *The Romance* tries to establish a certain familiarity between the audience and those exotic people living in faraway areas: intertitles are used to relate the images shown to the audience's own experience ("Fashions are various in the different sections of Eskimo land, just as they are elsewhere."), and cultural context (cf. Geller 2004, 101 on the use of well-known nursery rhymes and Robert Service's northern ballads in the intertitles).

The history of the HBC is presented as a story of progress, but there are also continuities between past and present: the ship still only visits remote areas once a year. Even if the lives of the Inuit are no longer unaffected by modern inventions, they are presented to the audience as non-modern people, closer to the dogs in their behaviour than to white people, living in a world "governed by 'natural rhythms' and 'unchanging traditions'" (Jordan 1995, 282). The focus of the film – with very few exceptions – is not on individuals; in general, large groups of people are filmed, and they are well aware of the presence of the camera.

Although material about *The Romance's* premiere is limited, it seems that the film was a success, at least in connection with the other anniversary activities of the HBC in Western Canada. The film medium brought images from faraway areas closer to its audiences, making them familiar with this vast country of Canada; a country that became smaller because of the film, allowing the audience to travel to remote areas without leaving their seats. By combining the older style of the travelogue with clearly staged, narrative sequences like the "film inside the film", one could even argue that *The Romance* invented a new type of film two years before *Nanook* premiered.

References

Bauman, Zygmunt. 2000. *Liquid Modernity*. Cambridge: Polity Press.

Barthes, Roland. 1980. "Rhetoric of the Image." In *Classic Essays on Photography*, edited by Alan Trachtenberg, 269–285. New Haven, CT: Leete's Island Books.

Berger, Sally. 1995. "Move over Nanook," *Wide Angle: A Film Quarterly of Theory, Criticism and Practice* 17 (1): 177–191.

Doane, Mary Anne. 1989. "The Economy of Desire: The Commodity Form in/of the Cinema." *Quarterly Review of Film and Video* 11 (1): 23–33.

Felski, Rita. 1995. *The Gender of Modernity*. Cambridge, MA: Harvard University Press.

Five Door Films. 2015. *The Romance of the Far Fur Country*. Published in conjunction with the DVD of the same name. A 28-page booklet featuring four contextual essays & suggestions for the classroom.

Geller, Peter. 2004. *Northern Exposures: Photographing and Filming the Canadian North, 1920–45*. Vancouver: UBC Press.

Gunning, Tom. 1997. "Before Documentary: Early Nonfiction Films and the 'View' Aesthetic." In *Uncharted Territory: Essays on Early Nonfiction Film*, edited by Daan Hertogs and Nico de Klerk, 9–24. Amsterdam: Stichting Nederlands Filmmuseum.

Jordan, Glenn. 1995. "Flight from Modernity: Time, the Other and the Discourse of Primitivism." *Time & Society* 4 (3): 281–303.

Marx, Leo. 1964. *The Machine in the Garden: Technology and the Pastoral Ideal in America*. New York: Oxford University Press.

Monod, David. 1986. "Bay Days: The Managerial Revolution and the Hudson's Bay Company Department Stores 1912–1939." *Historical Papers* 21 (1): 173–196.

Nagels, Kathrine. 2012. "'Those funny subtitles': Silent Film Intertitles in Exhibition and Discourse." *Early Popular Visual Culture* 10 (4): 367–382.

Peterson, Jennifer. 1997. "'Truth is Stranger than Fiction': Travelogues from the 1910s at the Nederlands Filmmuseum." In *Uncharted Territory: Essays on Early Nonfiction Film*, edited by Daan Hertogs and Nico de Klerk, 75–90. Amsterdam: Stichting Nederlands Filmmuseum.

Rotha, Paul and Jay Ruby. 1983. *Robert J. Flaherty: A Biography.* Philadelphia: University of Pennsylvania Press.

Ruby, Jay. 1989. "A Reexamination of the Early Career of Robert J. Flaherty." *Quarterly Review of Film Studies* 5: 431–457.

Shepard, David. 1974. "Letters to the Editor." *Film Comment* 10 (2): 59–60.

Schivelbusch, Wolfgang. 1979. *The Railway Journey: Trains and Travel in the Nineteenth Century.* New York: Urizen Press.

Filmography

The Romance of the Far Fur Country
Year: 1920/2015
Producer: Educational Films
Cinematography: H.M. Wyckoff, M. Derr
Length: 8 reels, about 120 minutes
Curatorial editor: Kevin Nikkel
Curatorial advisors: Chris Nikkel & Peter Geller
Contemporary score: Nathan Reimer
Colour & Tinting: Tony Wytnick
Sound effects: Kevin Gillen
Graphic elements: Patrick Neufeld & Anthony Nicholson
Additional titles: Mike Maryniuk
Graphic design: Doowah Design

Nanook of the North
Year: 1922/1976
Executive Producer: John Révillon
Director and producer: Robert J. Flaherty
Assistant: Captain Thierry Mallet
Editing: Robert J. Flaherty, Herbert Edwards and Charles Geib
Length: about 75 minutes
Restoration: David Shepard

Notes

[1] The film's intertitles often only refer to them as "Indians", "Eskimos", or "natives", or sometimes more specifically as "Cree Indians". I use "indigenous" as a general term and "Inuit" as a replacement for the film's "Eskimo".

[2] "H.B. CO. Motion Picture at Allen" in *Manitoba Free Press*, Saturday, May 1, 1920, p. 18.

[3] See Monod 1986 for more on the development of the Hudson's Bay Company Department Stores: "Since the 1890s, a demand had been growing within Canada for the HBC to expand its small saleshops in Winnipeg and Vancouver to satisfy the growing urban demand for consumer goods" (Monod 1986, 174).

[4] *The Heritage of Adventure*, the 90 minute UK version of the 1920 film, is available to today's audiences on the Five Door Films' DVD. According to Peter Geller, this version "does not appear to have been released in theatres" (Geller 2004, 208, note 39).

[5] See on this Doane 1989, 27: "If the film frame is a kind of display window and spectatorship consequently a form of window-shopping, the intimate association of looking and buying does indeed suggest that the prototype of the spectator-consumer is female."

[6] See Barthes 1980 for the relationship between linguistic and iconic message.

[7] Flaherty also describes the Inuit as both childlike and animal-like in *Nanook*, where both the puppies and the baby are described as a "young husky".

[8] See Five Door Films (2015, 15) for more about Wyckoff's reasons for using the Inuktitut syllabics: "the use of the 'Eskimo printed characters on the screen' would make the film 'appear more authentic'."

[9] Flaherty recorded BBC Talks, London, 14 Jun., 25 Jul., 5 Sep. 1949 (quoted in Rotha and Ruby 1983, 36). See also: "And the white man has a way of expression. His face reveals so much to a native. He can read your face like a book, while his face remains impassive" (Rotha and Ruby 1983, 13).

[10] One might argue that there were no railroads north of Edmonton in Western Canada and north of Churchill, Manitoba in Central Canada at the time these pictures were made. The total absence is nevertheless striking because the film also shows images from the cities.

THE DISASTER OF FRANKLIN: VICTORIAN EXPLORATION IN THE TWENTY-FIRST-CENTURY ARCTIC

ADRIANA CRACIUN

In September 2014 the John Franklin expedition made global headlines once again when Canada's then-Prime Minister Stephen Harper announced Parks Canada's discovery of Franklin's ship HMS *Erebus*, sought for nearly 170 years by three different countries and numerous private searchers. Found near the Adelaide Peninsula in a location suggested in Inuit oral histories, *Erebus* quickly ignited new international debates over the significance of disaster heritage, the nature of Arctic sovereignty, and the status of the Northwest Passage as an energy frontier. Then in 2016, the discovery of *Erebus'* sister ship HMS *Terror* by private searchers intensified the international debate over the ownership of the historic wrecks lying in Nunavut's waters (Craciun 2017).

Franklin's expedition had gained a new national dimension as a foundational episode in Canadian northern history and national identity since the serial searches and relic exhibitions of the nineteenth century (see Hulan 2002; Grace 2001). After confederation in 1867 and the legal transfer of the Arctic archipelago from Britain to Canada in 1880, the archipelago was designated the District of Franklin, acknowledging the central role of the Franklin searches in mapping and claiming it.[1]

The Canadian search that located *Erebus* in 2014 was the sixth in seven years, led by Parks Canada archeologists in collaboration with numerous agencies including the Canadian Coast Guard Service and the Canadian Hydrographical Service, and private sponsors Royal Canadian Geographical Society (RCGS), Shell Oil, a luxury cruise ship company, and a foundation funded by the co-founder of Blackberry, Jim Balsillie. Prime Minister Harper had taken a high profile interest in the multi-year searches and took center stage in announcing the discovery, describing it as "truly a historic moment for Canada" because "Franklin's ships are an important part of Canadian history given that his expeditions [...] laid the foundations of Canada's Arctic sovereignty" (Harper 2014). Harper's personal interest in the Franklin searches was part of the government's larger strategic use of the Franklin disaster as a public-facing component in its Northern Strategy, the comprehensive Arctic policy introduced in 2007. As part of its Northern Strategy, Canada prioritized new Arctic military and security projects, energy extraction development, science stations, and maritime regulations, all ostensibly working towards the end of strengthening its sovereignty and security in a rapidly changing Arctic. Of particular relevance in the Northern Strategy was the desire to define the Northwest Passage as internal historic waters, a claim denied by the USA and other nations that see the Passage as an international strait. Under the UN Convention of the Law of the Sea (UNCLOS) the distinction is a significant one: as internal historic waters, the Passage cannot be entered by foreign ships without permission from the Canadian government, whereas an international strait allows foreign ships transit passage. As the diminishing sea ice makes the Passage increasingly accessible to larger and more diverse kinds of ships, Canada faces new challenges in regulating traffic through the Arctic archipelago.

Franklin's shipwrecks are poised to play new roles in the Canadian government's approach to Arctic sovereignty, meriting the millions of dollars government agencies collectively spent since 2008 in locating *Erebus* and *Terror*. In 2010 we saw a trial run with the Franklin search that located HMS *Investigator*, one of several Franklin search ships that themselves became lost, described by Parks Canada as the "the most important shipwreck in history" (Martin 2010). When Parks Canada located Franklin's *Erebus* in 2014, it had to outdo this hyperbolic assessment of *Investigator* and so *Erebus* was immediately presented to the public as the lynchpin in Canada's historic sovereignty of the Arctic archipelago and of the Northwest Passage, and Franklin compared to Columbus. Government ministers of all ranks, Parks Canada staff, and the RCGS all chimed in together from the same talking points on Franklin's foundational role in Canada's Arctic sovereignty. But as Canada's territorial sovereignty is not in dispute (as opposed to the Northwest Passage waters), a British naval shipwreck's contributions towards its sovereignty are largely propagandistic.[2]

Public interest in Franklin is not inevitable, nor is it sustained by inertia – it has been repeatedly remade ever since he vanished, and it is the apparent continuity of this appeal, the so-called Franklin Mystery, that is used to obscure the different institutions, agents, and interests involved. In recent years Franklin has inspired novels suitable for literary prizes (Nadolny's *The Discovery of Slowness*), airport layovers (Cussler's *Arctic Drift*), and most recently major films and tv shows in the works (based on

Fortier's *The Proper Use of Stars* and Simmons' *The Terror*, respectively).[3] There have been operas, plays, poems, paintings, tea sets, and board games devoted to Franklin. So it is not surprising that the news of the *Erebus* discovery made global headlines, and that the Canadian government is anxious to start its own collection of Franklin relics, or perhaps even to raise and preserve the entire ship. Sweden did just that with its remarkable (but now threatened) restoration of the sunken seventeenth-century *Vasa,* which receives over one million visitors each year.[4] This enduring popular interest in Franklin specifically and in "celebrity shipwrecks" more generally (Gould 2000, 24), however, should not obscure the new dimensions of Franklin's significance that are driven more by the imagined future than the past.

In bringing on board Shell and the Royal Canadian Navy to help search for Franklin's ships, the Canadian government was expanding an existing link between the energy extraction industries, the military, and Franklin searches. Canada's largest Franklin search before the Harper searches was its "Project Franklin" in 1967, the centennial of Canada's confederation as an independent nation. Unusually, Project Franklin was a military exercise initiated and designed by a private citizen, an Ontario insurance broker and Franklin enthusiast named William McKenzie, who had been inspired by a popular 1961 Franklin book, *Island of the Lost*, by Paul Fenimore Cooper. McKenzie was determined to prove correct the Victorian theory that Franklin's expedition had failed because the British had been massacred by Inuit at "Omannak".[5] The Canadian Armed forces teamed up with geographer William Wonders of the Royal Canadian Geographical Society, an anthropologist from the National Museum, a journalist from the Canadian Broadcasting Service, and together with McKenzie they searched for the mythical "cemented vault" rumored to house Franklin's body and personal journals. While unsuccessful, Project Franklin did locate a debris field near O'Reilly Island that corroborated Inuit oral accounts collected by Victorian searchers which had placed one of Franklin's ships near Wilmot and Crampton Bay, where *Erebus* would be found in 2014 (Project Franklin, 1967). This public-private partnership had included academics, journalists, the RCGS, and three branches of the Canadian military – a coalition of public and private institutions brought together by one Franklin enthusiast's opportunistic timing in proposing the project just when the military was considering how to contribute to the Centennial celebrations.

As another successful "private-public partnership", Shell and the RCGS's collaboration with the Canadian government in the recent searches are on one level part of this longer entanglement of Arctic exploration with military and commercial interests, which began with the much longer tradition of early modern chartered companies that first developed British Arctic exploration. The Company of Cathay had sent Martin Frobisher with a flotilla of ships to Baffin Island in the 1570s in an effort to establish a mining outpost, a spectacular failure than resulted in a lengthy lawsuit and hundreds of tons of worthless ore being shipped back to England. Following the unsuccessful efforts of several early seventeenth-century chartered companies to send their ships through the Northwest Passage in an effort to establish a coveted northern trade route with China, the Hudson's Bay Company in 1670 received its royal charter, the longest in corporate history, and began its modestly successful resource extraction enterprise in the north (initially furs, then whales and walrus).

These early modern corporate bodies had their eyes on the bottom line, seeking exploitable routes and resources in competition with other European interests, but their similarity to the exploration culture of Franklin's day is superficial (Craciun 2016). The mercantilist cultures of exploration of the Company of Cathay and the HBC were incommensurate with those of Franklin's day, and of Shell today. Most significantly, their earlier notions of public and private were not comparable to those of the nineteenth and twentieth centuries, nor was there an early modern sphere of publicly-funded science with an increasingly fraught relationship with governmental and corporate influence. As we shall see below, the Harper administration's attempts to muzzle Canadian climate scientists is part of its larger efforts to politicize public knowledge (archaeological and historical in other cases) for its own interests. It is important therefore to distinguish what is new and specific to the twenty-first century incarnation of Franklin searches and commemoration, and to avoid relying on apparent continuities that can mask important breaks with the past.

The fate of HMS *Breadalbane*, the Franklin search ship lost in 1853 and discovered in 1981 in Lancaster Sound by a public-private search, can help us see the new course the *Erebus* and *Terror* are following in the twenty-first century. When the shipwreck hunter Joe MacInnis located the *Breadalbane*, he worked with Parks Canada, universities, and Big Oil, including two diving companies working with petroleum exploration. At the time, "Dome Petroleum said the information gathered by the expedition will have 'industrial applications'", i.e., for "oil pipelines" ("Arctic Dive" 1983). Dome was active in seeking Arctic pipeline possibilities in the early 1980s, part of a two-decade–long boom in oil and gas exploration in the High Arctic initiated by the 1955 Canadian Geological Survey expedition, called "Operation Franklin". Without Dome Petroleum's help, the *Breadalbane* search would never had had access to a $250 million WASP diving system that made possible the discovery. MacInnis went on to pioneer deep diving film work for James Cameron's *Titanic*, but the commercial savvy of his version of underwater archaeology was already evident in his unauthorized removal of the ship's wheel from *Breadalbane.* A fragile composite artifact, the *Breadalbane's* wheel had to spend 15 years in nitrogenated water awaiting an even more expensive conservation process for five more years, at great public expense (Carter 2003, 13-14; Woods 2011). It was this removal of a prized artifact that precipitated Parks Canada's inscription of the missing *Erebus* and *Terror* as a National Historic Site in 1992, leaving the placeless site in an unprecedented limbo as it awaited the discovery of the ships for decades.

The opportunities awaiting modern Franklin searchers were illustrated more graphically in MacInnis' advertisement for the Rolex Submariner chronometer. The 1985 ad boasts: "Searching for a shipwreck under the Arctic ice is not for ordinary divers. Or ordinary watches" (Figure 1). The Rolex Submariner was the iconic watch of James Bond, conspicuous in such films as *Goldfinger*, and in this ad Rolex and MacInnis linked the romance of contemporary adventurers with that of Franklin's day. Franklin's rescued chronometers had featured prominently in the 1850s relic displays and *ILN* representations, and remain among the few relics on permanent display at the National Maritime Museum (Craciun 2014).

Naval chronometers are both navigation instruments and highly individualized objects –each bears a unique number and signifies not only the status of its owner but, potentially, affective connections between officers (Phillips 2013). For example, some of the Franklin chronometers collected by John Rae in 1854 and currently in National Museums Scotland have had small circles cut from their cases and watch faces, which Rae gave as tokens to family and friends. This may be an appalling way to mistreat an artifact from our vantage point, but it illustrates the highly personal and emotive connection, to the Franklin relics in general, but specifically to chronometers and their evocative ability to signify time arrested. MacInnis and Rolex's commodification of this iconic exploration relic is a sign of how the powerful affective response to the Franklin disaster is continually put to new, larger-scale and more profitable uses.

The search for *Breadalbane* is thus a microcosm of the intertwined commercial, military and energy industry appeal of Franklin's legacy in our time. As the most northerly shipwreck in the world, located close to Beechey Island and its iconic Franklin graves, *Breadalbane* also became a tourist site along the increasingly popular Franklin disaster tours marketed

Figure 1. Rolex "Submariner" advertisement. *Skin Diver* 34.5 (May 1985) 25.

by northern cruise lines. In 2000, several adventure tour companies tried developing the *Breadalbane* National Historic Site into an under-ice submersible diving holiday, called "Ice Station Beechey",[6] which for

$10,000 would take tourists from the camp on the sea ice down to visit the well preserved wreck.

From 2012 to 2014 the *Breadalbane* wreck was absorbed into Canada's annual military High Arctic "sovereignty exercise", Operation Nunalivut. Operation Nunalivut's dives on the *Breadalbane* were staged from a sea ice camp resembling "Ice Station Beechey" (itself alluding to the cult cold war film, *Ice Station Zebra,* starring Rock Hudson). Asked in 2014 to comment on the *Breadalbane's* incorporation into military sovereignty exercises, Joe MacInnis replied, "I wasn't just searching for a shipwreck back then, it was a sovereignty issue — and it still is" (O'Connor 2014).[7] But searching for Franklin's ships in 1981 had little to do with sovereignty, and neither had Arctic exploration in Franklin's day of course. Because of the Franklin disaster's emotive and popular appeal, it is easily appropriated by private and governmental groups with little interest in the complex histories either of Franklin's disaster or of the Northwest Passage.

With the discovery of Franklin's ships we are on the verge of a new reinvention of the Franklin disaster. This reinvention evokes the transformative power of the Franklin myth in ways that resemble a celebration, indeed a fetishization, of exploration history, but is in fact hostile to questions of history. Michael Bravo has written eloquently about the urgent need to resist the emerging consensus that speaks of the Arctic solely in terms of future "opportunities" provided by climate change. This is "a classic recipe for amnesia, a focus on short to medium-term futures, with little or no sense of the past", writes Bravo, and it threatens to rob the Arctic, its peoples and places, of their histories, leaving in their place a dangerous nostalgia (Bravo 2017). The Franklin disaster's public presentation by the groups involved in the recent Canadian searches – from the various ministers, the Prime Minister, and Parks Canada, to the private partners like Shell and the RCGS – has been tightly coordinated from talking points, and what they share is precisely this language of Arctic opportunity to which they wish Franklin to contribute. While many (but not all) Victorians tried to imagine the lost expedition as martyrs to big science and to a national destiny of stewardship of the Passage, Harper's government wanted us to re-imagine Franklin's men as martyrs to Canadian Arctic sovereignty. Both instances represent instrumentalist interpretations of what was a complex human tragedy, but in our twenty-first century case this is more egregious because the stakes are higher and the historical distortion greater.

New to the twentieth-first century Arctic policy were the priorities of Harper's government that profoundly shaped the pursuit and planned preservation of this iconic artifact of exploration history before Harper's conservative government was crushed in the 2015 federal elections. Harper's version of *Erebus* as an "instrument of sovereignty" (Harley 1989, 10), a vision closely coordinated and reinforced across agencies in media coverage, was coeval with his government's heavy handed and politicized muzzling of government scientists. In 2006, Harper's government had introduced strict measures requiring that federal scientists seek prior authorization from the Privy Council before speaking to the media about "politically sensitive" research on global warming.[8] This was at the same time as his government gutted funding for sciences and particularly climate science, including defunding the Polar Environment Atmospheric Research Laboratory in the High Arctic.[9]

Similarly politicized interference with and defunding of publicly-funded archaeological research is directly connected to the ability of the underwater archaeology unit of Parks Canada to locate *Erebus*. In summer 2012 the Harper government slashed 80% of Parks Canada's archaeology and conservation staff nationwide. Franklin search staff were untouched, and indeed received major new equipment and government support for its project. These targeted cuts included the Canadian Museum of Civilization's curator of Arctic archaeology, Patricia Sutherland, who was dismissed just as her groundbreaking research on contacts between Norse and Dorset peoples in Baffin Island was gaining international attention.[10] The significance of what may appear as internecine civil servant strife is actually profound for how we understand Arctic histories, because the intense direction of public focus and funds upon Franklin is inseparable from the impoverishment of expertise and inquiry elsewhere. The elimination of archaeologists and scientists working throughout the Canadian Arctic, the closure and consolidation of regional archives, and the shut down of research revealing the multicultural richness of Arctic Canada five hundred years before the English arrived, together creates a climate of contempt for history.

The presentist and politicized version of history that Harper's administration adopted-- that Franklin's ships are "fundamental to Arctic sovereignty" – was emplaced at the expense of what multidisciplinary research shows – that this expedition became a *cause célèbre* because of a preventable disaster, and that in the history of Arctic discoveries Franklin deserves a minor role. In Harper's emergent narrative about Franklin's *Erebus*, the Arctic is in danger of appearing again as the empty, unpeopled wasteland that Franklin's British contemporaries imagined in the icescapes of the Arctic sublime. Seen again through this Victorian perspective, the Arctic's human histories began not with the arrival of the Dorset and Inuit thousands of years ago, nor with Norse traders one thousand years ago, but with the recent arrival of the ephemeral and unsuccessful British navy. In this version of Arctic opportunism, the Arctic is emptied of indigenous people and their multiplicity of histories, and reduced to an environment full of energy awaiting extraction. By crafting a northern policy around such an instrumentalist, anti-intellectual, and Anglocentric approach to the Arctic, Canada's government embraced Franklin's most dangerous legacy.

Erebus and the Franklin disaster were enlisted to play new roles in this scenario as part of heritage sites and educational materials that were to be developed in partnership with Shell Canada. Shell owns the largest number of oil and gas leases in the North American Arctic, and at the height of the Franklin searches from 2012–2015, it had continued moving forward on its plans to drill, until a disappointing test drill forced it to stop in 2015. Perhaps in an effort to popularize their controversial presence in the Arctic, Shell partnered with the RCGS and Parks Canada in order to connect the heroic associations of Victorian geographical exploration with today's controversial Arctic energy exploration. In effect, Shell was extending its version of "exploration" back in time: "Exploration has been important to the success of our business for over 100 years."[11] Shell's spokesperson had greeted the *Erebus* discovery by saying that Shell "looks forward to sharing the great story of the Franklin Expedition in schools throughout Canada".[12]

The RCGS and its Canadian Geographic Division (which publishes *Canadian Geographic* magazine) began a controversial partnership with the Canadian Association of Petroleum Producers (CAPP), the oil lobbyist group, in 2013. Together they created the "Energy IQ" and "Canada's Energy" curriculum and educational materials distributed to schools across Canada. With Shell the Canadian Geographic Division also created the "Classroom Energy Diet Challenge", to "help improve energy literacy" (Canadian Geographic 2015). With Shell and the Weston Foundation,[13] the RCGS had also been developing materials specifically about the Franklin expedition for Canadian schools. In 2014, the RCGS singled out Shell as its partner on these Arctic issues, promising "Through RCGS' partnership with Shell Canada, [to] create and deliver a multi-year national public and classroom educational program on the Canadian Arctic" (Partnering Agreement 2014). In partnering with these oil lobbyists and corporations in recent years, Canadian Geographic and the RCGS have come under increasing criticism for conflict of interest, including the possible failure to disclose financial sponsorship and allowing financial sponsors (Shell in particular) to alter editorial content (Brown 2015, parts 1 and 2). In its "energy literacy" materials in particular, Canada Geographic has been accused of allowing the oil lobbyist group "CAPP... to vet Canadian Geographic's educational content".[14]

The RCGS had brought Shell on board during Parks Canada's successful search in 2014, and the concern about the energy industry's influence over the educational materials produced by a non-profit group (Canadian Geographic) for public schools is equally applicable to the plans under development around *Erebus's* discovery. Shell's collaboration in searching for and educating about Franklin's ships had effectively attempted to rehabilitate their corporate image at a critical juncture (the renewal of drilling plans in 2014) by highlighting the continuities between Arctic exploration in the "heroic age" and in our own age of climate crisis and oil spill disasters. During the "Mission *Erebus* and *Terror* 2015" led by Parks Canada that August, Shell and the RCGS disappeared from the list of Parks Canada's search partners. In April 2015, a high-profile controversy had erupted alleging that a senior journalist at the *Toronto Star*, Paul Watson, had been effectively censored by editors while working on a story about the RCGS's role in the Franklin search. Watson, a Pulitzer-Prize winning journalist, resigned his position at the paper in protest over the paper's quashing his story on John Geiger (CEO of the RCGS) and his contested role in search for and discovery of *Erebus* (Geiger and the *Star's* editors were former colleagues).

The *Toronto Star* and Geiger denied Watson's allegations, and it remains unclear how or if this controversy was related to the disappearance of the RCGS and Shell as partners in Parks Canada's 2015 continuing search for *Erebus* artifacts and for *Terror*.[15] Shell's earlier interest in *Erebus* as an opportunity to generate educational materials about its vision of the Arctic, and CAPP's interest in similarly shaping geographical educational materials for public schools, signaled a worrying direction that the RCGS' influence could have taken in commercializing the history of exploration.

The Harper government's heavy handed rewriting of the Franklin disaster as the foundational event of his government's new version of Arctic history thus had worked in tandem with private energy interests to fuse the history of hydrocarbon extraction with that of sovereignty claims. Such a presentist approach to exploration history is particularly attractive to those who see the Arctic as a space in which to pursue future opportunities, including sovereignty as a "security commodity" (Dittmer et al 2011, 210; Dodds 2011). To be sure, centuries before Shell's "100 years" of exploration, the Company of Cathay, the Hudson's Bay Company, and the British Admiralty all placed great (but distinct) economic value on their respective future Arctic plans. But theirs were not

Arctics defined as energy frontiers by large-scale, transnational industrial operations like Shell's. Their precarious and ephemeral footholds in Arctic places, their encounters with Inuit, and their share of political clout in geopolitics, were on scales incommensurable with those of Shell, Gazprom, or ExxonMobil.

When speaking of the Arctic as energy frontier in our moment of anthropogenic climate crisis, it is important to distinguish not only the interests of the present bearing down upon it but also the difficulties specific to our time and alien to Franklin's day. One key feature in our era of grappling with anthropogenic climate change is identified by Mike Hulme as "climate reductionism", where "the future is reduced to climate" (2011, 264): in "this new variety of the determinist fallacy", Hulme writes, "the complexity of interactions between climates, environments and societies is reduced", and both human histories and futures impoverished as a result (2011, 247). Conceived through the lens of prescriptive natural sciences, the future appears "free of visions, ideologies and values" (Hulme 2011, 265). The Arctic is particularly vulnerable to such reductionism in part because of the historically limited humanistic scholarly engagement with it, and the corresponding success with which the Arctic has been increasingly defined as environment, energy frontier, and "abstract space (security commodity)" rather than "occupied place (homeland)" (Dittmer et al 2011, 210). "In the Arctic", writes Michael Bravo, "peoples, livelihoods, languages, and ecosystems contain within them the resources of long memories, in ways that are closely tied to the dynamics of frozen states" (Bravo 2017). According to Bravo, in attempting to empty out these long memories, human and nonhuman, we are at risk of reducing the Arctic to its economic futures. Current instrumentalist uses of exploration's histories, i.e., the transformation of Franklin into a martyr for Canada's Arctic sovereignty and/or a precursor for Shell, do not enliven those histories and places so much as consign them to obscurity.

We have often heard that the discovery of *Erebus* "solved one of Canada's greatest mysteries", so that at last we will know the truth of "what really happened" on the expedition (Harper 2014). But those details are not mysterious, merely unknown. Demystification of Franklin's legacy is welcome, but what we are now seeing is its further obfuscation. Boasting that the 2014 Franklin search was "the largest and most technologically advanced" (Geiger 2014) while failing to mention the fatal role played by the Franklin crew's large numbers, unwieldy ships, and "advanced" but ill-suited technology, reduces these recovery efforts to the proliferation of more ignorance.

Likewise the selective acknowledgment of the significance of Inuit knowledge in the searches for *Erebus* and *Terror* reveals the narrow range – geographical accuracy – of indigenous knowledge that is being incorporated to the exclusion of other dimensions. Detailed Inuit oral histories of what happened to Franklin's ships and men (and of much earlier English ships) had been collected by Victorian searchers Charles Francis Hall and Frederick Schwatka, had continued to circulate in Inuit communities to this day, were recorded in the Igloolik Oral History Project, and were again compiled in print in important studies by David Woodman and Dorothy Eber. The Inuit accounts' veracity had been debated in Victorian Britain and resisted by authorities (though by no means the entire public) because they confirmed the Victorians' worst fears – that the expedition had unraveled into chaos and cannibalism.

Spokespersons for the recent Canadian searches regularly invoked the significance of this long record of Inuit histories for locating *Erebus* and *Terror*. (Indeed the previous large-scale Canadian search, Project Franklin in 1967, had also searched this area around O'Reilly Island intensively based on Inuit histories and found numerous ship artifacts and underwater debris indicating the nearby presence of a wooden ship.) Ryan Harris, chief underwater archaeologist in the Parks Canada searches, described how the location of *Erebus* "validates, in many ways, their [the Inuit's] testimony, which, in some quarters, had been disputed" (in Watson 2014). Virtually every other search agency spokesperson echoed this point that the *Erebus* search "validates" Inuit oral histories of the ships' probable locations. The high profile involvement of then-Environment Minister Leona Aglukkaq (an Inuk from Gjoa Haven who also chaired the Arctic Council) in these acknowledgments of Inuit knowledge gave credibility to the Franklin searches' engagement with Inuit communities: "I have great pride in knowing this discovery will help Canadians learn how valuable Inuit traditional knowledge is" (Aglukkaq 2015, x).[16]

European admiration for Inuit geographical accuracy predates the Franklin voyages – indigenous knowledge was one of the most highly prized artifacts that Europeans collected. Franklin's Victorian searchers had eagerly sought out elders to question about possible survivors and ship locations, and to provide them with relics from the wrecks, which they then exhibited in London museums. "The native as mapmaker is a powerful myth" in the history of exploration according to Michael Bravo, "one of the elements which collectively constitute the ritual of the explorer's arrival and the mastery of the indigenous guide" (Bravo 1996, 21). Seen in this larger historical frame, the high profile "vindication" of Inuit accuracy and knowledge appears as a conventional component of the larger narrative of discovery, appropriation, and settlement that the Harper government nostalgically revived. Beyond the use value of Inuit knowledge of the ships' locations, where else have the Canadian searches shown room for or interest in Inuit perspectives on, and potential critiques of, the value of the searches themselves? For too long Canadian and British authorities have framed the issue solely in terms of the truth of Inuit knowledge, rather than in terms of how Inuit occupancy of the Arctic should entrust them as partners in decision-making.

The discovery of *Terror* in 2016 by the Arctic Research Foundation (ARF) highlighted the relationship of truth to trust, and why popular focus on the former has obscured the importance of the latter. On board the Arctic Research Foundation search ship *Martin Bergmann*, hunter Sammy Kogvik of Gjoa Haven told searchers that several years back he had found what may be part of a ship's mast sticking out of the ice in Terror Bay. Believing in him, Captain Adrian Shimnowski rerouted the *Martin Bergmann* to Terror Bay – they located the *Terror* within hours. When asked why he did not alert Parks Canada searchers, but told the head of the private Arctic Research Foundation instead, Kogvik said: "I trust him.... But I don't trust Parks Canada because they like to keep everything secret" (in Sorensen 2016). The priority dispute swirling around Parks Canada's *Erebus* discovery, and the Harper government's heavy-handed politicization and sensationalization of that project, had damaged Parks Canada's credibility with many Nunavummiut.

This break in trust also appeared inside the public-private partnership of Parks Canada with the ARF when the *Terror* discovery was announced in September 2016: the ARF's press release claimed independent credit for the discovery and did not even mention Parks Canada, its lead partner, while Parks Canada's follow-up press release similarly did not recognize the ARF. The ARF had also bypassed Canadian media and gave the global story to the UK *Guardian* via Paul Watson, the journalist who lost his job at the *Toronto Star* following his claims of being censored while investigating Geiger and the RCGS' roles in the Parks Canada searches. This hints at a breach reminiscent of the Victorian disputes that dogged Franklin searchers for decades as they squabbled over prize money and public credit.

Today what still needs vindication is not the truthfulness of the Inuit stories, but the trustworthiness of the other parties involved. Despite promises to rebuild relationships with the First Nations, Trudeau's new Liberal government stumbled when it came time to address how and by whom the Franklin ships would be preserved and commemorated. Trudeau's government upheld one of Harper's controversial campaign promises: a commitment to spend $16.9 million to build a new "Franklin Centre" in Gjoa Haven (and a further $1.5 million per year for upkeep), devoted to Franklin's legacy, and to showing "to the world the importance of Inuit Qaujimajatuqangit" ("Aglukkaq" 2015). But Gjoa Haven already has a museum devoted to Inuit traditional knowledge (Inuit Qaujimajatuqangit, or "IQ") – the Nattilik Heritage Centre (built in 2013 for $3 million) – and it had not been consulted at the time of the Conservatives' initial pronouncement before the election. According to the Inuit Heritage Trust, they would have preferred that a Franklin commemorative center be incorporated into an expanded Nattilik Heritage Centre instead of occupying a separate space (Zerehi 2015). Jack Arawak, an Inuit NDP candidate opposing Aglukkaq in the election, went further and argued that "Honouring somebody who's a failure I don't think is a good idea [...] I mean he failed at first at finding the Northwest passage, and secondly, failed at surviving in the North when he could have survived by using the expertise from the North" (in Zerehi 2015).

Building a monument to this failure and hubris, and then presenting it without consultation as a "contact zone" of cooperation between European exploration and Inuit knowledge, was indeed the Harper model for pursuing the Franklin searches. But as Arawak noted, Inuit knowledge encompasses much more than geographic accuracy. Louie Kamookak, an Inuit oral historian and Franklin search partner with Parks Canada, made a similar call for a broader understanding of Inuit knowledge: "For us Inuit it means that oral history is very strong in knowledge, not only for searching for Franklin's ships but also for [the] environment and other issues" ("Franklin" 2014). "Inuit traditional knowledge," writes Bravo, "often articulated by elders for researchers, is based on a moral understanding of the place of humans in their environment" (Bravo 2009, 272). This ethical, relational component of Inuit knowledge has been absent in the Franklin searches' celebration of Franklin's wholly antithetical way of relating to the Arctic as a hostile, empty, and lifeless space to be conquered. Thus, respect for and engagement with Inuit knowledge/IQ is not demonstrated by planning to build, without consulting those entrusted with preserving Inuit culture, a multi-million memorial to a Eurocentric disaster that was predicated on the superiority of British technology and (mis)understandings of the Arctic.

Instead of reconstructing a vestige of Victorian imperial hubris, if one were serious about how the Franklin disaster and its ships could serve as a truly hybrid contact zone for diverse public understandings of the Arctic's mutiplicity of histories and possible futures, one could imagine an innovative museum or cultural space shared with or developed in collaboration with the Nattilik Heritage Centre and Inuit Heritage Trust. Franklin and his ships could take their appropriate places in relation to the more expansive and more resilient Inuit culture with which they came into intermittent contact in the Arctic. The extravagant scale of the ships, the expedition, the loss of life, the searches, and the proposed funding of the Franklin Centre are all part of the same problem, and all of them, not just the location of the ships, can benefit from Inuit knowledge about how to live more gracefully in a dynamic Arctic.

Movement in this direction began in 2016 when Inuit groups revealed that they were being excluded from the closed-door negotiations between Britain and Canada regarding ownership of the ships. By July 2016, Inuit groups began to assert joint ownership rights over *Erebus* artifacts and refused to allow Parks Canada to remove more artifacts from Nunavut (Beeby 2016; Kassam 2016). The discovery of *Terror* in September exacerbated the situation, with Inuit groups' assertion of joint ownership going unanswered until Parks Canada was pressured to consider "options for co-ownership of the artifacts with our Inuit partners" (Beeby 2016).[17]

The status of the Franklin wrecks, artifacts and heritage sites remains complex. Under Harper's rewriting of the Franklin myth as the centerpiece of Canadian Arctic sovereignty, the party line had been that the British naval ships and their contents were unambiguously "Canadian". But according to a Memorandum of Understanding signed by Canada and Great Britain in 1997, while Britain granted Canada "custody and control" of the sunken ships and their salvage, it did "not waive ownership or sovereign immunity with respect to the wrecks themselves or their contents", retaining "any recovered artifacts identified by Britain as being of outstanding significance to the Royal Navy". According to international shipwreck law, military ships remain the property of the nation in question regardless of where they lie. And on top of that, an MOU does not have the same legal status as a binding contract, law, or a treaty.

There are additional overlapping agreements to untangle. There is the Nunavut Land Claims Agreement, Article 33 of which states that Canada and Nunavut "jointly own all archaeological specimens that are found within" Nunavut, including "explorers' cairns" and "explorers' documents". There is also a second MOU that Canada agreed with Nunavut in 2012, to "share resources and expertise in order to undertake research, search, preservation" of Franklin's ships. This 2012 agreement had tried to fold Franklin into the internal Canadian framework that Harper preferred, but fell short of engaging Inuit groups adequately. Parks Canada's *Terror* press release eventually signaled the welcome possibility that the "Wrecks of HMS Erebus and HMS Terror National Historic Site will be the first cooperatively-managed national historic site in Nunavut" (Parks Canada 2016).

The Canadian government, their captains of industry partners, Parks Canada, and their British and Inuit partners have only begun their transformation of *Erebus* and *Terror* into exhibits, educational materials, documentaries, souvenirs, (new Rolexes?) and new heritage sites. Thus far British imperial nostalgia has played too large a role in the peculiarly Anglocentric Arctic that *Erebus* and *Terror* have been imagined by its discoverers to inhabit. "We are inheritors of the British Arctic exploration legacy," said John Geiger, president of the RCGS: "What was once British is now Canadian" (in Oved 2014). Coauthor with Owen Beattie of the important Franklin study *Frozen in Time,* Geiger played a large public role in media coverage of the Franklin discovery, in enlisting the involvement of Shell and the luxury cruise line in the search, and in the ensuing media dispute alleged by journalist Paul Watson. This Canadian nationalist reappropriation of Franklin's ships would consolidate further the histories of exploration that emerged in the Victorian age of empire. It also threatened to reinscribe public museums and publicly-funded research along neocolonial lines strikingly out of step with twenty-first century scholarly and museum practices.

The ships and their contents are artifacts of an encounter stretching back two centuries, and extending into the future.[18] Their stewardship and presentation should reveal the meeting up of British, Canadian and Inuit histories. Acts of private appropriation, in contrast, obscure the manifold contingency of the past even while claiming to recover it. Writing of the nostalgic monuments to Capt. James Cook erected by British colonists in New Zealand, Nicholas Thomas observes that such a "devotional attitude toward history [...] blocks our view of the past" (Thomas 1999, 107). Such monuments are self-serving, and obscure the "risk and possibility that was vital" to such voyages and in their place create a "sense of inevitability" (Thomas 1999, 107). "An engagement with history may not enable us to anticipate the future", writes Thomas, "but it should make the past less predictable" (Thomas 1999, 9).

Franklin's continuing fascination is fascinating in its own way and has transformed what is imagined about (and increasingly pursued in) the Arctic. But that is why we should look before, after, and around this disaster, and resist what Jerome McGann describes as "the pressure of the consuming present, where illusions of sovereignty are fashioned from truthiness and memories of convenience" (McGann 2015, 217). The more unpredictable become our histories of Arctic exploration and Franklin in particular, the less useful they will be for the sovereignty and energy extraction exercises of our present.

The profuse artifacts of the Franklin disaster, now including that most charismatic technology of exploration, the ship, do not exist as precious "relics" self-evidently and throughout time, awaiting discovery and reverential display in our museums. They can only be only given a "relative existence" and visibility as relics and artifacts worthy of a Franklin Centre by the selective labor of those of us who come later (Latour 1999, 252). Rather than hagiography, this labor should be that of genealogy, as Foucault defined it: genealogy "disturbs what was previously considered immobile; it fragments what was thought unified; it shows the heterogeneity of what was imagined consistent with itself" (Foucault 1984, 83). Moreover this genealogy should expand beyond the narrowly nationalist lineages envisioned in the Stephen Harper years (as in the Lady Franklin years) and out into the transnational, relational, and indigenous communities of the dynamic Arctic. "What was once British" in the Arctic is not only not Canadian, but it was never British to begin with. Instead, *Erebus*, *Terror* and their artifacts present an "opportunity for a new relationship to be on display",[19] and it is that relationship that is at stake in the twenty-first century Arctic. If we reintroduce Franklin into these larger and longer histories of resilient Arctic places and peoples, his significance, like the course of *Erebus* and *Terror*, appears neither inevitable nor self-evident.

References

"Aglukkaq re-announces Franklin Centre for Gjoa Haven." 2015. *Nunatsiaq News*, Sep. 14. Accessed online Nov. 7, 2015.

Aglukkaq, Leona. 2015. Foreword to *Franklin's Lost Ship: The Historic Discovery of HMS Erebus* by John Geiger and Alanna Mitchell. New York: HarperCollins.

"Arctic Dive opens the way for oil hunt." 1983. *New Scientist* (June 23): 863.

Beeby, Dean. 2016. "Inuit press claims for co-ownership of Franklin artifacts." *CBC News*, Jul. 24. Accessed Sep. 2.

Bravo, Michael. 1996. "The Accuracy of Ethnoscience: A Study of Inuit Cartography and Cross-Cultural Commensurability." Manchester: Manchester Papers in Social Anthropology.

—. 2009. "Voices from the Sea Ice: the reception of climate impact narratives." *Journal of Historical Geography* 35: 256–278.

—. 2017. "Cryopolitics and Frozen States." In *Cryopolitics: Frozen Life in a Melting World*, edited by Joanna Radin and Emma Kowal, 27–58. Boston, MA: MIT Press.

Brown, Jesse. 2015. "The Sad Story of Canadian Geographic." Parts 1 and 2. Jul. 21 and 22. *CANADALAND*. Accessed Nov. 3.

Canadian Geographic. 2015. "Classroom Energy Diet Challenge." Accessed Nov. 2. http://energydiet.canadiangeographic.ca/

Carter, Kim. 2003. *Preserving Our Marine Heritage,* 2nd ed. Ontario: Save Ontario Shipwrecks.

Craciun, Adriana. 2012. "The Franklin Mystery." *Literary Review of Canada* 20 (4): 3–5.

—. 2014. "The Franklin Relics in the Arctic Archive." *Victorian Literature and Culture* 42 (1): 1–30.

—. 2016. *Writing Arctic Disaster: Authorship and Exploration.* Cambridge: Cambridge University Press.

—. 2017. "Of Shipwrecks and Sovreignty." *Ottawa Citizen*, July 14.

Dittmer, Jason, Sami Moisio, Alan Ingram, and Klaus Dodds. 2011. "Have you heard the one about the disappearing ice? Recasting Arctic geopolitics." *Political Geography* 30 (4): 202–214.

Dodds, Klaus. 2011. "We are a northern country: Stephen Harper and the Canadian Arctic." *Polar Record* 47: 371–382.

Eber, Dorothy. 2008. *Encounters on the Passage: Inuit Meet the Explorers*. Toronto: University of Toronto Press.

Foucault, Michel. 1984. "Nietzsche, Genealogy, History." In *The Foucault Reader*, edited by Paul Rabinow, 76–100. New York: Pantheon.

"Franklin find proves 'Inuit oral history is strong.'" 2014. CBC News, Sep. 10. Accessed online Nov. 7, 2015.

Geiger, John. 2014. Radio Interview on *Maritime Noon* (CBC radio). Sep. 17.

Gould, Richard. 2000. "Beyond Exploration: Underwater Archaeology after the Year 2000." *Historical Archaeology* 34 (4): 24–28.

Grace, Sherrill. 2001. *Canada and the Idea of North*. Montreal: McGill-Queen's University Press.

Ha, Tu Thanh. 2016. "Discovery of HMS Terror heats up Battle over who controls artifacts." *Globe and Mail*, Sep. 13.

Harley, Brian. 1989. "Deconstructing the Map." *Cartographica* 26: 1–20.

Harper, Stephen. 2014. "Statement by the Prime minister of Canada announcing the discovery of one of the ill-fated Franklin Expedition Ships" (Sep. 9, 2014). Accessed Nov. 6.

Hulan, Renée. 2002. *Northern Experience and the Myths of Canadian Culture*. Montréal: McGill-Queen's University Press.

Hulme, Mike. 2011. "Reducing the Future to Climate." *Osiris* 26: 245–266.

Kassam, Ashifa. 2016. "Inuit argue for say as Canada and Britain decide fate of HMS Terror Wreck." *Guardian*, Sep. 16.

Latour, Bruno. 1999. "On the Partial Existence of Existing and Nonexisting Objects." In *Biographies of Scientific Objects*, edited by Lorraine Daston. 247–269. Chicago, IL: University of Chicago Press.

Lewis, Tanya. 2012. "Gorgeous Decay: The Second Death of the Swedish Ship *Vasa*." *Wired* Sep. 3, 2012. Accessed Sep. 1, 2016.

Manasan, Althea. 2015. "FAQ: The issues around muzzling government scientists." CBC, May 20. Accessed Oct. 31.

Martin, Don. 2010. "Canadians Discover Long-Lost Ship 'Fundamental' to Arctic Sovereignty." *National Post,* Jul. 28.

McGann, Jerome. 2015. "Truth and Method: Humanities Scholarship as a Science of Exceptions." *Interdisciplinary Science Reviews* 40: 204–218.

Memorandum of Understanding Between the Governments of Great Britain and Canada Pertaining to the Shipwrecks HMS *Erebus* and HMS *Terror*. 1997.

Memorandum of Understanding Between the Government of Canada and the Government of Nunavut Respecting Cooperation and Coordination for Research, Search and Preservation Pertaining to the Shipwrecks HMS *Erebus* and HMS *Terror*. 2012.

Noik, Sherry. 2015. "Canadian Geographic denies oil lobbyist influenced its educational materials." *Yahoo New*, Jul. 22.

"Nunavut's Franklin artifacts caught in jurisdictional dispute." 2016. *Nunatsiaq Online* (Mar. 14). Accessed online Mar. 14 2016.

O'Connor, Joe. 2014. "Eerie new underwater footage shows historic Arctic shipwreck – the northernmost known – in fresh detail." *National Post*, Apr. 25. Accessed Sep. 2, 2016.

One Ocean Expeditions. 2014. "Victoria Strait Expedition Invitation."

Oved, Marco Chown. 2014. "Biggest search party yet for Franklin's lost ships." *Toronto Sun*, Jul. 27.

Parks Canada. Jul. 2015. Application for New Permit. Nunavut Impact Review Board.

Parks Canada. Sep. 26 2016. "Government of Canada Confirms Wreck of HMS Terror and Deepens Collaboration with Inuit in Nunavut through co-ownership of Franklin Artifacts."

Partnering Agreement Between Parks Canada Agency and The Royal Canadian Geographical Society. 2014.

Pharand, Donat. 2007. "The Arctic waters and the Northwest Passage: A final revisit." *Ocean Development & International Law* 38 (1–2): 3–69.

Phillips, Eoin. 2013. "The Economization of Time in the Pacific." Presented at the "Oceanic Enterprise" Conference, Huntington Library, San Marino, CA. Jan. 26.

Pringle, Heather. 2012. "Vikings and Native Americans." *National Geographic* 222.

Project Franklin. 1967. "Final Report." Glenbow Archives, Calgary.

Royal Canadian Geographical Society. Jul. 17, 2014. "2014 Search for Lost Franklin Vessels Targets New Location" (press release).

Royal Canadian Geographical Society. Sep. 9 2014. "Franklin Expedition Wreck Discovery" (press release).

Sisler, Julia. 2014. "Research Cutbacks by Government Alarm Scientists," CBC News, Jan. 10. Accessed Oct. 31, 2015.

Sorensen, Chris. 2016. "HMS *Terror*: How the Final Ship Was Found." *Macleans*, Sep. 14. Accessed Sep. 14.

Spears, Tom. 2016. "News of HMS Terror discovery didn't reach Parks Canada searchers." *Ottawa Citizen* Sep. 26, 2016. http://ottawacitizen.com/news/local-news/after-weeks-of-near-silence-parks-canada-confirms-arctic-wreck-is-hms-terror

Thomas, Nicholas. 1999. *Cook's Sites: Revisiting History*. Photographs by Mark Adams. Dunedin, N.Z.: University of Otago Press.

Watson, Paul. 2014. "Historic Moment as Franklin Wreck Found." *Toronto Star,* Sep. 10. Accessed Sep. 10.

—. 2016. "Ship found in Arctic 168 after doomed Norwest Passage Attempt" *Guardian*, Sep. 12. Accessed Sep. 12.

Weston Foundation. 2015. Northern Committee's website, accessed Nov. 9. http://www.westonfoundation.org/our-initiatives/science-in-canadas-north/

Woodman, David. 1991. *Unraveling the Franklin Mystery: Inuit Testimony*. Montreal: McGill-Queen's University Press.

Woods, Allan. 2011. "Exploring the Wreck of HMS *Investigator*." *The Star*, Jul. 9.

Zerehi, Sima Sahar. 2015. "Franklin Centre announcement catches Inuit Heritage Trust off guard" CBC News, Sep. 14. Accessed online Nov. 7.

Notes

[1] According to Donat Pharand, Canada did not make specific sovereignty claims to the entire archipelago until 1973, when it began to argue for the use of territorial baselines (Pharand 2007).

[2] The "Wrecks of the HMS *Erebus* and HMS *Terror* National Historic Site" (esp. if inscribed as a World Heritage Site), on the other hand, could play a role in helping to control access to the Northwest Passage (see Craciun 2012).

[3] Director Jean Marc Vallée (*Dallas Buyer's Club*) announced the day after the discovery of *Erebus* that he will direct a major studio film version of *The Proper Use of Stars*, about the Franklin disaster.

[4] Sweden's restoration of the sunken 1628 *Vasa* was renowned for its technical innovation, but in 2000 it was discovered that the ship is deteriorating at an alarming rate (see Lewis, 2012).

[5] David Woodman untangles the confusion over the Adam Beck massacre account regarding "Omannak" (an Inuktitut name applied to numerous different places, and turning up in conflicting nineteenth-century Franklin searches (Woodman 1991).

[6] An ad in *Sport Diver* (Jan/Feb 2000, p. 112) listed the price for the Zegrahm Deep Sea Voyages "Ice Station Beechey" holiday as US$ 9,980. A press notice in 1998 for the same adventure noted an "optional four-day extension to visit the North Pole for $8,980 per person" (Jorge Sidron, "Operator Offers Arctic Submersible Adventures," *Travel Weekly* Oct. 28, 1998). Deep Ocean Expeditions offered a similar "Submersibles to the *Breadalbane*" excursion for the same price.

[7] Concern over Canada's control of the Northwest Passage began in 1985, with the transit of the US icebreaker *Polar Sea* without permission, prompting an international incident and the controversial implementation of baselines designating the entire archipelago as Canadian waters.

[8] For a good summary of the most significant instances of this government censorship, see Manasan 2015.

[9] After PEARL had its funding eliminated in 2012, a different government agency was able to secure $5M in 2013 for an additional five years of operation. By comparison, Harper's "Franklin Centre" dedicated to preserving and displaying *Erebus* was promised $16.9M for five years plus 1.5M thereafter for operation, which Trudeau's government then supported.
For an overview of these cuts, see Sisler 2014.

[10] Patricia Sutherland's paradigm shifting research about a Norse colony at Nanook, Baffin Island, was profiled in *National Geographic* in 2012 (Pringle 2012). Her controversial dismissal was widely reported on in Canadian media.

[11] Shell spokesperson in RCGS Jul. 17, 2014.

[12] RCGS Sep. 9 2014.

[13] The W. Garfield Weston foundation "is providing $300,000 in funding to Parks Canada, ArcticNet, and Québéc-Océan to support scientific research at the site of the *Erebus* wreck" as of Sep. 2015; the Weston "foundation's funding enabled One Ocean Voyager", the luxury cruise ship, "to join the search" in 2014 (Weston Foundation 2015).

[14] Brown 2015, Part 2. *Canadian Geographic* denied these charges (Noik 2015).
[15] Despite the RCGS not being listed as a 2015 search partner, Parks Canada's 2015 Nunavut Impact Review Board permit application noted "a partnership between Parks Canada and the Royal Canadian Geographical Society... [for] the development of lesson plans and resource kits for distribution to teachers in all provinces and territories" (Parks Canada Jul. 2015). In Nov. 2015 I contacted both Ryan Harris (PC underwater archaeologist and PI on the NIRB application) and the RCGS media relations spokesperson for confirmation that this aspect of the RCGS-PC partnership was ongoing but neither replied to my emails.
[16] Aglukkaq herself had become unpopular in Nunavut, given her support of the Nutrition North program, her failure to acknowledge the environmental consequences of energy extraction while Environment Minister, and her lack of support of aboriginal rights groups like Idle No More despite also being Aboriginal Affairs minister. She came in a distant third in her riding (constituency) in the 2015 election, a crushing loss that many saw as part of a wholesale rejection of Harper's Arctic policies.
[17] The Inuit groups include the Kitikmeot Inuit Association and Nunavut Tunngavik Inc. (see "Nunavut's Franklin" 2016; Ha 2016). At the time of going to press, the ownership negotiations were ongoing.
[18] I borrow the term from the excellent "Artefacts of Encounter" (2010–2013) project at the Cambridge University Museum of Archaeology and Anthropology.
[19] Cathy Towtongie, president of Nunavut Tunngavik Inc. (Jul. 2016), quoted in Beeby 2016.

BETWEEN THE PRIMORDIAL, THE MODERN AND THE POST-MODERN: THE CHUKCHI WRITER YURI RYTKHEU

AUDUN J. MØRCH

In this chapter, I examine a selection of novels and short stories by the Chukchi writer Yuri Rytkheu (1931–2008). A native of the small village of Uelen on the easternmost tip of the Chukotka peninsula, Rytkheu experienced an almost primordial Arctic lifestyle during his childhood years. Modernity came to the Chukotka only with the Bolsheviks and, when Rytkheu was sent to school, he was among the first Chukchi to receive an education. A gifted boy, he was later allowed to go to Leningrad for further studies.

The Soviet Union was a totalitarian state with strict literary censorship that, to a varying degree during different periods, not only defined what was not allowed, but even demanded that all work should meet the requirements of Socialist Realism. Works written during the Soviet period, from the early 1930s onwards, always to a certain degree reflected the official, Soviet understanding of their themes. In this chapter I do not describe how Rytkheu complied with Socialist Realism (an interesting theme in itself), but rather, I try to understand and develop a picture of the modernization of the Soviet Arctic as seen from the point of view of the indigenous inhabitants of the area: the Chukchi, who had to make the transition from a near primordiality to the extreme, Soviet form of modernity in little more than ten years.

I have selected works that I find to be a good representation of the primordiality of the Chukchi lifestyle and modes of thinking, their first confrontations with the modern, their modernization and eventual transition to a post-modern state. The works are for the most part selected from Rytkheu's later, post-Soviet period, although a few earlier works of particular interest are also included.

Primordiality, modernity and postmodernity

If "primordiality" is understood as a lifestyle connected with Mesolithic up to Neolithic cultures, then most people would connect the notion with an *era* in the history of human culture, just as we connect modernity and postmodernity with eras. These notions represent lifestyles and modes of thinking that correspond to periods that can be fixed in time, although by no means accurately. The experiences of an indigenous people like the Chukchi, however, makes apparent the inherent eurocentricity of these terms. For an indigenous people, who have to make an almost instant transition from the primordial to the modern, modernity and postmodernity will be different. It is beyond the scope of this chapter to undertake a thorough investigation of such complex concepts as primordiality and modernity, to say nothing of postmodernity, but I will briefly discuss the concepts to clarify how I understand them.

The primordial, as defined by Webster's dictionary, means "constituting the earliest stages, original". The Arctic cultures of the Chukotka Peninsula were basically Mesolithic. The Chukchi had contact with modern European culture in the form of the Russian authorities, traders and others, including American traders and, in one famous instance, the Norwegian explorer Roald Amundsen's North East Passage Expedition of 1918–1925. Nevertheless, their own lifestyle was not modernized until the late 1930s and in some instances only after WWII. This is not to say that the lifestyle of the Chukchi before WWII was the one they had had since time immemorial. It was not. Research has shown that they were hunters of wild reindeer, and, to some extent, other land mammals (such as elk and wild sheep) until the seventeenth century. Neighbouring peoples, such as the Koriak and the Lamuts, had changed to reindeer herding a long time before, and eventually most of the Chukchi also chose to make the change. However, a part of the nation had already adopted the lifestyle of another neighbouring people, the Inuit. Thus the Chukchi nation came to consist of two separate groups: nomadic reindeer herders who lived on the tundra, and sedentary sea-mammal hunters, who lived on the coast (Vdovin 1965, 9).

The Chukchi retained an important and typical aspect of what could be called the primordial *Weltanschauung*: they understood their entire world as consisting of living entities. Animals, plants, inanimate objects such as rocks and even man-made objects had spirits. In addition, their space was inhabited by evil spirits, such as the *Kelet*, and goblins, such as the *Rekken*, carriers of disease (Bogoraz 2013, 2, 1–45).

Modernity, the second key term in this chapter, could be regarded as the second important step in a development that started with the Renaissance. The beginning of the Renaissance is usually fixed in time to Columbus' discovery of America in 1492. The date is arbitrary, but his discovery is nevertheless a seminal event of the era, and a typical one. Columbus' discovery meant an expansion of space. The Renaissance opened doors to new spaces and new, alternative truths and the ways in which people were thinking about the world changed correspondingly.

The beginning of modernity could, again somewhat arbitrarily, be fixed to a year; for instance, 1829, when George Stephenson's steam engine *The Rocket* reached a speed of 60 km per hour. From the outset, modernity has been connected with speed. Since prehistoric times, man had moved through space at more or less the same speed. Thus the relation between a certain space and the time it would take to move across it was more or less fixed. Now, as a result of technology and industrialization, this changed completely. Zygmunt Bauman has said that modernity meant that time and space *parted*, but the point seems rather to be that speed as such (the speed of almost everything) started to accelerate (Bauman 2000, 8). Still, Bauman's ideas about what he calls solid modernity, which was later superseded by liquid modernity, illustrate the acceleration of all processes that is typical of the modern era.

Thus, if the Renaissance meant an expansion of space while time became measurable (as clocks – invented in the late Middle Ages – became common), then modernity, contrastingly, meant a contraction of space relative to time. The general acceleration of time, which marked the beginning of modernity, is closely related to the idea of utopia. Acceleration meant the improvement of life. New technology kept saving time that could be used, for instance, to obtain education or create art. The utopian belief in continuous progress is palpable for instance in Chekhov's *Three Sisters* (1900) and it is also typical of modernity. One character, for instance, voices the belief that humankind itself is becoming ever more noble, cultured and non-violent as a result of improved education and progress in general. He seems convinced that we are on our way to utopia.

Postmodernity, as described by Frederic Jameson, is marked by an immense diversification; the absence of any unifying or even remotely common trait to the phenomena of our time which are, nevertheless, designated as postmodern. The only thing that can be said with certainty is that the postmodern is different from, and later than, the modern. Frederic Jameson seems almost at a loss to say anything affirmative about what postmodernism *is*, yet he confirms that we need the term since modernity has come to an end. He also talks of postmodernism as a cultural and experiential *break* (Jameson 1991, xiii). I want to suggest that the end of modernity as a coherent, unified era occurred in 1945, not only with the victory over Nazi Germany, but even more with the explosion of the first atomic bomb, the event that all of a sudden implied the possibility of the destruction of the world – of both time and space – within the foreseeable future. Thus, I understand modernity as an essentially utopian belief in human progress, whereas postmodernity represents a break, and basically an end to that belief. Postmodernity begins when nobody believes in utopia.

Soviet modernity

Familiar terms designating eras, such as antiquity, the Middle Ages, the Renaissance and even modernity are Eurocentric, and often problematic in relation to cultures in other parts of the world. It needs to be stated that Russia does not quite fit the general, Eurocentric understanding of time and eras. During the Middle Ages, *Rus* (the medieval term for Russian city states including Kiev, its largest and most important city) was fast becoming an integrated part of Europe when, in 1240, it was conquered by the Mongols. For about 240 years, most of Rus/Russia was part of the Golden Horde and isolated from Europe. In 1480, the rule of the khans was finally broken, but Russia remained isolated from Europe. As it rapidly took over large parts of the former territories of the Golden Horde, it might be said that Russia became a Byzantine empire rather than a European country. Only with the reforms of Peter the Great, some two hundred years later, did Russia begin to approach Europe. Yet, Peter's reforms only superficially made Russia a part of Europe. Importantly, Russia never experienced the Renaissance, the period which defined Europe more than anything else.

What the Bolsheviks did in 1917 was to set out, for the first time, to thoroughly modernize all of Russia, including the peasant class and every nation living within the territory of the former Russian, now Soviet, empire. The Bolsheviks denied that their ideology was utopian, claiming "scientific Marxism" to be entirely different from utopianism. Yet that, of course, is exactly what it was. It is impossible within the scope of this article to discuss the utopianism of their many programmes. Some were catastrophic; some were necessary (such as the elimination of illiteracy). Suffice it to say that Bolshevism was a particular and in every respect extreme form of modernity.

Modernity on the Chukotka

It seems that the Chukchi and the Russians encountered each other for the first time in 1644. In 1647, the Anadyrsk Fortress was built on the Anadyr River, and the first hostilities between the Chukchi and the Russians probably started soon after this. The Russians went on a military campaign against the Chukchi for the first time in 1729:

> For many years the reindeer-Chukchi in the vicinity of Anadyrsk fort had been a menace to the Russians, abducting them and their children, frequently besieging the fort itself, and harassing the foraging parties of Cossacks who, in order to get food, had to go down-river to the fishing grounds and points where migrating reindeer crossed the Anadyr. (Forsyth 2000, 145)

The Russians had the misfortune of running into a large Chukchi war party, which was on its way to raid its neighbours, the Koryaks. The Russians were defeated, and their commander killed. For almost forty years, the Russians waged an unsuccessful war against the Chukchi. In 1764 they finally gave up, and left the Anadyrsk Fortress. It was simply too difficult and costly to wage war against a tough and highly mobile enemy over such enormous distances and under the most adverse climatic conditions imaginable (Forsyth 2000, 149). After this, however, relations between the Russians and the Chukchi quickly improved.

In 1788, a Russian trading post was established on the Anyui River, where a market was arranged every year. Chukchi would come to the market and present the Russians with animal pelts, which the latter chose to regard as *yasak*, or the tax paid in pelts that the peoples of the Arctic had been forced to pay to Russian authorities everywhere else. The Chukchi, however, did not regard this as a tax, but as a friendly exchange of gifts. They highly valued the metal goods they received in return (Slezkine 1994, 106). The Orthodox missionaries at Anyui similarly thought that they were doing quite well. Many Chukchi allowed themselves to be baptised during market days, but again, from the Chukchi point of view, this was a friendly gesture to please the priest. Importantly, when they were baptised, they received tea, sugar and tobacco (Bogoraz 2013, 73). According to Vdovin, no Chukchi would ever allow himself to be baptised without payment in tobacco and metal objects, but afterwards he would just walk away and forget about it (Vdovin 1965, 255).

If Rytkheu's fictional account is accurate, the Chukchi also liked the metal crosses they received from the priest, for they could be worked into splendid fishing hooks. For this reason, some Chukchi allowed themselves be baptised again and again, but it never occurred to them to abandon their shamanistic beliefs (Rytkheu 2011, 85–86). Thus, although many were baptised, the Chukchi were never Christianized, as Vdovin (1965, 256) and Forsyth (2000, 151) both agree. Nor were they, of course, Russified. As far as the coastal Chukchi were concerned, the American influence may have been more significant than the Russian, as some Chukchi found work aboard American whaling ships and learned the English language during the latter part of the nineteenth century (Slezkine 1994, 107). However, although the Chukchi had learned to value commodities such as tea, sugar, tobacco, metal objects and even rifles, which were bought from the Americans, neither their lifestyle, nor their shamanistic beliefs nor their *Weltanschauung* in general were ever altered as a result of their contact with Russians or Americans, except for the consumption of alcohol. Strong drink had an immediate and devastating effect on the Chukchi (as on many other indigenous peoples of the Arctic) when they were exposed to it. Accounts given by participants of Amundsen's *Maud* expedition suggest that by the early 1920s some Chukchi communities on the coast were already suffering from this evil, whereas the Reindeer-Chukchi of the Tundra were less exposed (Amundsen 1921, 257–372, 398–421).

The collectivization and *dekulakization* programme of the agricultural sector in the Soviet Union was the most radical, utopian and massive modernization campaign the world had seen up to then. This programme included the small peoples of the Arctic and their livelihoods. It seems that the coastal Chukchi and Inuit hunting communities made the transition in a quick and surprisingly undramatic way. This was due to the fact that these communities were structured as fairly egalitarian collectives to begin with. By the end of the 1930s, close to 100% of the coastal sea-mammal hunting "industry" was reported to have been collectivized. With the Reindeer-Chukchi (and other reindeer-herding peoples), the situation was rather the opposite. Reindeer herds were privately owned, and the owners were obviously reluctant to give up their possessions. The largest owners were branded *kulaks* by the authorities and subjected to persecution. Some fled far into the immense tundra with their herds, and escaped collectivization until well into the 1950s (Forsyth 2000, 338–340).

To the Bolsheviks, the struggle against "backwardness" and illiteracy was scarcely less important than collectivization. It was necessary to win, and, indeed, to modernize the "hearts and minds" of the younger generation. During five years of schooling, the young learned to read and write both Russian and their native language. Among the nomadic Reindeer-Chukchi and other reindeer-herding peoples, parents only reluctantly sent their children to school, as the children were needed at home. For the coastal Chukchi, this was less of a problem. Also, because of their trade with the Americans, they had discovered the value of literacy (Forsyth 2000, 338). At one point, when a literate Chukchi boy died, he was buried not with a spear or a bow and arrows, but with his pen, pencil and a sheet of paper (Slezkine 1994, 238, 245–246).

Finally, a crucial element in the modernization of minds was the eradication of traditional beliefs. The most efficient way of achieving this was to eliminate the influence of the shamans, who, apart from treating illnesses, were the main carriers of the indigenous peoples' culture and inherited knowledge of their world. Shamans were vilified, humiliated and persecuted (Slezkine 1994, 226–228). The last Chukchi shamans probably disappeared during the 1950s as the last private reindeer owners were forced into collectivization. By this time, the world of which the shamans had unique knowledge had all but disappeared.

Primordiality

Yuri Rytkheu's *Teryky* (1979) was first published together with *When the Whales Leave* (1977) in a book titled *Sovremennye legendy* [Contemporary Legends] (1980). This book was republished in the volume *Polar Circle* (*Poliarnyi krug*) (1983). *When the Whales Leave* has its roots in a well-known etiological Chukchi myth. It develops into an allegory about the Chukchi nation and modernity (Mørch 2014).

Unlike *When the Whales Leave*, *Teryky* is structured much like a modern Russian or Soviet short novel (Rytkheu 1983). To be sure, one thing *Teryky* has in common with *When the Whales Leave* is that there is no mention of Russians or other white people. It is not possible to say anything with certainty about when the action supposedly takes place. Tools and weapons are not made of metal, but of flint and bones, which suggests the action takes place before the Chukchi started trading with the Russians. However, European-made tools were obviously rare during the seventeenth and eighteenth centuries. Thus, in principle, the action could be taking place any time between somewhat less than two hundred years ago and ten thousand years ago, were it not for the fact that the coastal Chukchi seem to have adopted the lifestyle of the Inuit relatively recently. The story defies any attempt to fix it in historical time. Mythological elements are absent, yet the scene is essentially primordial.

The story concerns a young hunter, Goigoi, who has recently married, is happy and in love. However, one day when hunting on sea ice, he runs out of luck. The ice he is on breaks off, creating a large ice floe that starts drifting into the open sea. Yet Goigoi does not die. He has killed a seal and a walrus, and therefore has an enormous amount of food. He can collect fresh water from pools forming on the ice. He can keep warm by using the skin of the dead seal as a sleeping bag. Months pass. Goigoi still does not die and even though the meat of the seal and the walrus has now gone (some of it was stolen by a polar bear), he is able to catch birds. What makes him hold out and not give up is his love for his wife, Tin Tin, whom he wants to see again. She loves him just as much, and it occurs to her that she can communicate with Goigoi telepathically. She knows he is alive. Goigoi's icefloe hits land, and he is able to go ashore. However, upon seeing his own image reflected in water, he is terrified: he is no longer a man, but has turned into a *teryky*.

According to Rytkheu, in Chukchi belief, the *teryky* is a kind of shapeshifter, a man who has turned into an animal as the result of having been isolated from human company for too long. Rytkheu's account is neither fantastic nor realistic. On the one hand, a *teryky* is a supernatural being, close to our idea of a werewolf, but otherwise the story is quite realistic. Indeed, it may be read as an argument for the non-existence of the supernatural *teryky*. If a man is to survive for almost a year on an icefloe, it seems probable that he will lose his human appearance to the extent that he is hardly recognizable. It is even known that malnutrition can cause *hypertrichosis*, or abnormal hair growth on the body. Thus, there is nothing indisputably supernatural in the story. It is rather a realistic story of life during an era when many natural phenomena were understood as supernatural, when nobody doubted the existence of spirits and monsters in nature.

Goigoi's wife Tin Tin recognizes him in spite of his transformation, but his brothers kill him, even though they too seem to *understand* that he is Goigoi. Since he has become a *teryky*, the law requires them to kill him. The law also prevented the brothers from looking for Goigoi while he was lost. If his destiny and hence the will of the gods was such, they had no right to interfere.

Teryky shows that Rytkheu was capable of writing masterpieces. It is a moving love story that also represents an experimentation with genre. This is something between a Russian short novel and a Chukchi legendary tale, an interplay between the realism of modern Russian literature and the fantastic elements of Chukchi tales. It should, however, be pointed out that Chukchi folktales are often brimming with realistic detail (see e.g. Bogoras 2007). Indeed, it would seem that Rytkheu draws just as much upon Chukchi legends as upon Russian and Soviet literature.

The old ways and laws of the Chukchi are portrayed as markedly cruel and irrational, perhaps the only element that is recognizably *Soviet* in

Teryky. Also, it would seem that it is the lack of any mention of Russians or Americans in *Teryky* and *When the Whales Leave* that makes it possible to call these stories "legends", thus creating an alibi for disregarding the traditional requirements of Socialist Realism.

These two stories give the reader an impression of how an indigenous people like the Chukchi conceive of the past. While Europeans look back at history, a long line of events that can be dated more or less accurately, the larger part of the indigenous past is indefinite. Events are difficult to fix according to a temporal system, although there is a beginning, which is the etiological myth, and a more recent past consisting of the memories of living people. In the first part of the novel *Poslednij shaman* (2004) (*The Chukchi Bible*), Rytkheu returns to this particular temporality in a more consistent way.[1] The foreword states that Rytkheu himself was able to trace his family back for about eight generations (Rytkheu 2011, viii–ix).

Modernity

In an earlier novel, *Magic Numbers* (1986), Kakot is a shaman who transforms into a modern man. The novel was written on the eve of *glasnost*, but still during the late Soviet era. Rytkheu does not deviate conspicuously from the principles laid down by Socialist Realism, which included among other things the presence of a "positive hero", usually a member of the Communist Party. Rytkheu does, however, provide an interesting twist on the required positive hero. Such a character is present, but he is neither a Soviet citizen, nor a communist. He is Roald Amundsen, the Norwegian explorer.

In accordance with Soviet ideology, the shaman must, by definition, be a negative figure. Yet, in this novel, the shaman is a "good" person, although shamanism as a phenomenon is bad. When Kakot's wife dies and he has proved unable to save her life, he decides to leave his call as shaman and flee to another part of Chukotka. In the novel, a shaman cannot leave his calling to become an ordinary person. He must continue as a shaman or commit suicide.

This harshness seems to be historically incorrect. Bogoraz met former shamans, who claimed to have lost their powers and quit (Bogoraz 2013, 2, 109). Since Kakot proved unable to save his wife, he could obviously have done this. It is somewhat disturbing that Rytkheu at this point seems to portray the Chukchi as more barbaric than they were, and one gets the impression that the author, even at this relatively late date, finds it necessary to do this in order to appease Soviet censorship.

Some years later, a search party from his original village finds Kakot. The runaway shaman, however, finds sanctuary aboard Roald Amundsen's ship, the *Maud*, which is frozen in the ice just outside the Chukotka peninsula. Thus, the former shaman is transformed into a cook on a Norwegian ship. He is washed, his hair is cut and he receives European clothing. Most importantly, he learns mathematics and discovers that the power of numbers is greater than the magic of the Chukchi shaman. The power of ancient magic is replaced by modern science.

The *Maud*, frozen in the ice, represents a small island of Norway and Europe isolated in a great Arctic wasteland. Yet isolation is not absolute, for after a while the Chukchi contact the crew. The expedition coincides with the Russian Revolution and civil war, so the indigenous population is still basically untouched by European culture. The ship, therefore, becomes an island of modernity; frozen, as it were, in a sea of primordiality. At the same time, modernity approaches the Chukotka from another direction, as the first representatives of Soviet power arrive.

Kakot was a historical person who was employed by Amundsen. The explorer does not mention him directly in his book about the *Maud* expedition, but does refer to him several times in his diaries (Bomann-Larsen 2011, 294–296, 330, 467). Kakot even appears in the film about the expedition that was shot between 1922 and 1925.[2] There is nothing, however, to suggest that the real Kakot was a shaman. In the foreword to *The Chukchi Bible*, Rytkheu writes that his grandfather, Mletkin, a well-known shaman, had been the inspiration for Kakot, as well as for Rinto in *The Wanderings of Anna Odintsova*. Thus it seems that Rytkheu chose to endow Kakot with some of Mletkin's traits, including his profession.

The novel *Interkontinentalnyi most* [The Intercontinental Bridge] (1989) was written during the *glasnost* era. For the first time in history, Russian authors enjoyed freedom from censorship. Yet this was still the totalitarian Soviet Union, and Rytkheu seems to have been cautious. Times could still change, and the country could very well have returned to harsher rule, which became obvious during the failed *putsch* in 1991. Nevertheless, the novel is different from anything Rytkheu had written before. Surprisingly, it is a kind of utopia. The action takes place in the future, probably around 2020 (Rytkheu 1989). Historically, outlandish plans to build a bridge across the Bering Strait have existed since the 1890s.

The Intercontinental Bridge raises questions about genre and sub-genre that should be mentioned at this stage. Firstly, Rytkheu's caution, which I just mentioned, may have genre-related implications. Throughout his career as a Soviet writer, Rytkheu's caution made him more or less abide by the principles set by Socialist Realism. As I have argued elsewhere (Mørch 1998, 185), according to Mikhail Bakhtin's definitions of the novel genre, the Socialist Realist novel barely qualifies as a novel proper. It cannot celebrate a leader, an ideology or a political system without losing genre integrity since the nature of the novel genre is to question and parody power. Rytkheu's Soviet novels never do this, at least not directly. Arguably, Rytkheu's early works are Socialist Realist novels rather than novels.

The question of genre, or sub-genre, is important with respect to the discussion of modernity. Utopias and dystopias are hallmarks of modern thinking and of modernist literature. *The Intercontinental Bridge* is neither a utopia nor a dystopia in a strict sense, although one Russian scholar has called it "anti-utopian science fiction" (Kolupaeva 2005, 138).

Rytkheu's text seems to express utopian yearnings in some respects, while at the same time taking an anti-utopian position in certain areas. It seems to be an example of what the American Slavist Gary Saul Morson has called a "metautopia", a work that treats the utopian question without taking a clear position either for or against it (Morson 1981, 142–185). Diffuse and relative, the term "metautopia" comes across as post-modern.

The idea of the intercontinental bridge itself is clearly utopian. The bridge, which is going to stretch across the Bering Strait, uniting the Chukotka peninsula with Alaska, will also unite the USSR and the USA in a common quest for worldwide peace and cooperation. The bridge is being built with advanced technology, another trait of many utopias. The Soviet Union itself seems to have reached a perfect state – as seen from the point of view of the author. For instance, on the Soviet side everybody in charge, both politically and technically, is Chukchi or Inuit, representatives of the indigenous local population. The Russians are somewhere in the background. Also, the old scourge of these peoples, alcohol, is gone. Along with tobacco, it has been banned all over the country, a reflection of Mikhail Gorbachev's battle against alcohol at the time the novel was written. On the American side, in contrast, things are not so good.

This is, in a way, a sign of Rytkheu's caution. On one level, he obviously wants his novel to appear to be a dig at the American system. On another, he is also criticizing the present state of affairs in the Soviet Union. It also remains unclear which side the novel itself is actually on. On the one hand, the idea of consolidating peace and cooperation through a bridge can hardly be seen as anything but laudable. On the other, the author keeps one question alive: what about the indigenous peoples? Will they have to be the ones paying the price? The ending is anything but utopian: an American Inuit blows up the bridge, although the circumstances surrounding this action are unclear.

This ultimately "metautopian" novel is a sign of Rytkheu's transition from a modern to a post-modern writer. In this, he is aligned with the principal trend in Russian literature. It remains somewhat ironic that the Soviet Union, for which Rytkheu predicted a great future, ceased to exist a mere two years after the novel was published.

The Post-Modern

The Last Shaman (*The Chukchi Bible* in English) was written in 2000. By this time, Rytkheu is no longer a Soviet writer but can write what he wants. He tells the story of the Chukchi people, and of Mletkin, his own grandfather, who was known as a great Chukchi shaman. This is arguably Rytkheu's most important novel. The foreword, which I have already cited, is an important key to Rytkheu's works. He explains that he has used members of his own family and other historical figures as inspiration for his characters, although he has felt free to use his imagination when information was scarce. Nevertheless, he believes that his own version of the Chukchi past is more accurate than those found in scholarly books, although his information cannot be fixed in historical time. His information has been passed down through generations by people who themselves were part of the events, not by "some Cossack, who could not tell a Chukcha [singular form of "Chukchi", rarely used in English] or an Eskimo from a tundra beast" (Rytkheu 2011, viii). This wry comment is an example of a theme Rytkheu never dared to touch during Soviet times: Eurocentrism or, more accurately, *russocentrism* and racism.

The novel itself is a great expansion of the themes and motifs appearing in *When the Whales Leave* and some other works. At the beginning, Rytkheu goes back to the etiological myth of the creation of the world by a raven. As a matter of interest, indigenous peoples on both sides of the Bering Strait share different variations of this myth, which suggests that it dates back to ancient times (see e.g. https://en.wikipedia.org/wiki/Cultural_depictions_of_ravens). Revisiting motifs from *When the Whales Leave*, the author goes on to tell the saga of the Chukchi and his own family. When he reaches the point where characters and events begin to be known from Russian historical sources, he starts to use the dates from those sources. It appears that the author is living in two cultures simultaneously, the indigenous and the modern. The dual-cultured understanding of time expressed in the first part of *The Chukchi Bible* seems to be unique. Like the shamans of Chukchi tradition, Rytkheu is in a certain sense a shapeshifter, a man with two identities. As an author, he also emerges with two different identities, arguably a postmodern quality.

The second part of the novel tells the story of Mletkin, Rytkheu's grandfather, who was also a kind of shapeshifter. As a young man, he takes hire aboard an American whaling ship, learns English and experiences modern life the American way. When the whaling ship is destroyed by fire, Mletkin finds himself stranded in San Francisco. In America, he meets an old acquaintance from home, Vladimir Bogoraz, the ethnographer, who introduces him to American colleagues and persuades him to represent the Chukchi at the Chicago World Fair of 1893. Mletkin resents being presented as an exotic savage and feels humiliated and enraged when he understands how deeply even his friend Bogoraz is steeped in racist ways of thought. Nevertheless, he gets to demonstrate his shamanic powers and frighten people a little at the fair.

Back in San Francisco, Mletkin finds employment in a hospital. As a shaman, he feels he is essentially a health worker. He wants to learn as much as possible about modern medicine, adding this to his own traditional knowledge. With the money he earns, he buys a large set of surgical instruments – and a barometer. After all, as a shaman he is also a meteorologist. Back in Chukotka, he is involved in a duel, and kills the man who has married his betrothed. From then on, he continues his work as a hunter and shaman. After a while, Chukotka becomes Soviet Chukotka, and in the end Mletkin is shot in the back by a Soviet commissar. The story seems to be historically accurate, although it is difficult to tell which details may have originated in the author's imagination. There is, however, no reason to doubt that Mletkin knew Vladimir Bogoraz well. Although Rytkheu does not mention whether Mletkin ever met Roald Amundsen, it seems likely that he did. In his book about the North East passage, Amundsen relates how, at one point, he received valuable information from a group of Chukchis, one of whom spoke English, because he had worked aboard American whalers and had even spent three years in San Francisco. This was almost certainly Mletkin (Amundsen 1921, 244).

For the most part, Rytkheu is reluctant to endow Mletkin with any magical powers – even his ability to dodge bullets when fired at from in front is described as a simple trick. There is no real magic, for when Mletkin is unexpectedly shot from behind, he dies like any other person would. Bogoraz, who witnessed shamanistic rituals and shamans at work, saw things he could not explain rationally. It seems that skilled shamans were also great illusionists, who, among other things, mastered *ventriloquism* to an amazing degree. They could "splendidly combine ecstasy with cool blooded deception" as Bogoraz puts it (Bogoraz 2013, 2, 126). For Rytkheu, however, it is important to demonstrate that the shaman is basically a professional with great integrity who meets certain needs. However, the way in which Mletkin enthusiastically embraces technology also implies that in a modern Chukotka there will soon be no need for shamans. Modern medicine is more powerful on most counts than the shaman's art. There is no need for a shaman to travel in spirit to find the walrus or reindeer if an aeroplane can perform the same job better. Weather forecasts are available from the radio service. Yet Rytkheu certainly also makes the point that the shaman, being flexible, or fluid, to use Bauman's expression, is readily able to master modern life. In Rytkheu's depiction, the shaman is already modern. The author makes it seem as though this ancient profession has been characterized by traits related to modernity since time immemorial.

Mletkin's death in the book is highly symbolic. Although the shaman's art has, in some respects, been made redundant by modern inventions, the last shaman of the coastal Chukchi is nevertheless the carrier of a vast amount of unique knowledge about everything from traditional medicine and astronomy to advanced illusionism; in short, of Chukchi culture. When he is shot in the back by a communist commissar, that bullet kills more than just a man.

Skitaniia Anny Odintsovoi [*The Wanderings of Anna Odintsova*] (2003) contains what is perhaps an even more radical instance of shapeshifting than the Chukchi shaman becoming a modern health worker (Rytkheu, 2003). In this novel, a young Russian ethnographer comes to Chukotka to do fieldwork. She is ambitious, and wants to gain more knowledge about Chukchi life than any other European before her, Bogoraz included. To attain this goal, she marries a young Chukchi and goes with him to become a reindeer herder's wife in the interior of Chukotka. Rinto, her father-in-law, is a shaman, and the relationship between him and Anna becomes a central theme in the novel. *The Wanderings of Anna Odintsova* is a very dramatic novel, and certainly one of Rytkheu's best. The action is set in the late 1940s, when traditional reindeer herding and sea mammal hunting was forcefully collectivized, as Russian agriculture was during the 1930s. The last remnants of the traditional lifestyle are all but annihilated, and those who offer resistance are killed or sent to the gulags.

The novel returns to the ethnographer as a literary motif and the theme of ethnography as a problem. Bogoraz, called Veip by the Chukchi, makes several appearances in *The Chukchi Bible*. In *The Intercontinental Bridge*, we hear about a female American ethnographer who coolly falls in love and marries an Alaskan Inuit in order to acquire a deeper knowledge of the natives than anyone before her. She leaves him when it is time to go back to civilization and finish her dissertation. Anna Odintsova's initial plan seems to be similar, but her fate turns out quite differently. The author makes the point that, although these ethnographers have done a lot of valuable work, the discipline itself is Eurocentric, and perhaps even ultimately racist.

This problem is treated extensively in Johannes Fabian's *Time and the Other*, and it seems that Rytkheu would have agreed on many counts. As long as indigenous peoples and cultures are regarded as primitive, and positioned as being at a stage equivalent to some point in the prior development of the Western world, there is no escape from this. Ethnography takes for granted that *it* is more advanced and sophisticated than the culture it studies (Fabian 2014).

Anna Odintsova, having married a Chukchi boy, finds it difficult to gain acceptance from her new family, especially her father-in-law, who is a shaman. However, his new daughter soon impresses him. She learns the Chukchi language to perfection. She learns to do all the work required of a Chukchi woman. Having been a Russian woman, a *tangitanka*, she is transformed into a Chukchi woman, or a real woman. (The word the Chukchi use for themselves literally means real man or woman, real human being.)

The time comes for the shaman to teach a successor. His sons do not seem to be up to the task, so he chooses Anna. Rytkheu does not explain how Rinto the shaman comes to this conclusion. There have been many female shamans, so it is not Anna's womanhood that is the problem, but the fact that she is Russian. Yet Anna has demonstrated her power to *transform*. This could be the reason why she is chosen. The power of transformation in itself is the crucial point. As an example of this, their proven ability to transform was exactly why homosexuals and transsexuals were sometimes chosen to become shamans (Bogoraz 2013, 2, 130–136). As a matter of interest, among the Chukchi, homosexuality and transsexuality were by no means unheard of. Harald Ulrik Sverdrup, who wrote an ethnographical essay on the Reindeer-Chukchi, includes several encounters with a certain Chukchi woman, who, although beautiful, is unusually tall and speaks in a deep voice. He then learns that she is not really a woman, but rather a "soft man". Sverdrup realizes that this "woman" is actually a homosexual man dressed in women's clothing and is appalled (Amundsen 1921, 339). From the perspective of the present day, the "primitive" Chukchi, with their relaxed attitude towards homosexuality, thus come across as far more modern than the Norwegian explorer.

The powers of the old shaman and his young apprentice, however, prove to be too weak as Soviet power closes in on them. Being the last reindeer herders to resist collectivization, they are eventually spotted from an aeroplane, hunted down like wolves and arrested. Most of the family is killed, not as the result of punishment, but by an avalanche as they are being taken to the administration centre to be judged. Anna, however, survives and is able to make one last transition by fleeing over the frozen Bering Strait to Alaska, where the author claims to have met her in her old age.

The Wanderings of Anna Odintsova contains many of the elements recognizable from other works by Rytkheu. One of these is obviously the clash between the indigenous and the Russian, between the primordial and the modern. Another recurring set of themes is passionate love, unlawful lust, sex and jealousy. In *The Wanderings of Anna Odintsova*, there is more of this than usual. Atata, an Inuit NKVD commander, has been madly in love with Anna for a long time. By the end, Atata has killed most of the last free Reindeer-Chukchi and the rest, Anna included, are in jail. Now the two of them decide to flee together over the frozen Bering Strait to Alaska. During their flight, Anna gets her revenge. She kills Atata and reaches Alaska alone.

Both *The Chukchi Bible* and *The Wanderings of Anna Odintsova* strike the reader as post-Soviet in the sense that they contain elements that would have been unimaginable during Soviet times and certainly before *glasnost*. Such elements include favourable depictions of shamans and their art and, contrastingly, unfavourable depictions of Russians, Soviet power and the highlighting of racism as a problem in the Soviet Union, not only in the West. It is also hard to imagine a book concerning the details of Anna's training to become a shaman being published in the Soviet Union.

Within the context of Russian literature, "post-Soviet" is sometimes seen as synonymous with postmodern. Rytkheu's post-Soviet novels are for the most part traditional in terms of form. An exception is *The Chukchi Bible*, which, with its particular, gradual evolution from myth to an alternative historical reality, might suggest associations with more typical postmodern Russian writers.

Conclusion

Concepts like primordiality, modernity and postmodernity are always problematic, because they encompass many different phenomena believed to be typical of a certain age or era. In relation to Rytkheu's work, these concepts have to be rethought. The term primordial is certainly not accurate. In *The Chukchi Bible*, Rytkheu demonstrates, for example, that the enormous increase in speed, which we associate with modernity, did not represent anything new to the Chukchi. They were already familiar with the idea of such speed since, in his state of ecstasy, the shaman was able to cover enormous distances in a very short space of time, for instance to obtain necessary information before a hunting trip. Furthermore, the shaman Mletkin has no problem functioning in America during the 1890s. While his mind and ways of thinking are primordial in some respects, it is resolutely modern in others. Chukchi time is by no means an equivalent of the European past.

When modernity in its extreme form, Bolshevism, is forced on the Chukchi, a completely different picture emerges. In this setting, modernity does not represent speed, newness or utopian hopes for the future. Paradoxically, Chukchi primordiality contained elements that Europeans would think of as modern. An example of this is the acceptance of homosexuality, and another is the basically ecological worldview implied by their animistic beliefs (Mørch 2014, 20). Thus, when modernity is introduced through force, when the Chukchi are modernized according to Bolshevik ideas, this brings about a break, a termination, rather than an advance: in other words, it is a postmodern experience for the indigenous people. The Chukchi thought of the world as a living organism, within which every being and object possessed a spirit. With the introduction of schools, where atheism and materialism were guiding ideologies, space itself was, in a way, destroyed. It had been alive, but was now pronounced dead. Indigenous myth and memory were replaced by history. It meant the termination of a world.

References

Amundsen, Roald. 1921. *Nordostpassagen*. [Northeast Passage.] Kristiania: Gyldendalske Boghandel.

Bauman, Zygmunt. 2000. *Liquid Modernity*. Cambridge: Polity Press.

Bogoras, Waldemar. 2007. *Chukchee Mythology*. Marston Gate: Forgotten books.

Bogoraz, V. (1934) 2013. *Chukchi*. Vol. 1. Leningrad: Izdatelstvo instituta narodov severa CIK SSSR.

—. (1939) 2013. *Chukchi*. Vol. 2. Leningrad: Izdatelstvo Glavsevmorputi.

Bomann-Larsen, Tor. 2011. *Roald Amundsen*. Oslo: Cappelen Damm.

Fabian, Johannes. 2014. *Time and the Other: How Anthropology Makes its Object*. New York: Columbia University Press.

Forsyth, James. 2000. *A History of the Peoples of Siberia: Russia's North Asian Colony 1581–1990*. Cambridge: Cambridge University Press.

Jameson, Fredric. 1991. *Postmodernism or, the Cultural Logic of Late Capitalism.* London: Verso.

Kern, Stephen. 1983. *The Culture of Time and Space 1880–1918*. Cambridge, MA: Harvard University Press.

Kjetsaa, Geir. 2004. *Tsjekhov. Liv og Diktning* [Chekov: Life and writing]. Oslo: Gyldendal.

Kolupaeva, Marina Viktorovna. 2005. "Problematika i zhanrovoe svoeobrazie romanov Iuriia Rytkheu 80-kh godov XX veka" [Problems and genre characteristics of Yuri Rytkheu's novels of the 1980s]. PhD diss., St. Petersburg: The Alexander Herzen Russian State Pedagogical University.

Mørch, Audun J. 1998. *The Novelistic Approach to the Utopian Question: Platonov's Čevengur in the Light of Dostoevskij's Anti-Utopian Legacy*. Oslo: Scandinavian University Press.

—. 2014. "The Chronotope of the Primordial: Yuri Rytkheu's *When the Whales Leave*." *Nordlit* 32: 17–32.

Morson, Gary Saul. 1981. *The Boundaries of Genre: Dostoevsky's "Diary of a Writer" and the Traditions of Literary Utopia*. Austin, TX: University of Texas Press.

Rytkheu, Yuri. 2011. *The Chukchi Bible*. Translated from the Russian by Ilona Yazhbin Chavasse. New York: Archipelago Books.

—. 1989. *Interkontinentalnyi most* [The intercontinental bridge]. Leningrad: "Sovetskii pisatel" Leningradskoe otdelenie.

—. 1983. "Teryky" in *Poliarnyi krug* [Polar Circle.]. Leningrad: "Sovetskii pisatel" Leningradskoe otdelenie.

—. 2003. *Skitaniia Anny Odintsovoi* [The wanderings of Anna Odintsova.]. Sankt Peterburg: Izdatelstvo zhurnala "Zvezda".

Slezkine, Yuri. 1994. *Arctic Mirrors: Russia and the Small Peoples of the North*. Ithaca and London: Cornell University Press.

Vdovin, I. S. 1965. *Ocherki po Istorii i etnografii chukchei* [Survey of Chukchi history and ethnography]. Moskva Leningrad: Izdatelstvo "Nauka".

Notes

[1] All references are to the 2011 English translation, which was entitled *The Chukchi Bible* rather than *The Last Shaman*, which would be a direct translation of the original title.

[2] *Roald Amundsen's* Maud *expedition 1922–1925*. This film can be obtained from the National Library of Norway.

EVERYDAY ARCTIC

SVALBARD IN AND BEYOND EUROPEAN MODERNITY

ANKA RYALL

With his 1967 lecture "Different Spaces" ("Des espace autres") Michel Foucault anticipated the spatial turn across the disciplines that opened up new ways of looking at history. If, as he argued, both history and time are now viewed in terms of "the simultaneous, of juxtaposition, of the near and the far, of the side-by-side, of the scattered" (Foucault 1998, 175), even places conventionally considered "outside time", such as locations across the circumpolar Arctic, become historicized. As Susan Stanford Friedman, among others, has pointed out (Friedman 2001; Friedman 2012), this spatialization of time has also been a challenge to the predominant narrative of modernity, which is based on the notion that metropolitan Europe makes history through progress and innovation while the rest of the world represents a permanent periphery that imitates the centre(s) and therefore always lags behind (Blaut 1993). Instead, when every geographical location is seen as constituting its own centre, there is a multiplicity of evolving versions of modernity (cf. Gaonkar 1999).

A spatial view of history is the backdrop for my reading of two narratives of Svalbard, a High-Arctic archipelago located approximately midway between Norway and the North Pole. I use the texts to explore what European modernity looks like from the perspective of its most marginal geographical space, without an indigenous or permanent population, surrounded on all sides by the Arctic Ocean. Before the Svalbard Treaty of 1920, this group of islands, first discovered and named Spitsbergen by a Dutch expedition in 1596, was a *terra nullius* – an uncharted and unclaimed free-for-all for hunters, sealers, whalers, trappers, prospectors and coal-miners of many nations. It was a male stronghold where, until very recently, women were interlopers, and where people never raised families or settled for life. Until the late 1940s, when aeroplanes began flying over in early spring to drop bags of mail to the residents of the largest settlement, Longyearbyen, it was completely closed off by ice from the rest of the world between October and May. In hindsight those mail drops, announced on Norwegian national radio, might have been the most obvious sign of Svalbard's incipient modernity.

Now, in the twenty-first century, Svalbard has been incorporated into both the Norwegian welfare state and global modernity through technology, research, international tourism and daily flights throughout the year. Even so, this far-off archipelago with its distinctive and challenging treeless landscape, 60 percent of which consists of glaciers, its three-month polar night, lack of roads and tiny transient population, in most respects still remains a place apart. It therefore comes as no surprise that in literature Svalbard is consistently represented by way of its difference from the rest of Norway. The archipelago is clearly an Arctic version of what Eric Prieto calls an *entre-deux*, or an in-between (Prieto 2013).[1] No longer a pristine Arctic wilderness nor yet a location that has been thoroughly subsumed into European modernity, it must be located between these two categories and understood as an "emergent" form of place (Prieto 2013, 2).

Arguably, Svalbard's transition to modernity began with early twentieth-century coal-mining and continued with the gradual demise of hunting and trapping – now strictly controlled and limited to species like the Arctic fox and Svalbard reindeer. The negotiation of this historical in-betweenness is the topic of this chapter, in which I specifically consider how emergent modernity is textually represented.

The emergence of modernity in Svalbard

Encompassing expedition narratives, travelogues, memoirs and journals, as well as fiction, Norwegian Svalbard literature often represents the archipelago as both a contrast to and mirror image of the mainland. This duality is particularly noticeable in texts from the mid-twentieth century when in the wake of the Second World War, the rapid rebuilding and expansion of settlements and mines began transforming the archipelago. The two very different memoirs that I discuss in this chapter both belong to this period. Both were published in 1955, each was its author's only book, and both are modern classics of Norwegian Svalbard literature. In both, Svalbard is a site of Arctic subsistence living as well as of an emergent European modernity.

According to Prieto, literature on emergent places tends to "oscillate between utopianism and nostalgia or mourning" (2013, 2). *Svalbard var min verden* [Svalbard was my world], written by the trapper and hunter Arthur Oxaas (1888–1974), gravitates towards the latter. It is a retrospective of a historical era that was coming to an end with the Second World War – an era when Svalbard, with the exception of Norwegian and

Russian coal-mining company settlements, lay outside of European modernity. From the 1890s until 1941, when civilians were evacuated from Svalbard by Allied Forces, the archipelago was a hunting ground for about 400 Norwegian over-wintering hunters and trappers (Drivenes and Jølle 2006, 140). Oxaas' career spanned almost the whole of this period, and his book belongs to a non-fictional genre – the High-Arctic trapper's memoir – that gained a brief popularity during the mid-twentieth century in Norway, not coincidentally at a time when there was very little hunting and trapping left in Svalbard. Its title promises an insider's account of the archipelago, based on the experiences of a lifetime, as well as a nostalgic recollection of a vanishing way of life.

Liv Balstad (1915–66), in contrast, looks ahead and towards a bold new future for the archipelago. Arriving in Svalbard in 1946 as the fiancée, soon-to-be wife, of the first governor (*sysselmann*) after the war, Håkon Balstad (1904–64), she stayed for nine years and by her own account left with "a stack of papers", the manuscript of her book *Nord for det øde hav* [North of the desolate sea] (Balstad 1955, 428).[2] As its title indicates, this memoir brings to bear a southern outsider's perspective, from which the Svalbard archipelago is defined not only as a marginal place (cf. Shields 1991), but literally – if we understand the Arctic Ocean as a boundary to the north – as a place beyond the margin. It recounts the challenges of living in Longyearbyen during the period of reconstruction following the devastations of the war. Well-received and widely read and debated, it was translated into several languages including English (in 1958, as *North of the Desolate Sea*, a version unfortunately based on an abbreviation of the original).

Whereas Oxaas' narrative covers almost half a century, ending with a final post-war over-wintering, Balstad's starts almost exactly when his concludes. When the trapper makes a brief appearance towards the end of her book, as a visitor to Longyearbyen before returning to the mainland for good, he is held up as an admirable example of a dying breed. "What Svalbard can give in return for toil and trouble and loneliness and darkness and cold, does not meet the expectations of modern people in a welfare state," she notes in explanation for his decision to retire (Balstad 1955, 345).[3] Her own focus is on those expectations, a utopianism that is thwarted but not undermined by the uphill struggle to incorporate Svalbard into a post-war Norwegian welfare society and transform Longyearbyen from a coal-mining company town into a modern municipality.

Both books may be read for the insights they provide into transitional Svalbard, and both have been and can be used as historical sources. But, as I have indicated, my aim is different. As Prieto points out, literary representations can help us gain a deeper understanding of those uncategorizable places that he characterizes as *entre-deux*. The reason is not that literary representations are imitative and should be "judged in terms of accuracy and conformity to the original", but that, as creative and performative acts, they are "part of the process that brings places into being *as* places" (Prieto 2013, 11). Significantly, both Oxaas and Balstad approach the *entre-deux* of Svalbard through narratives of coping with familiar everyday activities in a challenging environment. That context itself works as a form of defamiliarization (Felski 2000, 90; Felski 2002, 609; Haugdal, this volume), since during the isolation, darkness and extreme climate of the polar winter, activities that are usually routine and prosaic can easily turn into matters of survival.

A central theme for both Oxaas and Balstad is the civilising practice of homemaking in the dual sense of literally building a home – in Oxaas' case many temporary homes – and managing a home. It is therefore helpful to view their narratives in terms of a Heideggerian emphasis on the relation between the material practice of building and the human experience of dwelling. In his 1951 lecture "Bauen Wohnen Denken" ("Building Dwelling Thinking") Heidegger emphasized the essential interdependence of building and dwelling: "For building is not merely a means and a way toward dwelling," he writes, "to build is in itself already to dwell" (Heidegger 1971, 348). He supports his argument with a discussion of the etymology of the word *bauen*, which is derived from the Old German *buan*, meaning to dwell. Moreover, he notes that *bauen* and *buan* are connected to the verb *bin* as in *ich bin*, I am. In other words, to *be* is to dwell: "The way in which you are and I am, the manner in which we humans are on the earth is *buan*, dwelling" (Heidegger 1971, 349). How do Oxaas and Balstad describe building, dwelling, *being* – in the sense of immersing themselves – in an emergent place like Svalbard?

Building and dwelling in the Svalbard wilderness

Arthur Oxaas' *Svalbard var min verden* could be used as a down-to-earth illustration of Heidegger's threefold concept of building, dwelling, thinking. Much of his memoir deals with the practice of building and maintaining hunters' huts, the more or less primitive shelters representing the material basis for surviving in an inhospitable environment. In this way the book reflects Oxaas' sense of belonging to that specific landscape and of being in that world. The same can be said about his over-wintering diaries. Of the nine that have been preserved, two – which unfortunately only exist in copies – contain sketches of ground plans of hunters' huts.[4]

Even in his diaries, which are brief and factual like log books, Oxaas makes it clear that Svalbard is where he feels at home – in contrast to Bjørnøya (Bear Island), where he only spent one winter because it was "a living hell" (Oxaas 1916).[5]

Oxaas opens his memoir with his first voyage to the Arctic from his home town of Tromsø in northern Norway in the summer of 1902, when he is only 14 years old and has just finished school, and concludes it with the return from his 29th and final Arctic over-wintering in 1946–47 after an enforced break during the war years. Apart from the single winter on Bjørnøya and another on the island of Jan Mayen, his time was spent in Svalbard. Unlike his most famous Norwegian contemporary, Hilmar Nøis, who over-wintered nineteen times at his station in the Sassen Valley, Oxaas initially had no permanent base. For the first part of his career as a hunter and trapper, he did not return to the same place from his summers on the mainland. Before spending six winters at Flathuk, on the very northern tip of Spitsbergen (the main island in the archipelago), followed by nine at Kapp Wijk at the entrance to Dickson Fjord, he tended to winter in different locations, where he established himself in the autumn either by building new huts or repairing existing ones. Indeed, he became one of the foremost builders of hunters' huts in Svalbard, and the work of building, repairing and furnishing huts constitutes a recurrent theme in his memoir.

As was the common practice, Oxaas had one main hut or station and a few minor ones where he could stay overnight when out and about hunting reindeer and polar bears and tending his fox traps. He usually wintered with only one companion or alone and often saw no one else for several months, mainly because the stations were situated so far apart. Hunting and trapping in Svalbard was never a common occupation. Even during the 1930s, the peak years of Norwegian fox trapping on the archipelago, only an average of 25–35 persons over-wintered each year (Rossnes 1993, 8). A map in the geologist Anders K. Orvin's survey of settlements and huts in Svalbard in 1939, showing the scattered locations of buildings and dwellings along the coast, gives a poignant sense of the hunters' isolation (Figure 1).

Oxaas participated in the construction of his first hut in 1906, at the age of 18, ahead of his first over-wintering in Svalbard. It was located on Halvmåneøya (Half Moon Island), a small island off the south-east coast of Edgeøya (Edge Island), known as a good hunting ground for polar bears. Another reason for choosing this site was the amount of driftwood from Russia to be found nearby, but the over-wintering team he belonged to had also brought materials from the mainland. In his memoir, Oxaas describes the building of this first hut in detail:

Figure 1. Few and far between. Anders K. Orvin's map in "The Settlements and Huts of Svalbard" (1939).

> The timber frame was quickly raised, and it did not take long to clad the walls with one and a half inch boards. The whole house was 24 square metres, and attached to it we built a 4 x 4 metres shed for supplies. Inside the over-wintering house we put up four bunks along one of the end walls. In the corner by the shed we placed a little stove with an oven. On each long wall there was a window. Under one there was room for a table and under the other a kitchen cabinet. When we had brought in a couple of benches, the house was ready for occupation. […]
>
> We were now into September, and it was urgent to finish all the external work before winter arrived. The first thing we did was to lay a thick stone wall around the house. We used moss as padding between the wall and the rocks. In addition we covered both roof and walls with birch bark and peat. (Oxaas 1955, 23)[6]

Based on Oxaas' description, the interior was obviously extremely simple, containing only the most essential – clearly homemade – pieces of furniture. Little was done to make it cosy or comfortable.

Oxaas spent the winter on Halvmåneøya with two other inexperienced young hunters. Their leader, who over-wintered in a another newly built hut in Dianabukta (Diana Bay) further north on Edgeøya itself, had to teach them the ropes before he left: "First we learned to bake bread, then to use the spring-guns, the traps and the poison" (Oxaas 1955, 24).[7] (This was twenty years before the use of poison was prohibited.) The leader's priority is clear and significant. Domestic proficiency was obviously of equal or greater importance for survival than hunting and trapping skills – apart from the ability to handle a gun, which in Oxaas' narrative seems to have been taken for granted. In an article about his female contemporary, Wanny Woldstad, Marit Hauan argues that what makes Woldstad's Svalbard diary "feminine" is that she does not distinguish between indoor and outdoor tasks such as baking and shooting seagulls (Hauan 1992, 37). However, as Oxaas' memoir and diaries show, neither did the men. A note in his Jan Mayen diary on 14 December 1917 that he has "made waffles and skinned 2 blue foxes" (Oxaas 1917–18), is a not uncommon linking of domestic and hunting activities.[8] The Arctic wilderness was clearly no place for a gendered division of chores.

When Oxaas returned for a second over-wintering two years later (1908–9), he was hired as leader of a group of four hunters/trappers. They wintered on Kapp Lyell in Bellsund, where they built a hut that has now disappeared. Oxaas writes nothing about it in his memoir, except that the construction was swiftly accomplished. Likewise, while his first over-wintering is described from month to month, he writes little about daily life the second time around. Clearly, after describing his first hut and over-wintering in detail, he does not find it necessary to repeat himself. Instead, the emphasis in the rest of the book is on unusual events and over-winterings that were exceptional, either because he had a particularly congenial companion or (very rarely) the opposite, spent several winters in Svalbard without returning to Tromsø in the summer, or over-wintered alone.

At least seven of his winters in Svalbard Oxaas spent by himself, miles away from other people, the first (in 1912–13) in a small and primitive hut in Hornsund. It was the site of one of the many dramatic incidents that enliven Oxaas' narrative, when he returns from a hunting expedition in February to find that visiting polar bears have more or less wrecked it. Attracted by his store of polar-bear meat, they have destroyed the interior, torn off part of the roof and broken the stovepipe:

> I told myself that I after this would be careful not to go too far from the house. Now I had experienced how intrusive polar bears can really be. I could simply risk being without a roof over my head another time. The small hut with the small boat stove was the only place for miles around where one could survive here in the polar winter. Well, I rolled up my sleeves and started clearing up. First I had to mend the stove pipe to be able to get heat back in the room. But before I began repairing the roof, I had to carry out some loads of snow and assorted trash that covered the floor inside. I shovelled and swept until it finally began looking like the floor of a house again. And by the time I had been on the roof and nailed down loose boards and replaced broken planks with new ones, night had fallen […] (Oxaas 1955, 89)[9]

This episode is representative of Oxaas' attitude throughout – as both narrator and protagonist of his own narrative. He deals with the issues of subsistence and survival from day to day in a thoroughly practical, unsentimental and straightforward manner. He never dwells on loneliness, tedium or fear, but takes things as they come. Only occasionally – as in the reference to his small hut as "the only place for miles around where one could survive here in the polar winter" – are there indications of the desperate precariousness of his existence.

It is often taken for granted that Arctic writing is not only primarily male-authored, which is certainly the case, but also strongly masculinist in orientation. "Heroic, rugged individuals, usually male, people the literature set in the north," as Renée Hulan has observed (Hulan 2002, 99). In many respects, Oxaas' memoir fits the bill. However, I would like to suggest that Arctic masculinity is considerably more complex than has been acknowledged. Nils Magne Knutsen's attempt to define the Svalbard hunters' memoir as a genre is a case in point. Using a heroic stereotype as a template, he labels it "Arctic life as a robinsonnade". Over and over again, he writes, we get new stories "about a man standing on a deserted beach with some cases of equipment and a gun, and then he builds an existence and takes possession of the area as absolute ruler" (Knutsen 2013, 184). But Oxaas and his colleagues were neither castaways nor sovereigns. They had come north as migrant workers, in the hope of making a modest living at a time when there were few opportunities on the mainland, especially in northern Norway, and – as in the episode above – there is little sense of conquering masculinity in their memoirs. Instead, the wreckage of Oxaas' hut by polar bears shows how easily nature (or natural beings) reclaims the ground that the visitors had attempted to make their own.

As Silje Solheim Karlsen has convincingly argued, Knutsen makes the mistake of conflating narratives of polar expeditions and trappers' memoirs (Karlsen 2014, 7). Although the former may perhaps be read as "masculine self-heroising" (Knutsen 2013, 187), trappers' memoirs usually emphasize more mundane qualities, such as competence, experience, adaptation, collaboration and hard work, indoors as well as outdoors – all needed for survival during polar winters. If the trappers sometimes escape life-threatening situations – caused by violent storms, ice or polar bears – they represent themselves as lucky rather than heroic. Oxaas is certainly not alone in highlighting the everyday struggle of making the Arctic wilderness into a dwelling place in the Heideggerian sense.

A good example of this is a chapter dealing with three years – summers as well as winters – that Oxaas spent in Svalbard without his customary summer visits to the mainland. It opens with the following statement: "It may sound strange, but it is actually possible to get used to a lack of civilized surroundings" (Oxaas 1955, 171).[10] Svalbard, in other words, belongs to the category of places Kevin Hetherington ironically calls "the badlands of modernity": lacking civilized amenities, it is a "marginal space in relation to modern societies" (Hetherington 1997, viii). As Oxaas notes, the absence of civilization may not be ideal, but neither is it entirely negative; one can simply get used to it. In fact, he goes on to say that during this period (1928–31) he never felt the urge to return to the mainland. He spent the winters at the same station, at Flathuk, where he had over-wintered since 1924, felt happy and became familiar with every rock and spot. His attachment is described in Heideggerian terms: "Even if nature is bleak," he writes, "and it can seem desolate and barren up there during the cold winter months, one somehow grows attached to the place and becomes fond of it" (Oxaas 1955, 171).[11]

Arctic homemaking

A few years later, however, Oxaas seems to have had a change of heart about the value of civilization. During the summer of 1936, at home in Tromsø on holiday, he became engaged to Anna Andreassen (1890–1975). Back in Svalbard at the end of August, he made a note in his diary that signals a momentous life change: "At midnight the Lyngen arrived [...].[12] My fiancée came along to stay here during the winter. Perhaps this is a blunder, – time will tell" (Oxaas 1936–37).[13] He was 48 and had spent over twenty winters as a hunter and trapper in Svalbard, alone or with other men; she was two years younger and a widow. Yet, after what seems to have been only a brief acquaintance they were prepared to spend the winter together in a ramshackle hunter's station in Kapp Wijk, at 78°34' north, many miles away from their nearest neighbour – and with no possibility of leaving until the following summer.

Oxaas' apprehension is easy to understand. As a photo of him in front of the Kapp Wijk hut shows, the accommodation he had to offer was far from luxurious (Figure 2). But three weeks later he notes that his fiancée has settled in well: "Anna is busy taking care of the house, and well content" (Oxaas 1936–37).[14] After New Year there was a torrential downpour, and he had to dig a ditch around the hut to prevent the water from rising over the floor. "Hope this is the last year I have to use this hovel," he commented in his diary. "Everything is destroyed, our clothes rot, and one becomes quite disabled by gout." He added that one side of the house had also been sinking several degrees: "Yes now it is less pleasant here." But soon after, their life was obviously back to normal: "Anna baked today. She works hard every day and is very content" (Oxaas 1936–37).[15] Only the many diary entries registering how Anna copes and adapts to her new circumstances indicate Oxaas' initial concern and eventual relief.

Nothing in Oxaas' memoir or diaries suggests that he had been foolish to invite Anna to share his life in Svalbard. She, likewise, seems not to have regretted her decision to come. Interviewed in a local Tromsø newspaper after their retirement in 1947, she compared living conditions in Kapp Wijk favourably with those on the mainland, in spite of the long dark winters, which the routines of "daily work" helped her endure ("En av dem" 1947). In his memoir, Oxaas concludes that his marriage had led to "a complete change in habits and routines", adding that it was "a change for the better. Now everything was in order. Food, clothing and the house were capably taken care of. Somehow there was more civilization up here with a woman in the house" (Oxaas 1955, 190).[16] As his diary notes indicate, Oxaas connects “civilization” with the establishment of a gendered division of tasks – even in the context of an isolated Arctic hunting hut – conforming to the common pattern on the Norwegian mainland.

Figure 2. Arthur Oxaas in front of ”the hovel” in Kapp Wijk, photographed by Adolf Hoel in 1936. Courtesy the Norwegian Polar Institute.

As a builder, Oxaas was himself responsible for much of this change “for the better”. Soon after his fiancée arrived, he began the construction of a new and more solid house, describing the process in detail in his diary. The following summer, he and Anna got married in Longyearbyen, and after a trip to the mainland they returned to Kapp Wijk in September to finish the house. On 11 September 1937, Oxaas noted in his dairy that they had moved in:

> This pm we moved into the new house. All the time since we came I have had enough to do to finish the building. It has a double roof, floor, walls throughout. Panelling inside and outside, likewise painted inside. We feel like gentlefolk after moving from the hovel. The interior here is pleasant and practical. Large bright living room, bedroom next to it, a large room for storage, room for skins, expecting WC hall and a large shed with floor and window. I have also brought a radio that will now be rigged up, so it looks like we will now be able to keep up with everything, or anything, that happens outside our little world. (Oxaas 1937–38)[17]

Compared to his earlier huts, the new house was quite spacious. It had a ground plan of 5 x 8 metres (Reymert and Moen 2015, 337) and a separate bedroom like houses on the mainland. The old hut was converted into a shed. Most importantly, Oxaas' emphasis on the instalment of a radio – which required a 10-metre antenna to work – signals the incorporation of the new station into the modern world.

According to Odd Lønø, radios became quite common in Svalbard trappers' huts during the period 1935–41, although not everyone was able to receive a signal (Lønø 2014, 382). It is therefore more than likely that Oxaas would have embraced this modern technology even without a wife. His narrative makes it clear, however, that he viewed the radio together with his marriage and the building of a new house fit for "gentlefolk" as steps towards the incorporation of the Kapp Wijk station into the modern world. Implicitly, he considered himself more at home in the pre-modern Arctic wilderness as a lone hunter/trapper than as a married man.

A photo of the new station illustrates Oxaas' memoir (Figure 3). It is still intact and is now one of Svalbard's most important cultural heritage sites.

Figure 3. "We feel like gentlefolk after moving from the hovel." Photograph of the new station in Kapp Wijk, reproduced in Arthur Oxaas' memoir.

Although the house was vandalized after the couple left for good in 1947, and there are no surviving signs of Anna's contribution, her role is commemorated in her husband's memoir. In the chapter dealing with this building project – the final one of his career – Oxaas highlights the importance of her attention to homemaking, not only in the sense of everyday maintenance such as cooking, baking, cleaning and mending, but also in the sense of making the house into a welcoming domestic space through furniture, textiles and decorative objects. This is demonstrated in one of the illustrations, a photograph taken by Oxaas of his wife in the living room (Figure 4). While the caption tells us that she has "transformed the primitive hunter's hut into a comfortable home", the text itself explains how:

> Anna furnished the house with rugs on the floors and curtains over the windows. Between the sitting room and the bedroom she put up draperies instead of a door. With her feminine sense of the importance of small things, she made this house into a real home that made everyone thrive. (Oxaas 1955, 194)[18]

Figure 4. "Mrs Oxaas transformed the primitive hunter's hut into a comfortable home." Photograph reproduced in Arthur Oxaas' memoir.

The picture of Anna in the living room shows a modest 1930s interior that could have been anywhere in Scandinavia. It is the result, as Oxaas notes, of his wife's emphasis on comfort and her "feminine" attention to detail (cf. Schor 1987, 97).

However, Oxaas neglects to note his own contribution to the decoration of the living room, the hand-carved picture frames that we see on the wall behind Anna on the left. According to Oxaas' friend Jacob Jacobsen, who in August 1963 rescued three of the frames after vandals had demolished the interior of the Kapp Wijk house, they had been left there when Arthur and Anna returned to the mainland for good. Jacobsen found them in a cardboard box on the beach and restored them before presenting them to Svalbard Museum in 2002.[19] In a summary accompanying the frames, he explains that Oxaas carved them from driftwood during his winter on Jan Mayen in 1917–18, and that he had used them to display photos of the royal family cut from magazines and newspapers. The reason Jacobsen gave the frames to the museum was that he wanted others to know "what kind of man and artist Oxaas really was" (Jacobsen n.d.). Jacobsen also writes that Oxaas had a special gift for making his surroundings comfortable – exactly the special gift that Oxaas himself attributes to Anna.

After Anna's arrival in Svalbard, Oxaas not only physically replaced the primitive hunting hut with a proper small house, but also, in his diary, abandoned his gender-neutral account of everyday tasks. When narrating their domestic life together in his memoir, he does so in terms of a conventionally gendered spatial ordering. *His* work is outdoors, as builder and hunter/trapper, *hers* primarily indoors as homemaker and housewife – although he does mention that she participated in outdoor activities and contributed to the family economy by tending her own fox traps near the hut: "When the weather was good, she always went out to look after the equipment, and she was a very precise and capable trapper" (Oxaas 1955, 195).[20] The couple's division of labour suggests that they transposed to the Arctic wilderness the "strong male breadwinner model" in which married women were by definition homemakers that characterized Norwegian modernization during the mid-twentieth century (Åmark 2006, 324).[21]

As described in Oxaas' memoir, the living quarters he and Anna created together are clearly a version of Prieto's *entre-deux* – neither a typical hunter's hut nor entirely an ordinary 1930s Norwegian home, or more correctly a liminal space between *both*. Trying to explain why Anna decided to accompany him to Svalbard even before they were married, Oxaas surmises that she must have been motivated by a sense of adventure (Oxaas 1955, 190). In fact, her decision to move there with a man she hardly knew must have been a radical break with contemporary sexual norms, even if so-called "loose connections" were both more common and more accepted in Svalbard than on the mainland (Evjen 2006, 60–61). But, by fulfilling the daily routines of an accomplished housewife, as well as by transporting to Kapp Wijk objects representing domestic comfort at home, Anna as represented in her husband's writings signals the opposite – that there is *no* clear rupture between the Arctic wilderness and familiar everyday modernity, between her new life and the old. As Iris Marion Young puts it in her feminist critique of Heidegger's "Building Dwelling Thinking", "[t]he activities of homemaking […] give material support to the identity of those whose home it is" (Young 2001, 271). In the case of Anna Oxaas, this obviously meant a respectable female identity unchanged by displacement.

In Young's view, Heidegger's "division of dwelling into moments of building and preservation", as well as his privileging of "building as the world-founding of an active subject", reveal a male bias (Young 2001, 253). Even if he does underline that *bauen* "*also* means at the same time to cherish and protect, to preserve and care for" (Heidegger 1971, 349; original emphasis), according to Young he then "leaves preservation behind to focus on construction" (Young 2001, 255). The consequence is a devaluation of an important mode of meaning-making that is usually performed by women, coded as feminine and connected to traditional household tasks. Although Young warns against romanticising the activity of preservation, her argument for the revaluation of homemaking helps us to appreciate the equal contributions made by Anna to her and Arthur's mutual project of not only building, dwelling and place-making, but also of materialising an emergent modernity in the Arctic.

Integrating Svalbard into modern Norway

Turning from Oxaas' memoir to Liv Balstad's *Nord for det øde hav* means moving from the establishment of modernity in the form of a proper house and a gendered division of tasks in the practically uninhabited wilderness to Svalbard's most urban area, Longyearbyen. One of the illustrations in Balstad's book makes it clear that the designation "urban" was relative, however. The image has the ironic caption "Svalbard's metropolis" and depicts a scattered settlement of between 800 and 1000 inhabitants in moonlight during the dark season (Figure 5).

Figure 5. Longyearbyen, "Svalbard's metropolis," in the 1950s. Photograph reproduced in Liv Balstad's memoir.

Turning from Oxaas to Balstad also means ascending the social ladder, to the very highest echelon of Svalbard society.

Whereas Oxaas writes retrospectively about a period of his life that is over, Balstad's book reflects a developing perspective as she narrates her personal journey from innocence to experience. But even as she adapts to everyday life in an Arctic outpost, she continually reminds her readers of the utter *otherness* of Svalbard. Its enormous distance from both Oslo, where she used to live, and the traditional rural inland valley of Valdres in southern Norway, where she grew up, is not only geographical but also conceptual. All the same, Balstad has more than a little in common with Anna Oxaas. Like Anna, Liv Gullvåg as she then was, came to Svalbard as the fiancée of a man she had only recently met – in love but clearly also looking for adventure. Moreover, the book she later published about her experiences shows that, in much the same way as Anna, she tried to assimilate by becoming an Arctic homemaker.

As previously mentioned, Balstad's book recounts her nine years as governor's wife in Longyearbyen during the first decade of reconstruction after the devastation of the Second World War:

> The Germans had made a clean sweep, only three or four houses were left standing in Longyearbyen itself, so everything had to be rebuilt from scratch. [...] Everything revolved around the reconstruction. Only work mattered; what it meant to *live* was hardly in anyone's thoughts, nor was

> there any time for it. Everywhere there were stacks of material, heaps of timber and boards and planks; cement, machines; a confusion seemingly impossible to disentangle. (Balstad 1955, 82–83)[22]

Among the buildings destroyed by bombing was the original governor's residence. Arriving in 1946, Balstad therefore moved with her fiancé into a dilapidated and draughty Swedish prefab barracks (*svenskebrakke*), with rats under the floorboards and derelict furniture, which housed all government services in addition to the living quarters of the governor and herself. Her memoir recounts how, at the beginning, she was able to find some humour in the situation – particularly in her futile attempts to be an efficient housewife in a primitive kitchen with an antiquated stove that eventually exploded like a volcano in clouds of smoke (Balstad 1955, 72) – if only because she believed her husband's promise that a new residence would be built soon.

Unlike Oxaas, the Balstads cannot simply build a new house. Instead, they are dependent on a centralized bureaucracy that in a period of post-war austerity has other priorities than the reconstruction and modernization of a small Arctic settlement. The prolonged struggle to get the government to accept the necessity for a new governor's residence is a cause of mounting frustration, a sentiment that informs much of Balstad's memoir. It is also a source of tension, if not outright conflict, with her husband, who comes across in her text as a self-effacing civil servant constrained by loyalty and a reluctance to spend government money. Although she declares that what is most needed for life in Svalbard is "the talent for adaptability and the art of resignation" (Balstad 1955, 175), she adamantly refuses to resign herself to living in the temporary barracks.[23] In fact, this statement introduces one of several chapters that are sharply critical of the Norwegian government's indifference to Svalbard's affairs.

However, both the Balstads resolutely oppose government plans to simply rebuild the pre-war governor's residence. "Times have changed," as the governor informs the Ministry in 1947 (Balstad 1955, 171).[24] Looking ahead towards a future Arctic modernity, Liv Balstad uses her narrative to stress the importance of transforming the archipelago from a primitive and neglected coal-mining outpost into a truly integral part of Norway. Only a governor's residence that is "more spacious, more modern and also more suitable for official functions" would suffice as a proper symbol of this development (Balstad 1955, 171).[25] Moreover, it would have to be a suitable home not only for the governor – who, before the war, was expected to be a bachelor (Balstad 1955, 171) – but also for the governor's family.

Via connections, Håkon Balstad finally enlists the services of a private architect, Eindride Slaatto, an acquaintance of Liv's from the Valdres Valley. After a three-week visit to Longyearbyen, Slaatto leaves with finished plans for a residence that will combine the functions of being the Balstads' home, the governor's office and an official symbol of Norwegian sovereignty. Dreaming over the architectual sketches, Balstad sees the residence as "Slaatto's poem":

> the simple lines expressed a proud dignity. The large timbered banqueting hall with a fireplace – "Liv, I shall bring the forest into your hall!" – the concrete tower that connected the building – the guest rooms, the kitchen, the living rooms, – for me it became not only a house, but also a poem. (Balstad 1955, 179).[26]

Balstad's allusion to poetry may refer to Slaatto's statement about creating a residence that in the treeless Svalbard landscape will remind her of the woods back home. However, the architectural drawings show a building that in its harmonious combination of modern international functionalism with elements inspired by traditional Norwegian architecture may easily be interpreted as a poetic composition.

When the governenor's residence is finally completed – in 1950, after Balstad herself has had to travel to Oslo to confront the Minister and demand action – she delights in the way the architect has kept his promise to her by including in his design a large sitting-room with timbered walls and an open fireplace. As Hege Skogvang (2004) has noted, the sitting-room, a brown one-story wooden building with a saddle roof, reminiscent of a traditional Norwegian log house, is one component in a complex of four very different buildings. It is connected by a portico to a three-story square tower with white-plastered concrete walls and a low pyramidal roof that constitutes the corner of an L-shape completed by a rectangular red two-story main building. Like Anna's homely touches to the Oxaas house, these components, which are all transposed from the vernacular architecture of rural Norway to the High Arctic, make the new governor's residence evocative of the *entre-deux* of an emerging Svalbard society (Figure 6).

Obviously, the connotations of an official building like the governor's residence are very different from those of the Kapp Wijk station. In a settlement like Longyearbyen, where buildings were (and are) primarily functional, it stands apart – literally, by being situated on a hill overlooking the town – as an Arctic equivalent of the royal castle in Oslo.

Figure 6. The Norwegian governor's residence in Longyearbyen, photographed in September 2011 by Erlend Bjørtvedt (CC BY–SA).

Like the name Svalbard, which first appeared in twelfth-century Icelandic annals and was popularly – though erroneously – believed to prove that the archipelago had been discovered by the Vikings centuries before the Dutch (Arlov 2003, 50–51), it constructs the archipelago as a native Norwegian space with roots in medieval history. More than just a dwelling place, the new residence symbolizes the full integration of Svalbard into Norway as a modern country, but it does so by harking back to and renewing long and venerable traditions.

Everyday modernity in the High Arctic

Balstad's background and role make her extremely sensitive to the spatialization of modernity. While Oxaas, at least until his marriage, rarely reflects explicitly on how Svalbard is being gradually incorporated into the modern world, this is a major theme of Balstad's book. As an observer of everyday life, she registers how Longyearbyen is being transformed from the wreckage of war into a modern Norwegian welfare society. This, she repeatedly underlines, means a society that accommodates not only male breadwinners but also women, children and families. In fact, as Ingrid Urberg has argued, she seems to have felt it as her special responsibility to record the contributions made by women and wives to Svalbard society (Urberg 2007, 187).

An important theme in *Nord for det øde hav* is the conflict of interest between the mining company and civil society. Modernity before the war had meant coal-mining. The population was predominantly male, and the few women over-wintering in Longyearbyen were either wives of coal-company functionaries or service personnel. While Balstad is well aware that coal-mining constitutes the economic basis of modern Svalbard, and that Longyearbyen is still – as it was before the war – primarily a company town, she argues that the future of the archipelago lies in the development of a viable civil society. She therefore applauds efforts at "normalising" life in Longyearbyen by providing facilities for families and children and improving connections with mainland Norway. Important examples of the latter are better radio transmissions and the annual mail drop onto the Longyearbyen sports field a month or so before the first ship can get through the ice. But because of the demand for coal in post-war Europe, the welfare of the miners always has the highest priority, and she complains that as a consequence there is a continual shortage of provisions for other inhabitants.

One of many everyday issues signifying Svalbard's difference from the mainland, and one of great concern for housewives and mothers, is the shortage of ice to melt for water during the winter. A lack of ice in the Arctic may seem absurd to outsiders, Balstad explains, but because the snow and ice in Longyearbyen is contaminated by sand and coal dust, blocks of ice have to be laboriously transported from a frozen reservoir in the Advent Valley, some miles east of the town: "First the tanks of the mess halls were filled, then the bath tubs of the coal miners, and then finally it was the turn of private households" (Balstad 1955, 74).[27] Babies have no precedence, as she discovers for herself after her son is born in 1947. It is therefore as a representative of Svalbard mothers in general that she exhorts government officials – whose children doubtless get daily baths in tubs easily filled with hot water from a tap – to come north to view the primitive conditions in Longyearbyen for themselves (Balstad 1955, 181).

As the mother of both a young child and an older son from her first marriage, who joins her after two years in Longyearbyen, Balstad makes her own experiences exemplary of both the challenges and the joys of bringing up children in a High-Arctic environment. She writes that children in Longyearbyen are well taken care of and enjoy the freedom of being outside in every kind of weather, even through the dark winter months. One of her illustrations, of two happy toddlers playing in the snow underscores this point. The caption "ruddy-cheeked children – *that* too is Svalbard" posits a modern family-friendly Arctic, even what may be called an Arctic pastoral (cf. Wærp, this volume).[28] At the same time, she argues that children ought to visit the mainland every summer to recuperate from the long polar winter. Moreover, she believes that they need to learn through actual contact that the world as a whole is different from Svalbard, "that there are trees and animals in reality and not only in fairytales" (Balstad 1955, 152).[29] Unavoidably, that reality is her own frame of reference throughout the text. Although the overall mission of her book is to promote Svalbard as an integral part of modern Norway, she clearly – in contrast to Oxaas – would never have called the archipelago "my world".

Like Oxaas, Balstad complies with the gender system of the mid-twentieth century. She defines herself as "a housewifely woman" (Balstad 1955, 187) and never explicitly questions this role.[30] Nor does she express much resentment at having to function as her husband's unpaid secretary – even, on occasion, as unpaid prison guard and radio announcer. Yet, there is an incipient feminist protest in her defence of "the Svalbard woman", whose important work in the community is usually overlooked. As she puts it, "nobody considers dealing with woman as a member of society. She seems to be taken for granted. Woman as a human being and citizen seems to be something that nobody has begun ticking off in questionaires and tables" (Balstad 1955, 304).[31] In a long digression from her personal narrative she attempts to remedy this by paying homage to some of the women who have braved the elements, over-wintered in Svalbard and thrived. There she also suggests that – contrary to received opinion – being a woman is no disadvantage in the Arctic: "Perhaps it takes longer before she puts down roots, and perhaps there are fewer women who actually do so. But once she finds the Svalbard melody, she becomes more attached to the land and the surroundings than the man" (Balstad 1955, 304–305).[32]

By highlighting the importance of a capacity for emotional connection rather than such qualities as mastery or physical stamina, Balstad accomplishes a rather startling reversal of Arctic stereotypes. But the utopian impulse of her book is clearly inspired by a vision of a Svalbard society in which the feminine values of relationships, family, homemaking and good childcare prevail. Although Oxaas is neither an advocate nor explicitly oriented towards the future development of the archipelago, in his final chapters on married life he implies that he shares this vision. Balstad's goals have not been achieved when she and her husband leave Longyearbyen for good in 1955, and in contrast to Oxaas she is equivocal about whether she herself found "the Svalbard melody".

However, the case that both authors make for a Svalbard *within* European modernity cannot be separated from the issues of gender, citizenship and welfare developments. The frontispiece of Balstad's memoir, a photograph of a group of school girls in Longyearbyen pointing out their home at the top of a globe, is the clearest indication of this. As an image of both modernity and gender equality it signals that, geographical distance apart, Svalbard in no way lags behind more metropolitan centres.

Acknowledgements

An earlier version of this chapter was presented as a keynote lecture at the International Association of Scandinavian Studies Conference at the University of Groningen in the Netherlands, on 11 August 2016. Many thanks to Janke Klok, who invited me, and to the conference participants. Thanks also to the members of the FemArc Network at UiT The Arctic University of Norway, whose feedback on an unfinished draft was invaluable, and to Heidi Hansson for her incisive and constructive critical comments.

References

Arlov, Thor B. 2003. *Svalbards historie* [The history of Svalbard]. Trondheim: Tapir Akademisk Forlag.

Balstad, Liv. 1955. *Nord for det øde hav* [North of the desolate sea]. Bergen: J.W. Eides Forlag.

—. 1958. *North of the Desolate Sea.* Adapted by Alice Hope and translated by Joan Bulman. London: Souvenir Press.

Blaut, J.M. 1993. *The Colonizer's Model of the World: Geographical Diffusionism and Eurocentric History*. New York: Guilford Press.

Drivenes, Einar-Arne and Harald Dag Jølle, eds. 2006. *Into the Ice: The History of Norway and the Polar Regions*. Oslo: Gyldendal Akademisk.

"En av dem som har bygd Svalbard legger opp etter 29 overvintringer" [One of those who have built Svalbard retires after 29 over-winterings]. 1947. *Tromsø*, 8 July.

Evjen, Bjørg. 2006. *Kull, karer og kvinnfolk. Longyearbyen 1916–1975: fra arktisk arbeidsplass til etablert industrisamfunn?* [Coal, men and women: Longyearbyen 1916–1975: from Arctic workplace to established industrial society?]. Oslo: Unipax.

Felski, Rita. 2000. "The Invention of Everyday Life." In *Doing Time,* 77–98. New York: New York University Press.

—. 2002. “Introduction.” *New Literary History* 33, Everyday Life (Autumn): 607–22.
Foucault, Michel. 1998. “Different Spaces.” Translated by Robert Hurley. In *Aesthetics, Method and Epistemology: Essential Works of Foucault*, vol. 2, edited by James D. Faubion, 175–85. New York: The New Press.
Friedman, Susan Stanford. 2001. “Definitional Excursions: The Meanings of Modern/Modernity/Modernism.” *Modernism/modernity* 8 (3): 493–513.
—. 2012. “World Modernisms, World Literature, and Comparativity.” In *The Oxford Handbook of Global Modernisms*, edited by Mark Wolleager, 499–525. Oxford: Oxford University Press.
Gaonkar, Dilip Parameshwar. 1999. “On Alternative Modernities.” *Public Culture* 11: 1–18.
Hauan, Marit. 1992. “Wanny har bakt . . . også har hun skudt 3 stormåser” [Wanny has baked . . . and she has also shot 3 great gulls]. In *Kvinnfolk på tvers*, edited by Elin Svenneby, 25–41. Oslo: Emilia.
Hetherington, Kevin. 1997. *The Badlands of Modernity: Heterotopia and Social Ordering*. London: Routledge.
Heidegger Martin. 1971. “Building Dwelling Thinking.” Translated by Albert Hofstadter. In *Poetry, Language, Thought*, edited by Albert Hofstadter, 347–63. New York: Harper & Row.
Hulan, Renée. 2002. *Northern Experience and the Myths of Canadian Culture*. Montreal: McGill-Queen’s University Press.
Jacobsen, Jacob. n.d. “Resyme” [Summary]. Unpublished manuscript. Svalbard Museum SVB 4846 B/C.
Karlsen, Silje Solheim. 2014. “Fangstmannsbiografien og heltelitteratur. Med et skråblikk på sjangeren: Wanny Woldstad, Henry Ette, Thorleif Bjertnes og Henry Rudi” [The trapper’s memoir and heroic literature. With a glance at the genre: Wanny Woldstad, Henry Ette, Thorleif Bjertnes and Henry Rudi]. *Nordlit* 32: 1–16.
Knutsen, Nils Magne. 2013. *Litteratur i æventyrland. Artikler om nordnorsk litteratur* [Literature in the land of adventure: Articles on North-Norwegian literature]. Edited by Henning Howlid Wærp. Tromsø: Angelica Akademisk.
Lønø, Odd. 2014. *Norske Fangstmenns overvintringer på Svalbard og Jan Mayen 1795–1973* [Over-winterings by Norwegian trappers in Svalbard and on Jan Mayen 1795–1973]. Edited by Sander Solnes and Per Kyrre Reymert. Longyearbyen: Stiftelsen Svalbard Museum.
Orvin, Anders K. 1939. “The Settlements and Huts of Svalbard.” *Norsk Geografisk Tidsskrift* 7: 315–28.

Oxaas, Arthur. 1916. "Dagbok Bjørnøya" [Diary Bear Island]. Unpublished manuscript. Norwegian Polar Institute.

—. 1917–18. "Dagbog for overvintringsexpedition paa Jan Mayen" [Diary of an over-wintering expedition on Jan Mayen]. Unpublished manuscript. Norwegian Polar Institute.

—. 1936–37. "Dagbok fra Kap Wijk, Isefjord (for dosent Hoel)" [Diary from Kapp Wijk, Ice Fjord (for Docent Hoel)]. Unpublished manuscript. Norwegian Polar Institute.

—. 1939–40. "Dagbok fra Kap Wijk, Isefjord (ført for dosent Hoel)" [Diary from Kapp Wijk, Ice Fjord (written for Docent Hoel)]. Unpublished manuscript. Norwegian Polar Institute.

—. 1955. *Svalbard var min verden* [Svalbard was my world]. Oslo: Aschehoug.

Prieto, Eric. 2013. *Literature, Geography, and the Postmodern Poetics of Place*. New York: Palgrave Macmillan.

Rossnes, Gustav. 1993. "Norsk overvintringsfangst på Svalbard 1895–1940" [Norwegian over-wintering trappers in Svalbard 1895–1940]. *Meddelelser* 127. Oslo: Norsk polarinstitutt.

Schor, Naomi. 1987. *Reading in Detail: Aesthetics and the Feminine*. London: Methuen.

Shields, Rob. 1991. *Places on the Margin: Alternative Geographies of Modernity*. London: Routledge.

Skogvang, Hege. 2004. "Sysselmannsanleggene i Longyearbyen på Svalbard – om arkitektur som suverenitetsmarkør" [The Governor's estate in Svalbard – on architecture as a marker of sovereignty]. MA Thesis, University of Tromsø.

Urberg, Ingrid. 2007. "'Svalbard's Daughters': Personal Accounts by Svalbard's Female Pioneers." *Nordlit* 22: 167–191.

Young, Iris Marion. 2001. "House and Home: Feminist Variations on a Theme." In *Feminist Interpretations of Martin Heidegger*, edited by Nancy J. Holland and Patricia Huntington, 252–88. University Park PA: Pennsylvania State University Press.

Åmark, Klas. 2006. "Women's Labour Force Participation in the Nordic Countries During the Twentieth Century." In *The Nordic Model of Welfare – a Historical Reappraisal*, edited by Niels Finn Christiansen, Klaus Petersen, Nils Edling and Per Haave, 299–334. Copenhagen: Museum Tusculanum Press.

Notes

[1] Thanks to Kristina Malmio, University of Helsinki, who alerted me to Prieto's text.

[2] "[…] en bunke papirer." All translations are my own.

[3] "Det Svalbard kan gi igjen for slit og savn og ensomhet og mørke og kulde, svarer ikke til det et modern menneske krever av et liv i en velferdsstat."

[4] Oxaas' over-wintering diaries are kept in the library of the Norwegian Polar Institute in Tromsø, and all have been digitalized. Three of them (for the years 1936–37, 1937–38 and 1939–40) were written for Adolf Hoel, the leader of the Norwegian Svalbard and Arctic Ocean Survey, the forerunner of the present-day Polar Institute.

[5] "En helvetes plads."

[6] "Selve reisverket kom opp i en fart, og det var snart gjort å kle veggene med [en og en halv] toms bord. Hele huset var på 24 kvadratmeter, og i flukt med det førte vi opp et proviantskur på 4 x 4 meter. Inne i overvintringshuset slo vi opp fire køyer langs den ene tverrveggen. I kroken mot tilbygget plasserte vi en liten komfyr med stekeovn. På hver langvegg var det et vindu. Under det ene fikk vi plass til et bord og under det andre et kjøkkenskap. Da vi så hadde fått inn et par benker, var huset ferdig til innflytting. […]

Vi var kommet ut i september, og det gjaldt å få ferdig alle utearbeidene før vinteren satte inn. Det første vi gjorde, var å legge opp en tykk steinmur rundt huset. Vi fôret dyktig med mose mellom veggen og steinene. Ellers kledde vi både tak og vegger med russenever og torv."

[7] "Først lærte vi å bake brød, dernest å bruke selvskuddene, fellene og giften."

[8] "Har […] stegt vafler og flaaet 2 Blaaræver."

[9] "Jeg sa til meg selv at jeg etter dette skulle jeg være varsom med å dra for langt vekk fra huset. Nå hadde jeg opplevd hvor nærgående isbjørn virkelig kan være. Jeg kunne simpelthen risikere å bli stående uten tak over hodet en annen gang. Den vesle hytta med den lille båtovnen var det eneste sted i mange mils omkrets, hvor man kunne livberge seg her i polarvinteren. Først måtte jeg forsøke å skjøte ovnsrøret, så jeg kunne få varme i stuen igjen. Men før jeg begynte på reparasjonen av taket, måtte jeg bære ut noen lass med snø og forskjellig avfall som dekket gulvet inn. Jeg spadde og børstet og omsider begynte det å se ut som stuegulv der inne igjen. Og da jeg hadde vært oppe på taket og spikret fast løse bord og erstattet brukne planker med nye, var kvelden falt på […]"

[10] "Det lyder kanskje underlig, men en kan simpelthen venne seg av med siviliserte omgivelser også."

[11] "Selv om naturen er karrig, og det kan virke ødslig og goldt der oppe i de kalde vintermånedene, gror en liksom fast på stedet og blir glad i det."

[12] From 1934 the ship *D/S Lyngen* had made regular trips between the mainland and Svalbard five or six times each summer.

[13] "Kl. 24 kom Lyngen hertil […]. Min forlovede kom da også med for at være her i vinter. Kanskje er det en dumhet, – man får se."

[14] "Anna beskjeftiger sig med huslige sysler, og trives bra."

[15] "Håber det er siste året jeg må bruke denne torvsjåen. Alt ødelegges, klærne rådner op, og selv bliver man aldeles maroder av jikt. [...] Ja nu er her mindre hyggeligt. [...] Anna bagte idag. Hun stræver vær dag og trives bare godt."

[16] "For meg betydde dette en hel omlegging av vaner og livsrytme, men en omlegging avgjort til det bedre. Nå ble det orden i sakene. Mat, klær og hus ble stelt av kyndige hender. Det ble liksom mer sivilisasjon her opp med en kvinne i huset."

[17] "I em flyttet vi in i nyhuset. Har hele tiden siden vi kom havt nok at jøre med at få bygget ferdig. Det har dobbelt tak, gulv, vegger overalt. Topslået in og utvendig, likeså er der malt inne. Vi føler os som goseiere efter at være flyttet fra torvsjåen. Her er da også triveligt og praktisk inredet. Stort lyst dagligværelse, soveværelse ved siden, et stort rum som proviantrum, skindrum, ventes W.C. gang og en stor sjå med gulv og vindu. Jeg havde også med et radioapparat som nu skal rigges op, så det ser da ut for at man nu skal få følge med alt, eller noe av det som forregår utenfor vår lille verden."

[18] "Anna utstyrte huset med filleryer på gulvene og gardiner i vinduene. Mellom stuen og soveværelset hengte hun opp nye portièrer istedenfor dør. Med sin kvinnelig sans for småtingenes betydning, gjorde hun dette huset til et virkelig hjem, som var egnet til å skape trivsel for alle."

[19] Thanks to Herdis Lien and Sander Solnes at Svalbard Museum for helpful information about Oxaas' picture frames.

[20] "Når været var bra, dro hun alltid ut for å se etter redskapene, og hun var en meget nøyaktig og flink fangstmann."

[21] Åmark notes that "the idea that married women should not work" was stronger in twentieth-century Norway than in the other Nordic countries (Åmark 2006, 310).

[22] "Tyskerne hadde gjort rent bord etter seg, bare tre-fire hus sto igjen i selve Longyearbyen, så alt måtte bygges opp fra grunnen av. [...] Alt dreide seg om gjenreisningen. Det gjaldt å arbeide; det å *leve* tenkte så menn ingen på, og ikke var det tid til det heller. Overalt lå det materialstabler, hauger av tømmer og bord og plank; sement, maskiner; et virvar som man ikke skulle tro var til å finne ut av."

[23] "Tilpasningens evne og resignasjonens kunst er noe av det som trengs mest for den som skal bo på Svalbard."

[24] "Tiden har blitt en annen."

[25] "[...] romsligere, mer tidsmessig og også mer representativt."

[26] "'Slaattos dikt' kalte jeg det i et poetisk øyeblikk – de rene, enkle linjene hadde en stolt verdighet i seg. Den store peisetua i rundtømmer – 'Liv, jeg skal bringe skogen inn i studa til deg!' – tårnet i mur som bandt bygningene sammen – gjesteværelsene, kjøkkenet, stuene, – for meg ble det ikke bare et hus, men et dikt."

[27] "Først ble beholderne i messene fylt, så badene til gruvearbeiderne, og så til sist kom turen til privathusholdningene."

[28] "[...] rødmussede barn – også det er Svalbard."

[29] "[...] at det er trær og dyr til i virkeligheten og ikke bare i eventyrene."

[30] "[...] en husmoderlig kvinne [...]"

[31] "[…] ingen finner på å befatte seg med kvinnen som samfunnsmedlem. Hun går liksom med i slengen. Kvinnen som menneske og borger er liksom noe som ingen har begynt å prikke inn på skjemaer og tabeller."

[32] "Kanskje tar det lenger tid før hun slår rot, og kanskje er det færre kvinner som overhodet gjør det. Men finner hun først Svalbardmelodien, knytter hun seg sterkere til landet og omgivelsene enn mannen."

MEDIATING EVERYDAY LIFE IN SVALBARD: HERTA GRØNDAL'S PHOTOGRAPHS, 1950S–70S

ELIN HAUGDAL

A woman in a spotted dress sits in a living room equipped with modern furniture, carpets and cushions, a houseplant on a woven tablecloth, pictures and lamps on the walls, and in the corner a radio cabinet and a rocking horse for the children (Figure 1). She seems confident and relaxed, dressed up and ready to make her home a showcase for the photographer. As this woman is portrayed with two small children in other photographs, we assume that she is a mother, and that this picture shows her family home.[1] The motif is arranged and hence constructed, but nevertheless a witness to time and place.

The picture was probably shot in Ny-Ålesund (Kings Bay), one of two Norwegian settlements in Svalbard, in the mid-1950s or early 1960s. The photographer is the Austrian Herta Grøndal (b. 1930). During half a century of extensive documentation of daily life in Svalbard, she took thousands of pictures of women, children and men at work, engaging in leisure-time activities or in their homes.

As a starting point for this chapter, I have chosen one of the most ordinary pictures from this photographer's archive, a picture that could have been taken almost anywhere in the western world during the middle of the twentieth century. Other photographs from Grøndal's archive, such as the pictures taken at a dramatic crevasse on Kings Bay Glacier, represent time, place and women in more exceptional and spectacular ways.[2] However, Herta Grøndal's photographic motifs seem to alternate between the ordinary and the exceptional, thus mediating the duality of everyday life in the northernmost settlement in the world. Her pictures show small Arctic towns at the dawn of its emergence as a modern family society, with an increasing presence of women and children. She depicts women and men in traditional gender roles, but her pictures also transform these roles, portraying harsh miners with intimacy, and solitary women at crevasses or on mountain tops.

Figure 1. A home in one of the Norwegian settlements in Svalbard in the mid-1950s or early 1960s. Photo: Tromsø University Museum – Herta Lampert, id: tsnf57615

Herta Grøndal worked as a photographer in Svalbard during the period from 1952 to 1974, and continued to visit the archipelago regularly until 2008. Her maiden name was Niedermayer and she changed her name to Grøndal after her marriage in Svalbard in the 1950s. During the 1970s, she moved back to Vienna, remarried there and took her second husband's last name, Lampert. A large collection of her photographs, recently donated to Tromsø University Museum, is labelled "Herta Lampert", on her instructions, but in Svalbard and as a photographer she is most commonly known as Herta Grøndal. Her pictures have been documented neither in detail nor at length, except for a few interviews and feature articles in Norwegian newspapers and magazines.[3] My perspective on her is through a gendered lens, as a photographer who was also a woman with a strong agency within Svalbard society, and my main focus is on her pictures of everyday life. I will explore how these pictures mediate, and also construct, the new identity of Svalbard as a place appropriate even for women and their children, thus challenging common ideas of the Arctic as male, heroic, uninhabited and inhospitable.

The connection between everyday life, modernity and gender has been discussed by a number of scholars from different disciplines, and with different perspectives. In this context, I find Rita Felski's article "The Invention of Everyday Life" (1999/2000) particularly useful, because of both her critical summary and her feminist position. Felski posits *time*, *space* and *modality* as the three constitutive categories of the everyday, and invites her readers to explore these categories phenomenologically, as they appear as *repetition*, *home* and *habit*. The term "everyday life" does not only describe the lives of ordinary people, of course, since "every life contains an element of the ordinary" (Felski 1999/2000, 16), but Felski recognizes that some groups and classes, such as women and workers, are considered more closely tied to daily life than others. This may also be true of Svalbard, even though, despite strong class divisions, the population to a large extent shares the same everyday conditions, the same reality.

The interest in everyday life and the significance attributed to it is a modern phenomenon, which emerged during the last part of the nineteenth century and into the twentieth. During this period, everyday life became visible in literature and art, as Felski points out, and she wonders about the reason for this "new visibility" (Felski 1999/2000, 16). I argue that the answer, at least when we look at the twentieth century, has to include the growth of photographic culture, the increased number of photographic cameras, the reproduction and dissemination of pictures, and the "photographic eye" of modern culture (Mayer 1946). Photography and photographs are entangled with the everyday life of modernity as a way of seeing, and thus thinking. Photographs affect the construction of self and others and have a strong impact on collective identity. Photographs are printed, put in family albums or hung on walls, they are sent as postcards, exchanged as gifts or sold to magazines, and in different ways disseminated and circulated throughout contemporary society. According to the photo theoretician Elizabeth Edwards, this important aspect of the photograph's social function or performativity relies on its materiality and relationality (Edwards 2004, 16; Edwards and Hart 2004, 3–4). The social functions of digitized images are of no lesser importance. Photographs have a strong retrospective function as "memory machines", boosted by photo-sharing on social media.

The prevalence of photographic devices during the mid-twentieth century coincided with social changes and modernization in Svalbard. Herta Grøndal's photographs, taken during the 1950s, 1960s and 1970s, represent these changes, primarily taking the ordinary lives of women and workers as motifs, and I explore how these photographs mediate repetition, home and habit in Longyearbyen.[4] However, Felski reminds us that the study of the everyday "beckons to us with the beguiling allure of the 'really real'" (2002, 607). Likewise, the study of photographic *representations* of the everyday is no less tantalizing, and therefore the pictures need to be constantly disputed as witnesses of truth and reality.

Herta Grøndal – photographer

More than once, Herta Grøndal climbed the steep mountainside nearby with heavy equipment, a tripod and large-format camera to take a panorama of Longyearbyen. In one of her most iconic images, she made the area known as Sverdrupbyen the main subject (Figure 2).[5] In Longyearbyen, which at the time was a strongly class-divided coal-mining company town, this framing is of the greatest significance. Former inhabitants refer to the existence of "two peoples", those who worked in the offices, and those who worked in the mine (Østlund 2016, 19). Sverdrupbyen and Nybyen contained the homes of the workers, while the higher and lower officials had their residences at Haugen and Skjæringa – positioned as peripheral in this picture.

A common meeting place, Huset ("The House"), was purposely erected in 1951 between the workers' housing area and the officials' homes. The town looks peaceful in Herta Grøndal's moonlight. Actually, most of her motifs are tranquil like this, whether she depicts landscapes or people. Hence, her photographs proclaim that the Arctic is a harmonious and good place in which to live, although the available written and spoken narratives about life and work in Svalbard tell us this was not always the case.

It was Herta Grøndal's adventures in the north of Norway that led the then 22-year-old woman to Svalbard and the mining town of Ny-Ålesund (Kings Bay) for the first time in August 1952 (Svarstad 2006). Equipped with curiosity and a simple Kodak camera, she only stayed for a short time. Back home in Vienna, she took courses in photography, then returned the following summer with a new and better camera to spend her first winter in Svalbard, in the other Norwegian mining town, Longyearbyen. She afterwards settled there, together with her husband, the

Figure 2. Herta Grøndal's iconic panorama of Longyearbyen from the late 1950s or, most probably, the first half of the 1960s. Colour photo: Tromsø University Museum – Herta Lampert, id: tsnd41023

miner Leif Archie Grøndal (1914–1991). Leif Grøndal was given assignments as a photographer – shooting for the mining company, Store Norske Spitsbergen Kulkompani (SNSK), documenting official events in Svalbard. He also had close ties with the Governor of Svalbard, Håkon Balstad, his family and network in both Longyearbyen and the Russian settlements. The more than 5000 photographs he left behind testify that he too had an eye for everyday life in the Arctic, not least for the lives of children.[6] Early in the 1950s, Herta Grøndal became his assistant, doing darkroom work. After the birth of their first daughter in 1954, she moved to the mainland and lived in Drammen, her husband's hometown, where she worked part time in a photographic laboratory while commuting regularly to Vienna. It must have been a stormy decade, with three daughters born in 1954, 1956 and 1959, a lot of travelling, and a broken marriage at the end. During the early 1960s, she stayed in Svalbard for short periods, and by the mid-1960s she had again settled in Longyearbyen, without her children. She lived there for nine years.

Even though Herta Grøndal was not trained as a photographer, but as a musician, she worked, was treated and behaved like a professional photographer.[7] Signs of professionalism included her top-notch technical equipment,[8] her use of a 6 x 6 medium format, and parallel shooting in black-and-white and colour even as early as the 1950s.[9] She continued taking black-and-white pictures even after colour film became cheaper – in part probably for aesthetic reasons, but also because she could develop the pictures herself rather than having to send them away. Even the fact that she put a stamp on the back of her prints, "Copyright Herta Grøndal Spitsbergen", shows her confidence as a photographer. During the period 1965 to 1974, when she was settled in Svalbard, Herta Grøndal was the local contact for the Norwegian Broadcasting Corporation (NRK). She worked both as a film photographer and as a reporter, and produced 16 mm films for television. She was hired by SNSK as well as by the Norwegian Foreign Ministry as a photographer and film maker (Hagevik 1992), and regularly delivered photographs to the mainland newspaper *Nordlys* as well as to the local paper, *Svalbardposten*. Moreover, she had many private commissions as a portrait photographer. During the period 1974 to 1985, she did not visit Svalbard at all,[10] but over the following years she visited regularly until 2008, normally for a month in the summertime and some weeks in the winter. While there, she was always taking pictures.

Although it is difficult to provide an overview of Herta Grøndal's photographic production in Svalbard, it may be categorized into three topics: nature (landscapes, wildlife and plants), cultural artefacts (houses, huts, industrial constructions and ruins) and people (work and leisure, and portraits). It is obvious that the latter category was more important during the first few decades, when she was more settled and integrated into Svalbard society, than in later years.

Herta Grøndal's photographic archive is split into two, the previously mentioned Herta Lampert collection at Tromsø University Museum (TMU), and a rather large collection managed by two of her daughters in Longyearbyen, Grøndal Photo, which also includes Leif Grøndal's photographs. Herta Grøndal also sold a larger amount of photocopies to

SNSK. Both the Grøndal pictures and SNSK's collection are deposited at Svalbard Museum.[11]

The collection at TMU, which is the primary source for this article, contains more than 12 000 diapositives and negatives in different formats. This large number of pictures was given to TMU in the autumn of 2014 and is now part of their digital photo collection. Several of Herta Grøndal's images are familiar to those connected with or interested in Svalbard; they have been used in magazines, news reports and books, they are reproduced on postcards and posters, and one of her famous polar-bear pictures is even used as a motif for hand-knotted rugs, the making of which was a popular hobby among Svalbard's inhabitants, men as well as women.[12] However, most of the pictures now being digitized and published in the Unimus system[13] have not been previously available. These digitized images still lack all kinds of metadata, and Herta Grøndal's envelopes and boxes only provide a very limited amount of information on time, places and persons.

At the present stage, this lack of information gives us an opportunity to look at the pictures themselves, exploring pictorial qualities such as composition, light, close up, perspective, the choice of motifs and the depicted people's gestures, gaze and relations – in an attempt to delay the search for the identification of historical persons, places and events. The images used in this article were chosen primarily because of their visual potential, as well as their capacity to provoke discussions about everyday life in the Arctic during the mid-twentieth century. Thus, it is not my intention to represent the full scope of the Herta Lampert collection.

Constructed rituals

A picture taken at Skjæringa, just below the governor's estate, shows a spatial structure of elements characteristic of Longyearbyen in the post-war period: the bell tower, the flagpole, the war memorial and not least the cableway trestles. These are all solid, human installations, framed by the unending landscape of barren tundra and mountains (Figure 3). The dynamic and temporal elements of this structure – the colourful Norwegian flag, the hanging polar bear skin, the shadow of the fence, the moving coal wagon, a child running and two women carrying a tub presumably full of laundry – are recorded by the photographic eye. It is springtime in the 1950s or early 1960s. The picture raises questions about the relation between this permanent spatial landscape and the different people who at various times have inhabited, used and exploited it.

Figure 3. This photo from Skjæringa includes the main motif groups of Herta Grøndal's photographic production: landscapes and animal life, constructions and artefacts, and people at work and leisure. Colour photo: Tromsø University Museum – Herta Lampert, id: tsnd41903

The Arctic society that Herta experienced for the first time in 1952 was quite different from the one she visited in 2008, the date stamped on her last photographs. In the early 1950s, the Norwegian settlements were expanding. The oldest mining town, Longyearbyen, was under reconstruction after the devastation inflicted during the Second World War, a development led by SNSK. In Ny-Ålesund as well, mining operations resumed after the war, directed by another company, the Kings Bay Coal Company. This company and the small community suffered great losses due to several major mining accidents in 1948, 1952 and 1953, in which a total of 43 people lost their lives, and more was to come. In Ny-Ålesund, mining operations ceased in 1962, after the deaths of 21 miners in a final accident, and only a small number of residents remained. In Longyearbyen, the mining industry survived the 1950s, and both the industry and the population were still increasing. It stabilized at around 1000 in the 1960s. By comparison, there were a total of 2000 inhabitants in the Russian settlements of Grumant, Barentsburg and Pyramiden.[14]

In this context, the most important change during the 1950s, 1960s, and 1970s was the increasing number of families living in Longyearbyen (Evjen 2006, 64).[15] Only the engineers and officials working for SNSK were allowed to bring their spouses, more for the purposes of recruitment than with the establishment of a family community in mind (Evjen 1995, 74; Arlov and Holm 2001, 33). Longyearbyen was a company town from 1916, when SNSK was founded, until 1989. This means that SNSK operated schools, hospitals and telephone installations, built roads and airfields, and decided everything, from working conditions to the curtains in the barracks and the price of beer (Arlov and Holm 2001). During the 1970s, there was a shift from company town into a government corporation, and the beginnings of a developing infrastructure and services that were more welcoming to families (St.meld. no. 9 1999–2000, 3.5). But the transformation of this small town from an all-male society to a family community was not completed until the 1980s.[16] Herta Grøndal's photographs are of course historical sources that illuminate the beginning of this social change, mediating new aspects of people's everyday lives during the post-war years – for women and children as well as for the men.

Her pictures of children from this period take an insider's perspective. One of the photographs in her archive, probably from the mid-1950s, depicts a score of children sitting on a pile of parcels (Figure 4). The narrative is quite simple. A ship is in town and the mail has been delivered – a big event for this small settlement, and the children have got a ride with the wagon on which the parcels are being transported from the harbour. The children sit waiting, forming a triangular composition that points above the railings and on to the picture's dynamic moment, where two men are unloading mail bags to the person on the quay. The children have turned their backs on this activity, and the photographer has chosen the children's viewpoint, their rituals, their patience and excitement, which is underscored by the photographic angle and focal range.

Although it is a tight group, the children's pale faces profiled against the black shipboard, with their gazes in different directions, show individuality.[17] One girl is looking straight into the lens, thus fixing the very centre of the picture and confirming an awareness of the photographer, who is viewing her at her own eye-level. It is probable that this photograph was shot by Leif Grøndal, or his brother Harald, and Herta Grøndal may have developed it in the darkroom.[18] In any case, the Grøndal photographers' motifs and perspectives reveal a set of seemingly insignificant, yet important, rituals constituting everyday life at 78 degrees north.

Figure 4. A Grøndal photo from the mid-1950s showing children of Longyearbyen taking part in the mail arrival event. This ship is most probably "Ingerfire" on one of her first trips in May 1954. Photo: Tromsø University Museum – Herta Lampert, id: tsnf56170

Mail delivery was an important event. During the winter, after the return of the sun, mail was carried to Longyearbyen by aeroplane.[19] "Soon the mail plane came to be the slender thread that connected us with the outside world through the long winter months," Liv Balstad, the governor's wife, writes in her book *North of the Desolate Sea* (1955). "It was the mail plane we waited for. It became a sort of measure of time" (Balstad 1958, 187).[20] The large number of pictures of these aeroplanes delivering mail, taken by the inhabitants of Longyearbyen during the 1950s and 1960s, indicates the importance of this event.

Mail is not the focus of a photograph shot by Herta Grøndal a decade later (even though we can spot a parcel on the ground under the fuselage), but the picture represents the tenuous connection with the world outside the Arctic (Figure 5). The photographer's position is now more that of an observer than an insider, positioned at a significant distance from the scene and the motif. At the very centre of the picture, in the intersection of perspective lines, is another photographer, a man with a camera. It is hard to tell if he is an inhabitant or a visitor, but his function in this photograph is clearly to reflect the photographer's own activity, or this Arctic destination as a photographic motif as such.

Figure 5. The airstrip in the Advent Valley, which was in use during the war and between 1959 and 1974. This picture must have been shot in the spring (March – May). Colour photo: Tromsø University Museum – Herta Lampert, id: tsnd41008

However, what really attracts attention in this picture is the colour stain: a child dressed in turquoise faux fur and red trousers, sitting on a sledge pulled by shaggy dogs. Together with two other children wearing fur hats, the little group forms a contrast to the lightly dressed flight crew. Entering from the bottom left corner, traditionally a significant placement of figures in western pictorial culture, this group of children and dogs is established as the real representatives of the residents of Longyearbyen.

The arrivals and departures of ships and planes – bringing mail, goods and people – embody time in Longyearbyen. Not following exact dates, but depending on weather conditions, and announced over national radio, these events represent longing, expectations and feelings, while structuring time and thought. When they occurred, these events triggered certain behaviours in the inhabitants, and seem to have served like rituals for both adults and children, as a lot of the Grøndal pictures illustrate. Expectations, feelings and actions obviously changed between the 1950s and the 1970s, but the ritual patterns connected to these arrivals and departures were nonetheless maintained.

Some of these habits or rituals were determined by external conditions, like the seasons, with periods of total darkness alternating with midnight sun, with ice on the fjords or melting ground in the summertime. Other repetitive patterns and rituals were regulated by the mining industry and work shifts, which were mainly controlled by the clock. Such culturally constructed rituals seem to be of immense importance in Arctic settlements. Traditions brought from the Norwegian mainland are maintained, representing "a little piece of Norway" (*Dagen* 1989), such as the National Day on 17 May with a children's parade, the midsummer celebration with bonfires, the "Christmas tree" party, or children's skiing races. Other cultural rituals have emerged specifically within the Svalbard social setting, such as the cultural and sporting exchanges with the Russians, and the sun party to celebrate the first rays in March after three months of darkness, or parties just for pleasure, like the so-called Shang-Po-Lar festive evenings in Sverdrupbyen, featuring music and dance. Herta Grøndal's archive is a valuable source of information about these different kinds of traditions and cultural rituals which defined the year's cycle for the inhabitants of Longyearbyen.

According to Felski, "repetition, understood as ritual, provides a connection to ancestry and tradition; it situates the individual in an imagined community that spans historical time" (Felski 1999/2000, 20). Because Svalbard has no permanent population, ancestry is hardly a relevant category there, and this makes the significance of an imagined community even greater. The rituals are constructed to "survive", to quote

Felski, who sees "the pragmatic need for repetition, familiarity and taken-for-grantedness in everyday life, as a necessary precondition for human survival" (Felski 1999/2000, 31). Repetition throughout the year seems to be of immense importance as a way of establishing Longyearbyen as a thriving, sustainable and normal society, with its own traditions and history. Some annual rituals contribute to the naturalization of this community as Norwegian, other routines and temporal repetitions are signs of modern civilization.

On the one hand, annual events and regular festivities may not fit the concept of the everyday. On the other hand, neither postal deliveries nor even shopping were everyday events in Svalbard. Furthermore, the photographs shot on anniversaries reveal the ordinariness of the life of the people. Herta Grøndal's photographs of common celebrations like the National Day offer some good examples. Shot outside Stormessa in the workers' housing area in Nybyen, at the monument outside "the House", or at the governor's estate at Skjæringa, Herta Grøndal's pictures display how the inhabitants took part in these annual rituals at the socially differentiated places in this Arctic town.

Even though she acted as a professional, Herta Grøndal's photographic practice has much in common with the amateur photographer's, which may be considered a *ritual* practice. The amateur photographer uses her camera in private or socio-cultural contexts that she expects to be memorable, thus confirming and reinforcing the meaning of those rituals (Bourdieu [1965] 1990, 31). The quotidian in such pictures is revealed in details, gestures, interactions between people, as it appears in singular motifs, or in the picture's background. However, Herta Grøndal also studied everyday life in the settlements of Svalbard in a more systematic and documentary way, as can be seen in her photographic series of the major workplaces, of the school, the regular leisure activities, and from everyday life in the Russian settlements.

Gendered places

Herta Grøndal's pictures of miners, smiling at the photographer despite their hard and dirty work digging deep into the mountains, have become iconic of Svalbard in modern times.[21] One of these pictures shows six men on the slope-train heading for a shift in the mine named 2b (Figure 6). Our gaze is drawn from their smiles to the mouth of the mine high up on the steep mountainside. They are rough workers equipped with headlamps and coffee in bottles for a shift in the dark mine.[22]

Figure 6. Smiling miners heading for their shift in Mine 2b in Longyearbyen, shot by Herta Grøndal probably in the 1950s or 1960s. Photo: Tromsø University Museum – Herta Lampert, id: tsnd56214

During the whole of the twentieth century, the major workplace in Svalbard was the mine. It was a highly gendered space that women were usually not allowed to enter. Yet Herta Grøndal obtained access on a commission from SNSK to document life and work in the mines, and she produced a large number of pictures of technical installations, of the miners' working processes and also many portraits of coal-black miners in straight photography style. Grøndal seems to be both open and objective in her photographic practice, recording rather than judging, but she is also affected by a gendered view. An example is her colour photograph of the lunch room in Mine 6, taken during the early 1970s. It may be accused of being a stereotypical example of gendered space, but it nevertheless show

how the miners made their "home" inside the mountain and decorated the bare, steel-panelled walls with posters of half-naked female bodies (Tromsø Museum – Herta Lampert, id: tsnd42119).

Not until the 1980s did women work "underground", most of them employed to clean and keep in order lunch rooms like this and other enclosures in the mine (Evjen 2006, 99–100). The majority of women were employed as cooks, servers and cleaners in the mess barracks, at least until the 1960s.[23] Herta Grøndal's photographs from the women's workplace are less spectacular and exotic than those of coal-black miners in narrow mine shafts, but even if the motifs seem at first sight to be of minor interest, they reveal details of everyday life that we might hardly notice.

One instance is a picture of women doing the dishes after a meal in Stormessa, the major mess hall for workers in Nybyen (Figure 7). In this well-lit room with white wall panels, there are three women and the hand of a fourth coming in with stacks of dirty plates. One woman is depicted from behind in full figure, wearing a blue work dress and black rubbers. She is mirrored in the frontally posed woman behind the rack, while the third is stretching over the conveyor belt.

The central element is the dishwasher, apparently a new device in this kitchen, compared to the rickety trolleys and chipped red-painted concrete floor.[24] The shiny dishwasher is indeed a sign of modern times, of work efficiency, routine, a new standard of hygiene – and also a sign of the upgrading of women's work in an Arctic context.

Photographs like this snapshot from the kitchen in Stormessa give the impression of being authentic glimpses into everyday realities. Close up, cropped and candid, the photograph conveys a snapshot aesthetic in which we can hardly insist on either the photographer's intentionality or the picture's sign quality. We read the picture as a glimpse, ignoring the study of details; for instance, how well the three women have done their hair, or the small pieces of jewellery they are wearing. However, the hand with ring finger and bracelet in the blurred lower right corner, pushing the trolley with dirty plates, is more than just a pictorial phenomenon which "happened" to be snapped. Seen through the lenses of everyday life and gender these details may be read as signs of femininity and of women's multiple roles in Svalbard society.

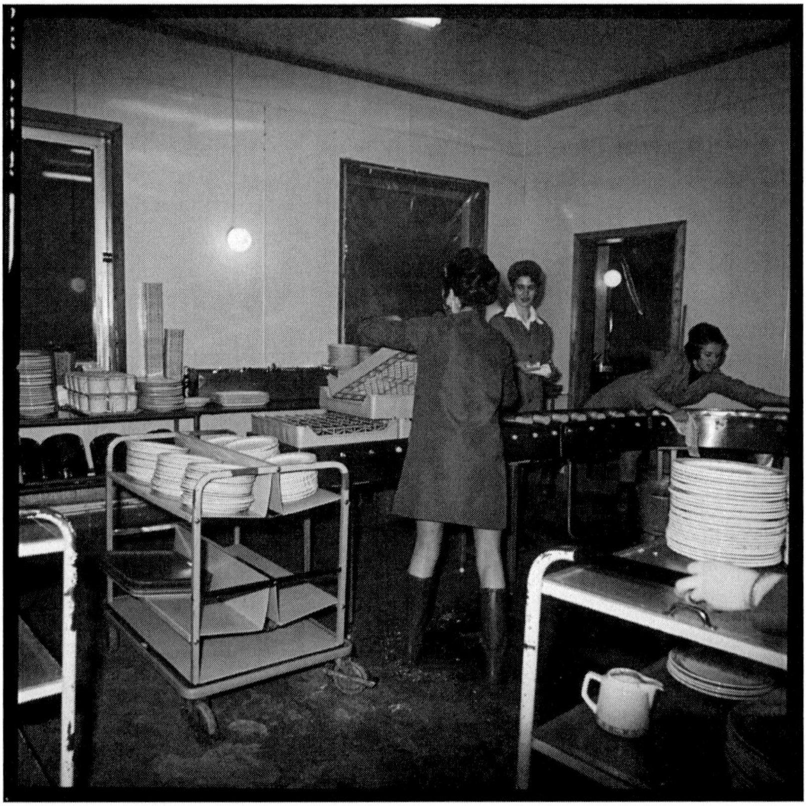

Figure 7. Women's work at Stormessa in Nybyen, probably in the late 1960s. Herta Grøndal's photographs from Stormessa as a working place also show men in the kitchen preparing meals. Color Photo: Tromsø University Museum – Herta Lampert, id: tsnd41455

Women are depicted in a variety of spatial contexts in Herta Grøndal's photographs during the period 1950–70, thus showing how everyday life involves spaces for different functions related to work, family, eating and resting, to recreation and socialization. In Longyearbyen, space is differentiated and restricted socially by gender as well as by class. Some social spaces are obviously privileged for men, others for women. Porcelain painting, for instance, comes across as a women's leisure activity (Figure 8). Other sources tell us that these painting courses were organized in Svalbard's church.[25] The church was a multifunctional place that was as important as a social meeting place as a religious house, and it has been described as a common home to the inhabitants of Longyearbyen over several decades.

Figure 8. Porcelain painting in Svalbard church in the years around 1970. Colour photo: Tromsø University Museum – Herta Lampert, id: tsnd41555

On this occasion, Herta Grøndal's picture shows women relaxing, as though they were at home. They are totally concentrated on the painting of plates and cups, forgetting themselves and their surroundings, not even distracted by the photographer's presence. The hard flash lights up space far beyond the women's attention, such as their legs underneath the table, thus betraying intriguing details of their absorption, like the slipped-off shoe.

The women in Herta Grøndal's motifs are not only seen indoors; her photographs depict women's outdoor life as well, either together with children or simply alone in nature. Some of her pictures taken outside the settlements were shot from sites that are not easily accessible, like glaciers and mountain peaks. They depict moments of female solitude in a magnificent landscape. Herta Grøndal often used wide-angle lenses to exaggerate depth, or placed a person in the foreground overlooking great Svalbard panoramas. Even without people, her panoramas bear witness to her own moments of solitude in nature. However, most of her photographs of everyday life show people interacting and socializing, whether at work or at leisure. Moreover, several motifs explore the social spaces that were available for interaction between the sexes. Explicit examples are photographs taken at parties in "the House" or on other premises in the workers' districts of Nybyen and Svedrupbyen. The seeming over-representation of people partying in Herta Grøndal's archive is explained by the photographer herself as a result of people's direct requests: "Take a picture of us, Herta!" (Moe 1999).

It is difficult to tell whether Herta Grøndal had been invited to photograph an informal group portrait of eight men and two women shot in the basement storage are in Stormessa (Figure 9). Everyone is lined up and looking straight into the camera lens, smiling, except the man in the middle with two female waitresses on his lap, who is being "served" a drink.

An almost identical picture was published in a feature article on Herta in 2006 with a caption calling this a bachelor party, which may explain the scene (Svarstad 2006). Additionally, the collection contains a colour photo from the same occasion, shot a bit before or after the women's presence. This colour photo displays a more chaotic situation, and reveals a telephone on the table in the background, thus telling us that this is not just a storage area, but a social space for men and women to hang out and to meet during or after work in the kitchen. Herta Grøndal's documentation of the bachelor party in the basement at Stormessa does not seem haphazard. She brought with her a medium-format camera and changed its film backs from black-and-white to colour.

Of course this kind of motif reflects the photographer's own gender, age, interests, social status and class affiliation. But Herta Grøndal's pictures also illustrate negotiations of gender, of masculinity and femininity, which can be considered in a more general discourse on gender and modernity. For example, her portrait of two men smiling and smoking in their narrow twin room, sitting neatly knee to knee on their made-up beds, with large-flowered curtains and artificial sprigs of flowers on the wall, displays a more vulnerable masculinity than her picture series from the mine.[26] Even some of her photographs capturing two rough hunters in a cabin in the wilderness are intimate close-ups, playing humorously with the hard and primitive conditions as well as masculine stereotypes (Figure 10).

According to an interview, Herta Grøndal lived with these men for a period of nearly six weeks. She depicted them inside the cabin and outdoors engaged in hunting-related work. They all came to know each other well (Moe 1999).[27]

Figure 9. Bachelor Party in the storage in Stormessa, c. 1970. Photo: Tromsø University Museum – Herta Lampert, id: tsnf57713

Figure 10. Hunters in their cabin in Hornsund in 1969. Photo: Tromsø University Museum – Herta Lampert, id: tsnd59341

Another intimate double portrait of two women sitting on a divan in a dorm borders on the intrusive (Figure 11). Dressed almost identically, portrayed half facing, with one hand on their laps, the other holding a long drink in a non-familiar way, they mirror each other, yet do not seem very close.

Figure 11. Women with drinks, probably in the women's barracks in Nybyen, Longyearbyen. Photo: Tromsø University Museum – Herta Lampert, id: tsnf57683

In this shallow pictorial space, the women are cropped at the knees and pushed towards the spectator. Personal details like the purse hanging in the upper corner, and the bedspread betraying the sleeping place and the pillow underneath, brings us even closer. The tapestry, however, is a liberating piece of kitsch which allows the gaze to escape. Representing a more fertile landscape than the Arctic tundra outside their barracks, such romantic motifs were popular in workers' homes during this period.[28] The digitized photograph allows us to zoom in on the tapestry, and we can see how the owner of the room has hung her wristwatch on this textile fabric, at the head end of her bed. She has arranged her belongings in the space, as an extension of her bodily habits, and created support for her routines, to paraphrase parts of Iris Marion Young's definition of *home* (Young

2000, 62). Herta Grøndal's digitized image reveals details embedded in everyday life that are not meant to be noticed in a photographic representation.

Home must be understood as far more than a private enclosure preserved for the conjugal family. To make a home in the Arctic, in a kind of exile, or at least with an awareness of this home's temporality, seems to demand flexibility, adaptability and willingness to use, inhabit and be satisfied with those varied spaces that are actually offered. Home is not a given entity, but is rather to be understood as the "production of place", according to Felski and others, not a "fixed point in space, a firm position from which we 'proceed' […] and to which we return in due course" (Felski 1999/2000, 22).[29] The woman in the spotted dress may represent our culture's inherent image of *home*, and the governor's wife, Liv Balstad, also writes in her memoirs of how her dreams of such a "real home" came true around 1950 when their new residence was completed (Balstad 1958, 29 and passim). But, to most people in Longyearbyen during the 1950s and 1960s, this kind of domicile was not achievable. Most of the men and women working in Longyearbyen lived packed together in barracks with at least one other roommate. Their homes were not private.

Herta Grøndal's photographs encourage questions about living conditions as well as social mobility between areas in Longyearbyen, on the basis of class and gender. She herself seems to have had access to both male-dominated places and women's spaces, as a participant and as an observer using her camera to document everyday life. It is interesting to compare her motifs with other photographs from the same period, particularly those taken by Erling J. Nødtvedt, pastor of Svalbard during the period 1951–60. His high-quality photographs are now part of Svalbard Museum's collections, and they serve as an important historical source of knowledge about Svalbard society during this decade. The motifs in Nødtvedt's collection are quite similar to those of both Grøndals: Leif and later Herta. They depict annual rituals and everyday life, the mining industry and its modern constructions. But Nødtvedt's own family, his workplace and home, are obviously powerful forces in his photographing practice. Herta Grøndal does not seem attached to home and family life in the same way. Her motifs demonstrate a wide radius and entry into a diversity of social spaces.

Herta Grøndal had access to the mine, to Stormessa, to the workers' barracks in Sverdrupbyen and Nybyen, and to the town's nightlife. Her short period as an employee of the governor's family during the 1950s, later on as organist in the church, and as a teacher at the school, provided her with a wide range of contacts and social contexts. She was given access to the Russian settlements and was allowed to take pictures there during the cold-war period, although under strictly controlled conditions. In addition, she joined hunters, scientists and groups of tourists travelling around Svalbard (Grøndal 2008). Moreover, her official photographic missions from the 1960s onwards granted her entry into places and occasions usually reserved for officials. Her immense collection of photographs reveals that she possessed the freedom to move between different social, gendered and geographical spaces.

Photography of habit

During the 1950s, life and work in Longyearbyen were quite well documented by SNSK's photographers, first and foremost the Grøndals, but also by a number of amateur photographers. "I bought a Christmas gift for myself, a camera, Kodak Retina 1B," a former employee of SNSK recorded as an important memory from 1955. The price was 465 crowns, he remembered, "a lot of money at that time when my salary was around 1900 crowns a month" (Østlund 2016)[30]. The affordability of new cameras and their small size made them a natural part of people's equipment on trips and vacations, and from the 1960s onwards the practice grew in accordance with the increasing number of individuals owning a camera, whether they were temporary settlers in Svalbard or people travelling there as tourists. Pictures were taken on special occasions to show friends and family what life in the Arctic was like, and to create memories that would last a lifetime. There is quite an extensive collection of these historical photographs in Svalbard Museum documenting the 1950s and the following two decades. In addition, there are clearly a great number of pictures in private albums, to judge from the large amount of sharing on social media.[31]

Although the imagery changes somewhat from the 1950s to the 1970s, the resemblances in the choices of photographic subjects are striking. Those equipped with cameras obviously participated in the same rituals and inhabited the same places; their pictures confirm a common grounding in time and space. Worth mentioning again is Pastor Nødtvedt's photographs recorded during the 1950s. Even though a lot of his photographs are of his own family, he also depicted public life, the miners and their workplaces.[32] Another example is the Norwegian miner Roald Amundsen, who took a number of photographs of everyday life during the first half of the 1950s. These seem very much like Herta Grøndal's on the basis of their motifs, as well as the relation they show between photographer and his/her subject. Both depict ordinary people – either dressed up for portraits, as parts of documentary series, or as faces in snapshots – and the habitual life of the Norwegian settlements in Svalbard during the mid-twentieth century.[33]

Obviously, both the Grøndals and the amateur photographers play a part in revealing and making explicit unconscious and unrecognized aspects of everyday life in these Arctic settlements. They direct our attention towards unnoticed phenomena, thus making them significant both for the contemporary spectator and for posterity. These pictures of everyday life expose *habitus*, which may be understood as the deeply embedded sets of largely unconscious dispositions that Pierre Bourdieu tells us "cannot be touched by voluntary deliberate transformation, cannot even be made explicit" (Bourdieu 1977, 94). Bourdieu nonetheless recognizes the potential in photography and photographic practice to uncover and illuminate aspects of habitus (Sweetman 2009, 491 and passim). And Felski reminds us that habitus is not simply understood as actions, but also as attitudes, as a "way of experiencing the world" (1999/2000, 26, 31), thus determining certain practices and informing even the smallest of actions and gestures.

Many of Herta Grøndal's photographs of everyday life are so ordinary, and some of their details so close to life, that they are hardly noticeable. However, none of the photographs should be mistaken for being pictures of "the 'really real'" (Felski 2002, 607). Bourdieu's sociological perspective usefully reminds us of the necessity not only of detecting the meanings which the photograph *claims* (the photographer's intentions included), but also of interpreting the redundancy of meanings which it *reveals* by being part of the visual culture at a certain time (Bourdieu [1965] 1990, 7). Thus, both snapshots and, in particular, composed and arranged motifs stand out as redundant and complex objects of analysis.

The picture of a woman cultivating flowers in a cold frame (Figure 12) may illustrate such complexity. She has probably been asked by the photographer to squat beside her home-made frame, both to show ownership and to fit into the composition. Her colourful apron matches the hardy pansies, and her knitted cardigan harmonizes with the snow-flecked mountain backdrop, thus contrasting with the plastic sheet and muddy foreground.

Figure 12. The wife of Mining Inspector Johnsen challenges climatic conditions with her pansies. The photograph is shot in Longyearbyen sometime between the mid-1960s and the mid-1970s. Colour photo: Tromsø University Museum – Herta Lampert, id: tsnd41968

The woman's pose, gesture and smile present us with a kind of holy shrine in the cold Arctic summer, a strong will towards growth, against all the odds, and the human devotion to beauty.[34] This motif may seem ordinary, but in its Arctic context, where sprawling flowers are sparse, it is exceptional.

First and foremost, this picture seems to confirm Felski's view of the concept of everyday life as vital and innovative, rather than something negative. Felski is critical of analyses that see women as representatives and "victims" of everyday life, as suggested in Henri Lefebvre's influential analysis in *Everyday Life in the Modern World* (1968). According to Lefebvre's understanding, women are tied to their homes, to repetitive tasks and natural cycles, to the unreflected routines of daily life – that is to say, to "the realm of the insignificant, invisible, yet indispensable" (cf. Felski 1999/2000, 17). The photographs of Herta Grøndal provide us with a different way of seeing and thinking about everyday life in the Arctic settlement. Several of her pictures illuminate the necessity to really *invent* everyday life in the Arctic during the post-war years and thus they demonstrate the vitality of the everyday. However, we may ask whether motifs like this should be recognized as everyday, or as occasional – also because the pictures reveal the possibility of women's "aesthetic" activity, which is the privilege of only a few. The woman who cultivates flowers in the Arctic summer is neither insignificant nor invisible, but her activity goes beyond necessity. So does Herta Grøndal's photograph of her.

Memory machine

The photographs taken by Leif and, later, Herta Grøndal occupy a dominant position within the visual culture of Longyearbyen during the 1950s, 1960s and 1970s. Looking into Herta Grøndal's archives, the large quantity of never-published pictures reinforce the position she worked so hard to occupy, as the visual documentarian of Svalbard throughout these decades. Several of her pictures were disseminated to and by her contemporaries, they were bought as portraits, framed to hang on walls, or used as postcards. The inhabitants of this small settlement in the Arctic were at the beginning of a new visual era in which they were able to see themselves through photographs. A literal example is a postcard from the Norwegian Polar Institute's picture archive, showing Herta Grøndal's prospect of Longyearbyen seen from the mountain Sarkofagen. The sender has marked his home with an X and written: "Here we live on the other side of the card."[35] Such pictures helped to form the inhabitants' self-awareness and self-understanding, thus giving them the potential not just to confirm, but also to critique, their own rituals, homes and habits (cf. Sweetman 2009, 494–498 and passim). Herta Grøndal's photographs, especially her postcards, have also played a pivotal role in shaping our general image of Svalbard, and as such they have contributed to Arctic iconography.

However, most of the photographs of Svalbard's everyday life have never been seen prior to this digitized collection, and the large number of pictures of people, both formal portraits and snapshots, that have now been published, make the digitized Grøndal archive and the Lampert collection a "memory machine". Even photographs of seemingly lesser importance are becoming part of the visual culture of Svalbard. In combination with photo sharing on social media, these images have the potential to function as reinforcements of a collective identity: not as milestones in a diachronic history, but rather as an episodic recollection of the past. This is partly due to how Herta Grøndal saved her photographs, in a very haphazard fashion. Many strips of negatives were cut apart and placed in different envelopes and folders.[36] In addition, photographs that obviously belong in the same series are located in various different photographic collections or archives. This makes it a challenge to interpret individual images. The subsidiary images contained on the same strip of negatives might offer another angle or display elements that were hidden in the main image chosen. Accepting, however, that the focus on a singular image removed from its sequence may very easily lead to interpretational error, we can nevertheless explore each photograph as an image *per se*, overriding representation, cultural knowledge and official narratives. This requires visual analyses that are tentatively blind to historical facts, and do not seek simply to confirm already-established truths.

The burning question in studying documentary photographs such as Herta Grøndal's, is to what degree the photograph's real meaning is connected to time, place and memory; for example, in the picture of the woman in the spotted dress. Might today's viewers claim that these photographic images carry meaning even without historical and local knowledge of "there and then"; that is, of Svalbard during the decades between 1950 and 1980? This question is highlighted in different ways in theories about photography's dual nature of being an imprint of historical realities and at the same time constructions of worlds, and it is highly relevant when it comes to the study of large collections, such as Herta Grøndal's.

The basic hypothesis of this article – that a female photographer taking pictures of women and children contributes to altering the dominant view of Svalbard life as harsh and masculine – becomes far more complex after surveying Herta Grøndal's thousands of images. It is obvious that the negotiation of gender implies the involvement of both sexes, and her images seem in many cases to get closer to the men's lives than the women's. Herta Grøndal's images of both men and women display a diversity of habits and manners, as well as a multitude of possible roles and identities. Her photographs reflect the dialogue taking place not only between the photographer and the captured subject, but also between the society and the social systems represented by it. "In this dialogic confrontation by means of gaze, gesture, and cameras there is room for agency," as Elizabeth Edwards puts it (2004, 12), thus actualizing the picture as more than mere representation or documentation of historical situations. Herta Grøndal's mediation of the repetition, homes and habits of Longyearbyen leaves the photographic image, not as a neutral medium, documenting "facts", but as a contested and often paradoxical site of gender, identity and modernity.

References

Amundsen, Birger. 2008. "Evighetens Herta" [The eternal Herta]. *Svalbardposten*, Mar. 7.

Arlov, Tor B., and Arne O. Holm. 2001. *Fra Company town til folkestyre. Samfunnsbygging i Longyearbyen på 78 grader nord* [From company town to democracy. Building the Longyearbyen community at 78 degrees North]. Longyearbyen: Svalbard Samfunnsdrift.

Balstad, Liv. (1955) 1958. *North of the Desolated Sea*. Adapted by Alice Hope, translated by Joan Bulman. London: Souvenir Press.

Bourdieu, Pierre. 1977. *Outline of a Theory of Practice*. Cambridge and New York: Cambridge University Press.

—. (1965) 1990. *Photography: A Middle-Brow Art*. Translated by Shaun Whiteside. Stanford, CA: Stanford University Press.

Dagen (1989). "Et lite stykke Norge." [A little piece of Norway], 20 Oct. 1989.

Edwards, Elizabeth. 2004. "Thinking materially/Thinking relationally." In *Getting Pictures Right: Context and Interpretation*, edited by Michael Albrect, Veit Arlt, Barbara Müller, Jürg Schneider, 11–23. Köln: Rüdiger Köppe Verlag.

Edwards, Elizabeth, and Janice Hart. 2004. "Introduction – photographs as objects." In *Photographs Objects Histories: On the Materiality of Images*, edited by Elizabeth Edwards and Janice Hart, 1–16. New York and London: Routledge.

Evjen, Bjørg. 1995. "An Arctic Society – A Laboratory: Longyearbyen, Svalbard, 1916–1975." *Acta Borealia* 11 (1): 73–81.

—. 2006. *Kull, karer og kvinnfolk. Longyearbyen 1916–75* [Coal, men and women. Longyearbyen 1916–75]. Oslo: Unipax.

Felski, Rita. 1999/2000. "The Invention of Everyday Life." *New Formations* 39: 15–31.
—. 2002. "Introduction." *New Literary History* 33 (4): 607–622.
"Fotograf Herta Grøndals reiser til Svalbard – En kvinnelig pioner." [Photographer Herta Grøndal's travels to Svalbard.] 2006. *Skinn* 1: 4.
Grøndal, Eva. 2008. "Fotograf Herta Grøndal" [Photographer Herta Grøndal]. *Polarboken 2007–2008*, 106–112. Oslo: Norsk Polarklubb.
Gustad, Ragnhild. 2008. "Herta Grøndal tok sitt første bilde av øygruppa i 1952" [Herta Grøndal Shot Her First Picture of the Archipelago in 1952]. *Nordlys*, Mar. 17.
Hagevik, Ellen Marie. 1992. "Herta – sommerfugl på vinterbesøk" [Herta – Butterfly on Winter Visit]. *Svalbardposten*, no. 10.
Larsen, Peter and Sigrid Lien. 2007. *Norsk fotohistorie. Frå daguerreotypi til digitalisering* [Norwegian history of photography. From daguerreotypi to digitization]. Bergen: Det Norske Samlaget.
Lokvam, Kjell. 1965. "Modig wienerinne i polarøde" [Brave Viennese in the Polar desolate]. *Kvinner og klær*, Feb. 3.
Mayor, A. Hyatt. 1946. "The Photographic Eye." *The Metropolitan Museum of Art Bulletin: New Series* 5 (1): 15–26. doi: 10.2307/3257398.
Moe, Kirsti. 1999. "Copyright Herta Grøndal, Spitsbergen". *Svalbardposten* no. 9.
St.meld. 1999–2000. No. 9, "Svalbard."
Svarstad, Asbjørn. 2006. "Kald kjærlighet" [Cold love]. *Memo*, May 4.
Sweetman, Paul. 2009. "Revealing Habitus, Illuminating Practice: Bourdieu, Photography and Visual Methods." *The Sociological Review* 57 (3): 491–511. doi: 10.1111/j.1467-954X.2009.01851.x
Østlund, Andreas. 2016. "60-årsminne – høsten 1955" [The 60-year remebrance – Autumn 1955]. *Svalbardposten* no. 2, Jan.

Other Sources

The Herta Lampert photo collection at Tromsø University Museum
The photo collection at Svalbard Museum
Archives of Biography and of Pictures at the Norwegian Polar Institute
Interview with Eva Grøndal, daughter of Herta and Leif Grøndal and owner (with Nora Grøndal) of Grøndal Photo, Apr. 29, 2015; correspondence Feb. 13, 2017
Interview with Sveinulf Hegstad, photo archivist at Tromsø University Museum, Aug. 31, 2015

E-mail correspondence with Herdis Lien, curator of Svalbard Museum, Nov. 16, 2015

Correspondence with Tone Nødtvedt, digitizer of Nødtvedt collection, Dec. 21, 2016; Feb. 13, 2017

Facebook, "Gamle Svalbard" ["Old Svalbard"] (public group, 5634 members as of Dec. 22, 2016)

Notes

[1] There are at least three more pictures of the woman in the spotted dress, one in the opposite corner of the living room, playing her guitar, another a classic posed family portrait with an infant on her lap and a well-dressed boy on the rocking horse. These pictures have no metadata. However, one copy (SVN-F 00071.tlf) exits in Svalbard Mueseum's collection together with pictures labelled "Kings Bay". It will perhaps be possible to identify time, place and person, but that is not my intention in this article. However, some of the information that I have come across is included in the body text, captions and notes.

[2] One of these pictures (Tromsø Museum – Herta Lampert, id: tsnf56210) was reproduced in a feature article on Herta Grøndal in the women's magazine *Kvinner og klær* in 1965 (Lokvam 1965).

[3] According to the Norwegian Polar Institute's biographical archive: Lokvam 1965; *Dagen* 1989; Hagevik 1992; Svarstad 2006; "Fotograf Herta Grøndals reiser til Svalbard – En kvinnelig pioner" 2006; Gustad 2008.

[4] The photographs from the Russian settlements seem to be divided between at least two archives. My main focus in this article is on the pictures from Longyearbyen.

[5] This picture was taken with a Hasselblad or Rolleiflex, and it is quite similar to the iconic picture taken with a Linhof Technika 13x18 diapositiv in 1966 (Grøndal collection). The vantage point is the tip of Mine 1b, which is exactly the same place from which Leif Grøndal shot his prospects of Longyearbyen some years earlier. It is sometimes difficult to identify the photographer (according to an interview with Eva Grøndal). Herta Grøndal herself, however, seems to have been very careful and concerned about the ownership of the motifs.

[6] Leif Grøndal probably gained his first assignment as a photographer in 1949 to cover the crown prince's visit to Svalbard. Leif also worked together with his brother Harald. Leif left Svalbard in 1961. His photographs are deposited at the Svalbard Museum.

[7] Female photographers were not uncommon, and photography was seen as a suitable profession for women. By 1910 there were 118 female photographers registered in Norway (Larsen and Lien 2007, 49).

[8] During the 1950s, Herta Grøndal used a Voigtländer Vito, a Zorki (a Russian camera), and later on a more expensive Rolleiflex camera, medium format 6x6 cm. She also borrowed a large format Linoftecnica 13x18 cm from her husband for shooting diapositives. From the end of the 1970s she used 135 format (Contax

camera). Additionally, she used Hasselblad 6x6 during the 1960s and 70s (and actually until 2008) as well as colour positive film. In later years, she used a digital Nikon camera, and the last time she was visiting Svalbard she brought with her "a Leica-camera, a Contaflex, and the Hasselblad." (Amundsen 2008). Some images were printed as postcards, see https://www.grondalfoto.no/postkort_s_h/index.htm. Interview Sveinulf Hegstad, TMU, Aug. 31, 2015, correspondence with Eva Grøndal Mar. 13, 2017, and various sources.

[9] Information from Hegstad 2015.

[10] Herta Grøndal moved back to Vienna, remarried and gave birth to a son.

[11] The Grøndal collection contains approximately 11 000 negatives, the pictures in the SNSK archive run to about 15 000 digitized images, plus some still not digitized. These collections contain contributions from various photographers, including Herta Grøndal (information from Herdis Lien, curator at Svalbard Museum, Nov. 16, 2015).

[12] Her panorama hung on the walls at SNSK's offices and other private and public institutions in Longyearbyen, in the church as well as on the pub, and in private homes, cf. Svalbard Museum's photo collection SVF 06549.jpg. Today Herta's pictures are blown up on walls in both *Huset* and the recent opened restaurant *Gruvelageret*.

[13] http://www.unimus.no/foto/#/search?q=lampert

[14] In the period from the 1950s to the end of the 1980s, the ratio between Russian and Norwegian citizens was quite stable at approximately 2:1, according to Statistics Norway, available at http://www.ssb.no/a/histstat/svalbard/tab3.html. The biggest Russian settlement after the war was Grumant, but this mining town was abandoned in 1961–1962.

[15] In the second part of the 19502, approximately 180–190 women and children were living in Longyearbyen, i.e. 13–14% of the total population. In 1960–61 the number of women and children was 161, out of a total of 702 inhabitants (Evjen 1995, table 4).

[16] The sociologist Ferdinand Tönnie's distinctions between *Gemeinschaft* and *Gesellschaft* (1887), generally translated as *community* and *society*, is highly relevant to Svalbard society of this period.

[17] Two people have identified themselves in this picture, and dated the photograph to 1953/54. Facebook group "Gamle Svalbard", posted by the author Jan. 6, 2016.

[18] There are more pictures shot by Leif Grøndal, most likely during the same mail delivery (cf. Grøndal Photo Collection and information from Eva Grøndal; cf. Svalbard Museum, SVF 08059.tif).

[19] The first postal flight to Svalbard was in the late 1940s. The small Catalina aircraft owned by the Norwegian Air Force did not land, but dropped its parcels on the tundra in the Advent valley, near Longyearbyen. In 1954 Leif Grøndal took a picture of this moment, and Herta was on it, amongst the crowd of curious settlers, cf. Svalbard Museum SVF 03125. On its way home, this plane crashed at Bear Island and two postmen were killed in the accident. The air strip in the Advent Valley was built by the Germans during World War II, but was not used after the war. The first Catalina aircraft landed in 1958. The first aircraft with passenger

landed on Apr. 2, 1959. Erling J. Nødtvedt depicted the event, Svalbard Museum, SVF 11937b. During the 1950s, the arrivals of post to Svalbard were announced on national radio.

[20] The Norwegian original, *Nord for the øde hav*, was published in 1955. Cf. Anka Ryall's chapter in this book.

[21] There exist more, and older, photos of miners on slope-trains. A photograph from the Svalbard Museum Collection, SVF 00180, shows 15 men with coal-black faces getting off a shift in Advent Bay, taken some time before an explosion in 1920. The picture is used in Svalbard Museum's exhibition to illustrate the region's coal mining.

[22] This picture was posted by the author on the Facebook group "Gamle Svalbard" Apr. 11, 2016; it received 172 likes, but nobody recognized the people in it. However, another picture from the very same point of view was taken about the same time by Erling J. Nødtvedt, pastor of Svalbard 1951–60. Nødvedt's picture shows his wife and children, together with another family, on the slope-train heading for the mine. According to the caption, this is a leisure-time activity: "It was apparently the first time there was a stroller up the mine" (See Svalbard Museum's Photo Collection, SV 08176, among several). The smiling family members on the trolley somewhat disturb Grøndal's photograph of the men and miners as a representation of everyday life in Svalbard.

[23] During the 1960s, the variety of workplaces for women increased, but all within care and service. "From 1965 women were also employed as a hairdresser and photographer," Evjen writes (Evjen 2006, 64; my translation), probably referring directly to Herta Grøndal.

[24] Stormessa was renovated in 1961.

[25] Pastor Samson Vik held classes in traditional decorative painting and his spouse Berit Vik held classes in porcelain painting in the sitting room of Svalbard Church during the years between 1963 and 1973. Facebook group "Gamle Svalbard", posted by Knut Mellerud Jan. 22, 2016.

[26] Tromsø Museum – Herta Lampert, id: tsnf56773. From the workers' barracks probably in Longyearbyen in the 1970s.

[27] The men are Einar Sletbak and Terje Ovlien.

[28] Cf. Svalbard Museum's photo collection, for example SVF 13762.tif.

[29] Felski is referring to Agnes Heller.

[30] Translated from Norwegian: "Jeg kjøpte julegave til meg selv, et fotoapparat Kodak Retina 1B som kostet 465 kroner, mye penger på den tiden da min lønning var på cirka 1.900 kroner i måneden."

[31] New digitized photos and albums are constantly being published on the Facebook group "Gamle Svalbard".

[32] Nødtvedt's collection, given to Svalbard Museum, consists of more than 3600 recordings from the 1950s, all of them digitized.

[33] Amundsen's album of negatives has been donated to Svalbard museum. Another 1950s photographer was the governor's assistant, Carl A. Wendt, who is represented by 456 pictures in Svalbard Museum's photo collection. Most are landscapes, but also included are some pictures from the governor's estate and

from the Russian settlement Grumant. *Svalbardposten* has published some high-quality photographs of life and work in Longyearbyen shot by Felix Eckert, a German student of mining, who worked in the mine during 1954–55. *Svalbardposten* nos. 24 and 28, 2013. http://svalbardposten.no/index.php?page=vis_nyhet&NyhetID=3548. Several German and Austrian mining students held internships at SNSK in connection with their studies. Apart from Herta Grøndal, only a couple of women – Cathrine Bjelland and Synnøve Margrethe Hansen – are registered as photographers in the Museum's photo collection. The photographic culture in and of the Soviet settlements during these decades remains to be explored.

[34] The picture was used as the cover illustration of the book *Hager mot nord. Nytte og nytelse gjennom tre århundrer* [Gardens towards the north: Utility and beauty through three centuries], edited by Ingebjørg Hage et al. (Stamsund: Orkana Akademisk 2015). See also Pastor Nødtvedt's photographic studies of the green and blooming window in his family's apartment in the church during the late 1950s. Svalbard Museum's photo collection, SVF 10651, among several. One of these pictures was printed in the Christmas issue of the magazine *Alle Kvinner* [All Women] in 1960 in an article based on a letter from Ranveig Nødtvedt, "Mitt blomstervindu mot polarnatten" ["My flower window towards the Polar Night"].

[35] Norwegian Polar Institute's archive of pictures. The post card is signed "E".

[36] According to an interview with Hegstad in 2015.

TRACING THE ARCTIC CRIME SCENE

CATHRINE BJERKNES

The Arctic is currently a subject of political importance, and is commonly portrayed as a place of struggle and conflict.[1] Moreover, the areas north of the polar circle have increasingly been portrayed in popular literature, where themes such as social problems, climate change and the exploitation of resources have been problematized. The Arctic has been especially popular within genres such as climate fiction (cli-fi) and crime fiction, in which the place itself plays a prominent role or the detectives are indigenous. There has, however, been little research on how the Arctic is portrayed in these works, and this chapter begins to fill the gap by stimulating a discussion on how popular literature or, more specifically, crime fiction, presents the Arctic. This is particularly important since the readers of these novels have rarely visited the Arctic themselves and therefore only know the region in its mediated and not its real state. The impression that readers form is thus shaped by the media, and especially popular culture. Consequently, the literary Arctic presented in novels influences how it is understood as a place and an idea, at the same time as new representations are shaped by contemporary changes in the region.

A persistent discourse over the last 200 years has been the Arctic as a place where human interaction with the landscape is marked by heroic masculinity. In such heroic discourses, the Arctic landscape is generally perceived as desolate and remote, and as feminine, to be conquered by the male heroes (Hansson 2010). This notion is gradually being contested within the mass market; for example, through crime novels featuring strong female characters who interact with the landscape instead of conquering it, and who participate in populated communities, either as visitors or natives.

The crime genre offers myths connected to the experience of modernity, since its inherent interest in society necessarily involves an exploration of the experience of modernity (Thompson 1993, 8). In a globalized world influenced by movement and transcultural exchange, the criminal landscape changes as well. This influence is also visible in crime novels set in the Arctic. In the wake of what is usually perceived as the global world's entry into the northern territories, the portrayed Arctic communities struggle with clashes between tradition and modernity (see for example Ekman 1993; Høeg 1993; Pettersson 2012, 2014, 2016; Truc 2015). But notions of tradition and modernity in the North have a tendency to be oversimplified, especially within the crime genre's limited conventions. Accordingly, these tendencies respond to myths about the Arctic rather than the reality of the numerous and contradictory areas north of the polar circle.

In this chapter, I examine *White Heat,* a crime novel published in 2011 by the British author Melanie McGrath, under the name M. J. McGrath. In an interview, McGrath suggests that she has quite a static understanding of the Arctic when she claims that "apart from the melting of the ice caps, which will impact on all of us, the Arctic is the last place of innocence in the world" (McGrath 2007, 6). In associating the Arctic with untouched nature and lack of progress, McGrath seems to be a naïve victim of touristic stereotypes. However, the novel is more complex than this statement implies and offers a multifaceted representation of the Arctic. As emphasized by anthropologist Hugh Brody, the Inuit "are not representatives of an ancient world that is slowly and painfully sinking beneath waves of modernity. Rather they are our contemporaries. They have lived, learned, and adapted for as many millennia as all other human beings" (Brody 1987, xiv). In other words, Arctic society has been exposed to the same processes of change and negotiation as the rest of the world, and must be understood through the same concepts.

Instead of focusing on the traditional conflict between modernity and the pre-modern world as a starting point for understanding the representation of the Arctic in *White Heat*, I use hybridity as a lens. The novel has a part Inuit, part white main character – Edie Kiglatuk – and hybridity is one possible entry into the text's complex depiction of her and the culture and life world she inhabits. Such a perspective can open up space for multifaceted and dynamic interactions between notions of tradition and modernity, facilitated by the hybrid characters in the novel, and can enable us to explore how the representation of the Arctic in *White Heat* reflects a contemporary reality, and not necessarily a collision between an old, unchanged world on the one hand and a new, complex and constantly changing world on the other.

In addition, I discuss the novel's portrayal of encounters between insiders and outsiders in the light of Frantz Fanon's Manichean allegory, an illustration of the dualist power relation between binaries such as good and evil in texts about natives. Even though Edie herself represents an encounter between Inuit and western values and mindsets, as well as between masculine and feminine norms, which is both complex and destabilizing, this complexity is not necessarily sustained when the encounters are between different characters. Together with the notion of hybridity, the Manichean allegory makes it possible to understand how *White Heat* both challenges and reinforces dichotomies and stereotypes.

Crime fiction with a historical background

Five years before publishing *White Heat*, McGrath, who was then primarily known as a travel writer, wrote the non-fiction book *The Long Exile* (2006). This explored the consequences of the Canadian government's forced relocation of 19 Inuit families in 1953, from the east coast of Hudson's Bay to the wasteland of Ellesmere Island further north. The government was hoping that an Inuit presence on Ellesmere Island would discourage Greenland, Denmark and the USA from claiming it. The Inuit families were assured that they would be able to go back to Hudson's Bay after three years, and they were promised better conditions and better supplies of animals and food than they would have if they stayed. These promises were never fulfilled. Instead, the Inuit families were abandoned in an alien and unknown environment. Forty years later, in 1993, the Royal Commission on Aboriginal Peoples asserted that the relocation was "one of the worst human rights violations in the history of Canada" (Government of Canada 2010). The government, however, continued to state that the Inuit families had been moved for humanitarian reasons and had volunteered to go (Stevenson 2006, 19). Not until August 18, 2010 did a public apology appear, cited on the Government of Canada's website: "We would like to express our deepest sorrow for the extreme hardship and suffering caused by the relocation," said then Minister of Indian Affairs and Northern Development, John Duncan. "We deeply regret the mistakes and broken promises of this dark chapter of our history" (Government of Canada 2010). When McGrath wrote her documentary account, this apology had not yet been published.

With this governmental wrongdoing as a backdrop, McGrath's documentary is also an account of Inuit relations with whites throughout modern history: their encounters with whalers, fur traders, missionaries and scientists. In an interview related to the publication of *The Long Exile*, McGrath stated that she had begun writing a novel set in the Arctic "because I realised how much [the Arctic] had impacted on me" (McGrath 2007). That novel is *White Heat*, and by employing the popular genre of crime fiction it was possible for McGrath to reach a larger audience. In *White Heat* she draws on her research for *The Long Exile*, and the novel can thus be read as an extension of this non-fiction project, especially since *White Heat* mainly takes place on Ellesmere Island, where the imagined descendants of the relocated families still live, in a fictional community of 70 inhabitants called Autisaq. The novel also fictionalizes some of the facts reported in the documentary. Specifically, the documentary follows Josephie Flaherty – the son of the filmmaker Robert Flaherty – and Maggie, the Inuit lead actress of his film *Nanook of the North*, while the main character in *White Heat* is the daughter of an unidentified white film aficionado and an Inuit woman called Maggie.

Establishing an Arctic crime scene

The title *White Heat* can have several meanings, such as a white glow, intense passion, wrath or energy, or a temperature so hot that it gives off a white light. The term is an oxymoron, a self-contradictory antagonism: the colour white is normally cold, but here it is warm; warmth is often associated with the colour red, but here it is white. White is also a metonym for snow and, as we know, hot snow melts. In this way, the title points to the Arctic landscape itself, suggesting climate change in the Arctic. Another meaning of the word "heat" is police, which leads the title to suggest a white police force, indicating white superiority over the Inuit. In the novel, the police are represented by the police constable Derek Palliser, who works for the High Arctic Police Force, but who is not necessarily superior to the Inuit or a representative of power. He is part Inuit himself and represents neither force nor whiteness, but seems to be trapped in a static hybrid position out of which nothing new emerges.

Edie is in her early thirties, a former polar bear hunter and a dried-out alcoholic now working as a teacher in the local school in addition to taking tourists (mostly white men) on guided trips. The book starts *in medias res* with Edie on one of these trips with two Americans, Felix Wagner and Andy Taylor. The story is narrated in the third person, primarily from Edie's perspective, and there is a conspicuous absence of authorial distance. Through free indirect discourse, the narrator's knowledge of Edie's psyche coincides with Edie's self-knowledge. The reader has immediate access to Edie's scepticism concerning the outsiders:

> Edie Kiglatuk mulled over why it was that the hunting expedition she had been leading had been so spectacularly unsuccessful. For one thing, the two men she was guiding were lousy shots. For another, Felix Wagner and his sidekick Andy Taylor hadn't seemed to care if they made a kill or not. (McGrath 2011, 1)[2]

This statement implies that Edie suspects the Americans are not ordinary hunters but have a hidden agenda. This leads her to speculate on their strange behaviour: "Maybe it was just the romance of the High Arctic they were after, the promise of living authentically in the wild with the Eskimo, like the expedition brochure promised" (1). It soon becomes clear that there is a gap between such touristic stereotypes, or the Americans' alleged expectations, and "real life" in the novel, brutally demonstrated when Felix Wagner is shot soon afterwards, a death the elders in the community quickly dismiss as an accident. Shortly after this, Andy Taylor comes back with another American, Bill Fairfax. Edie guides Fairfax on a problem-free trip, while her stepson Joe takes Taylor on an expedition that has a tragic outcome when Taylor disappears and Joe arrives in Autisaq in a state of delirium. A few days later, Edie finds him dead in his bed with a stack of pill boxes beside him. From this point on, she throws herself into a hunt for the truth, since she is convinced that Joe has not taken his own life as the authorities in Autisaq believe.

Edie's life is centred on a few important people, primarily her stepson Joe and Joe's father Sammy – her ex-husband. Edie's aunt Martie is also an important character, tough as Edie, and "large, at least by Inuit standards, with skin the colour of an heirloom suitcase and a voice like a cartoon train wreck" (19). Martie lives alone and occasionally works as a local pilot – when she is not drinking. As the murder plot starts to gather pace, Constable Derek Palliser – who lives a short flight away in the fictional village Kuujuaq – becomes one of Edie's very few allies, even though she is the one who unravels the mysteries surrounding the murders.

Reading hybridity

Although Edie is the daughter of an Inuit mother and a western father, "[i]n her traditional garb you wouldn't even know she had a *qalunaat* [non-Inuit] for a father. The small, slightly fierce woman staring back at her [in the mirror] looked one hundred per cent Inuit and she liked it that way" (112). She identifies first and foremost with her indigenous background, having been born and raised in an Inuit community, but she cannot fully escape her western side. As Kirsten Thisted (2008) points out in an article about Peter Høeg's novel *Miss Smilla's Feeling for Snow,* genes play a remarkable role, as in Danish literature about Greenland in general. Even though Smilla – who has an Inuit mother and a Danish father – has hardly spent any time with her father, her father's genes become apparent in her behaviour as early as the age of five (Thisted 2008, 334). This is also the case in *White Heat*, where Edie's *qalunaat* genes influence her in various ways, even though her father left her and her mother when Edie was only six years old.

Hybridity is a much-used concept within postcolonial theory, where it has been used to interrogate how human identity formations are shaped by the conflicts between the wider modern society and the indigenous population, gender conflicts and other important conflicts. Originally stemming from biology, "hybridity" signifies the cross-breeding of plant or animal species (Friedman 1998, 83). In postcolonialism, the term refers to different forms of cultural mixing and interactive exchange (Friedman 1998, 83), referring to the relationship between colonizer and colonized and suggesting a transference between them or from one to the other. Susan Stanford Friedman (1998) refers to three basic approaches to the politics of hybridity. The first regards hybridity mainly as an effect of forced oppression by a dominant group over a less powerful group, causing forms of deculturation or assimilation. The second mode suggests that hybridity destabilizes authority and displaces binaries, and that the hybrid "disturbs, intervenes, unsettles, interrogates" (Friedman 1998, 89) and so on by refusing to adapt to established categories. Whereas the first position locates agency in the dominant power and the second focuses on the hybrid dissident, the third position calls for a "locational thick description" where power flows "multidirectionally in the contact zone instead of unidirectionally" (89). In the last passage, Friedman adapts Mary Louise Pratt's term "contact zones", which Pratt defines as "social spaces where cultures meet, clash, and grapple with each other, often in contexts of highly asymmetrical relations of power, such as colonialism, slavery, or their aftermaths as they are lived out in many parts of the world today" (Pratt 1991, 34).

Homi K. Bhabha understands hybridity as transgressive. He discusses it in relation to the term "third space", a position that enables the articulation of in-between identities and positions which are "*neither the One [...] nor the Other [...] but something else besides,* which contests the terms and territories of both" (Bhabha 1994, 41, original emphasis). In addition, Donna Haraway's cyborg figure may be useful in a discussion about modernity and hybridity. Her cyborg manifesto was written in 1985, in an attempt to relate to a society that evaded categorization in modern terms. As Anne-Jorunn Berg points out, one of modernity's fundamental ideas is a notion of either/or, rendering mixed forms dirty and unwanted (Berg 2014, 113). Against this backdrop, Haraway introduced the cyborg as a figure that could disturb and interrupt. The cyborg is located in the connecting zone between the human and the non-human, the organic and the non-organic, between nature and culture. It also unites threads that are usually kept apart and perceived as essentially different (Berg 2014, 112). In this way, it is a hybrid figure that challenges dichotomies and disrupts binaries, and thus potentially transcends modern categorization.

For the purposes of interpreting *White Heat*, hybridity may be seen as something that displaces binaries or where power flows multidirectionally within a contact zone. McGrath presents Edie's personality as a hybrid blend of indigenous and western knowledges and traditions in a number of ways. Edie loves old American comedies (Buster Keaton and Charlie Chaplin) and wears a wristwatch, aspects that mark her as a modern North American individual. On the other hand, she feels at home in the Arctic, and prefers to be out on the land, aspects that are connected to her Inuit upbringing. But, in contrast to most Inuit, who find their identity as members of the group (Collignon 2006), Edie favours solitude. In this respect, she performs Inuitness in terms of a western ethos, and she also intermingles western and Inuit practices. This is illustrated, for example, by her way of drinking tea: "For herself, she preferred Soma brand English Breakfast, brewed with iceberg water, sweetened with plenty of sugar and enriched with a knob of seal blubber" (1). So, on the very first page of the novel, Edie is shown engaging in an inuitification of something purely English, as she translates tea-drinking into an Inuit practice. At the same time, the western influence on Edie is connected to modernity, as both wristwatches and movies are items that may represent modernity. The items may in themselves look innocent and neutral, but the role that modernity plays in the novel is never neutral.

Modernity and the Manichean allegory

Through Edie, the novel problematizes the relationships and differences between Inuit and *qalunaat*, or southerners, the changing landscape and the current status of the Arctic. This dual perspective allows both insiders and outsiders to be seen partly from the outside. But, as the author is not Inuit herself, the novel tells its story about the Inuit by presuming to speak for the other. This way of speaking for the Inuit no longer goes down well with the Canadian literary public. In Canada, the question of appropriating voice has led to heated debates since the 1980s, and for Native women the issue crystallized at the Third International Women's Book Fair in Montreal in June 1988. According to Laura Smyth Groening in the book *Listening to Old Woman Speak* (2005), "native women were concerned with the way non-Native versions of First Nations stories distort different cultures", and they requested Euro-Canadian authors like Anne Cameron to "move over" and provide the space "Aboriginal writers needed to define their own reality" (Groening 2005, 5). The appropriating of voice was considered to be a continuing invasion and violation by whites over the First Nations, and the debate grew over the following months (Groening 2005, 21). It is not accidental that it is a British author who is making use of this narrative mode in a novel not primarily written for Canadian readers.

This particular debate had a primarily gendered basis, and in feminist critique the debate has taken this same direction; articulating a concern about who can speak for whom, and whether being a woman confers the right to speak for any woman. In relation to McGrath's novel, the consequences of writing from an external perspective are more pertinent. McGrath has after all created a character who is only part Inuit, and therefore she does not fully ventriloquize in her novel. Groening attempts to analyse "specific problems embedded in the representation of Native people in non-Native writing" (Groening 2005, 15), and in the course of her analysis she introduces Frantz Fanon's term, the Manichean allegory. This allegory is inspired by Manichaeism, which bases its dualist cosmology on the struggle between good and evil. Abdul R. JanMohamed (1986) has developed Fanon's metaphor further, and argues that the "dominant model of power- and interest-relations in all colonial societies is the manichean opposition between the putative superiority of the European and the supposed inferiority of the native" (quoted in Groening 2005, 14). Terry Goldie (1989), however, disagrees with JanMohamed, since in his view contemporary texts often reverse this power relationship so that instead the indigene is depicted as superior to the non-Native. Leslie Monkman similarly argues that primitivism often represents a positive entity because "it is used to express the ways in which the Native serves as a teacher to the white man" (quoted in Groening 2005, 14).

This educational aspect is close to what I will argue is McGrath's agenda, but instead of portraying the indigene as a teacher to the white man, the novel potentially serves as a teacher to the white reader by representing an Inuit world in which the Inuit are superior to the white outsiders. In an interview with her publisher, McGrath was asked whether Westerners are disrespectful of the Arctic. She answers yes, and states that:

> [this] was very much the inspiration for [*The Long Exile*]. The Western view of the Arctic has been coloured by the pioneering, mastering, conquering explorer ethos and if we're going to stand any chance of stewardship of that land, then we have to learn to look at it in the same way as the Inuit. (Tucker 2007, 3)

McGrath clearly wants to present an Inuit perception of the land, a "translation" of Inuit life and perception, in order to teach readers to view the Arctic in a more nuanced way. Although this is a praiseworthy project in itself, the portrayal of whites in the novel is in danger of becoming just as dichotomous and binary as the old descriptions of Inuit or First Peoples traditionally were. In *White Heat*, the reversal of how westerners have been speaking for the Inuit can be exemplified by the way in which Edie describes the outsiders whom she guides early in the novel, as I touched upon earlier. In a rather judgemental way, she emphasizes their arrogance and unrealistic belief in their own skills in mastering the environment:

> puffed up egos, in the Arctic for the first time, laden down with self-importance and high-tech kit, thinking it was going to be just like the duck shoot in Iowa [...] Then they got out on the sea ice and things didn't seem quite so easy. If the bears didn't spook them, then the blistering cold, the scouring winds, the ferocious sun and the roar of the ice pack usually did the job. They'd stave off their fear with casual bravado and booze and that was when the accidents happened. (4)

As this passage shows, the outsiders are denigrated as lacking the essential qualities that matter in the Arctic. Here it is not Inuit who are considered inferior by westerners, but the opposite. There is no sign of the outsiders learning from the insiders, or vice versa; power does not flow multidirectionally, but rather unidirectionally, and the binary is kept in place.

This dichotomy could of course be the direct result of Edie's ambivalence, her disgust at these particular white "puffed up egos" (4), but the same tendency is found in the novel's crime plot itself, in which the gravest crimes are committed exclusively by outsiders. In addition, most of the outsiders function as representatives of modernity, understanding the Arctic not in terms of its immanent values, but as a place to exploit for their own benefit. Wagner, for example, "was trying to pin down the exact location of a gas reserve" (315). Bill Fairfax, another American, has got hold of a meteorite that can help to pinpoint this location, a meteorite his ancestor Sir James Fairfax, who – perhaps not coincidentally – shares initials and destiny with the famous explorer John Franklin, had traded with Edie's great-great-great-grandfather Welatok. This analogy points to the historical dimension of colonialism and brings the contemporary textual level into contact with historical power relationships. It becomes clear that the trading was not an equal affair; instead, Fairfax/Franklin tried to exploit Welatok to get hold of the meteorite.

Exploitation of natural resources is the main goal of nearly all the outsiders in the novel, and they pursue this goal without any respect for the locals. The outsiders are not the only criminals, however. Their main accomplice from the local community is the nurse Robert Patma who – it turns out – works and even kills for the outsiders in order to finance his drug and gambling habit. Nevertheless, he is a white man, and in that sense belongs to the outsider category. Edie's corrupt brother-in-law and Autisaq's mayor, Simeonie, is also bought off by the outsiders, allowing them to search for the gas reserve at Craig Island in exchange for money he can use in his political westernization project. Even Edie's aunt Martie and the old Inuit loner Koperkuj fall under the power of the criminals, due to their drug addiction. Since both drug and alcohol addiction and gambling are consequences of white colonialism, they reflect evil on more than a local scale. Likewise, they connect the novel's crime plot to the abuses of colonial history.

But western values are also connected to innovation and industrial progress, which may sometimes save lives, and Edie is a case in point. When she is captured by the evil scientist DeSouza and everything seems hopeless, salvation comes in the shape of Martie's plane appearing in the sky like a *deus ex machina*: "A spot appeared among the clouds, blooming then resolving into the familiar shape of a Twin Otter. […] It was Auntie Martie's" (365). This is not the only time that a symbol of modernity serves as Edie's rescuer; more often than not, thinking about old American movies helps her in difficult situations; for example, when she is captured by two Russians and her jaw is dislocated in a fall:

> The first thing was to ignore the pain. She had a method for this, something her father had taught her. She sat with her eyes closed and replayed the scene in *Safety Last!* […] until the pain lay somewhere between the laughter and she could at least think clearly again. (266)

In a situation where the pain is almost unbearable, thinking about silent movies becomes a method to channel it elsewhere. Another example shows how the movies also serve as learning tools: "Edie shook her head. You had something to hide, it was best to act mute. Something she'd learned from silent movies" (126). The silent movies play an important role in Edie's life and, since she was introduced to them by her father, they also represent a link to him. After Joe's death, the movies even bring her into contact with her own grief: "The screen flickered for a moment then the familiar face of Harold Lloyd appeared. Only then did Edie feel the tears come" (344).

In all these examples, the *qalunaat* influence upon Edie is used in a genuinely positive manner, and in this way modernity plays a more complex role in *White Heat* than just representing a dichotomous pattern.

This also goes for the way in which Inuit and western knowledge intermingle at the end of the novel, as the former loner Koperkuj tells Edie stories about the old days while she introduces him to silent comedy. Although this fits neatly into the novel's happy ending, this exchange of apparent opposites actually creates a third position as Koperkuj gives his favourite actor Buster Keaton the Inuit nickname Kituq (374). Trapped in his sick bed, slowly recovering from a near-fatal attack, American movies also serve as a medicine for him.

Hybridity and investigation

According to Ed Christian (2001), hybridity is an essential component of most postcolonial detectives: "They are blends of Western police methods and indigenous cultural knowledge, and their abilities are greater because of this hybridity. They have appropriated what is useful from the empire, but transformed it" (Christian 2001, 285). This is in line with Bhabha's view on hybridity, which is that, in the most positive sense, hybridity can be understood as an intermingling of opposites, creating something new, something in-between that transcends differences. This is not necessarily what happens in *White Heat*, however.

Edie can perhaps be best understood as the personification of this intermingling of the local and the global as she combines knowledge from her western education with intuition and skills achieved through her Inuit upbringing. The reason that she manages to expose two Russian criminals who call themselves Raskolnikov and Petrovich is that she knows the western literary classics: "for the only time in her life she found herself thanking the Canadian government for her ridiculous, southern-curriculum education. […] Raskolnikov and Petrovich were characters in *Crime and Punishment,* the murderer and the detective sworn to bring him down" (217). At the same time, her insight into global politics is perhaps slightly too impressive; for example, in a discussion with Derek she explains why the oil company Zemmer blames an explosion on Chechen Islamists:

> Maybe that way, they get the US Government pouring troops into oil-rich regions so they can suck the oil out of them. Mention the words Islamist, someone in Washington DC adds another few noughts onto the defence budget and no one complains. In any case, either Zemmer or the Russian guy found out they were being misled and wanted Wagner taken out and they got some local guy to do it. (317)

Such knowledge implies an interest in contemporary politics and an insight that is not necessarily unrealistic. Yet, as far as the reader knows,

Edie never follows the news or reads newspapers; she only watches old American comedies. Such knowledge of contemporary politics is therefore not entirely convincing within the novel's framework.

Edie's skills as an Inuit investigator are far more convincing. As an Inuit, Edie knows the environment to such an extent that it helps her "read" the bones she finds on her search for the missing Andy Taylor: "Edie had a familiarity with bones. If you were Inuit, you couldn't not. All her life she'd been flensing them of their flesh, chopping them to get to the marrow, to make soup or to give to the dogs" (142). Bones are not just neutral remains; they tell stories if only one is able to listen: "She picked out a fragment of skull, from the top back of the head, she thought. In the midst of it was a small hole, about the size of a nickel, almost perfectly round. The unmistakeable entry wound of a bullet" (143). Where everyone else would just see a human bone, Edie sees a human being who has been shot: the missing Andy Taylor. It is exactly this kind of knowledge that allows her to succeed as a detective:

> Edie's eye was drawn to an intense sparkle not fifty metres up ahead. Ice sparkled, snow sparkled, in the right conditions some rocks sparkled; fish skins sparkled, as did the hooves of musk oxen and caribou and the metallic parts of snowbies and komatiks, but nothing Edie could remember encountering before sparkled quite like this. (140)

Thanks to knowledge she has acquired through her Inuit upbringing, she finds microscopic objects hidden under the snow or on the ground. The passage above is an example of the interplay between Edie and the natural environment – the sun reveals an earring; her feet feel the ground more intuitively than her eyes and so on. Edie even invokes the dead man's bones: "*If you've got something to tell me*, she said to the bones, *now would be a good time*. She waited, but the bones remained silent" (168, original emphasis). Despite of the lack of an answer, this passage implies an empathy with the material environment – nature is not only an object to exploit, but something to interact with and listen to. Edie's knowledge of the land and environment is strongly connected with her body; it is physical and practical, not abstract and intellectual.

From a gender perspective, the quoted passage is problematic, however. The fact that it is a woman who shows such closeness to nature and the landscape borders on being a cliché, since women are associated with body, landscape and nature – as opposed to men, who stand for intellect, progress and culture. In line with this perceived connection between femininity and nature, Edie makes use of so-called traditional methods to uncover the truth. She begins an "Inuit search", which must be understood as a bodily investigation of the landscape, as opposed to the police constable Derek Palliser who – if he investigates at all – makes use of modern police methods or "common sense". In this way, the "Inuit search" becomes a feminine activity within the frame of the novel.

But there is also dissonance in the novel between so-called traditional mindsets and the consequences of modernity, in Autisaq and within Edie herself. Detection and investigation are normally associated with the western world, with reason, law and order. In Autisaq, detection is unwished for. The locals are instead used to taking the law into their own hands, as in the recipe for dealing with murders "the Inuit way; take [the lawbreaker] up to the mountains and, when he was least expecting it, push him off a cliff" (37). A similar example involves Edie directly when, at the end of the novel, she realises that it is Robert Patma who has murdered Joe: "In the few minutes since she had found the Saran Wrap in Robert Patma's kitchen, it was as though she'd been inhabited by some other, unfamiliar, self" (343). This other self actually considers taking the law into her own hands, "*contemplating murder*" (343, original emphasis).

Taking the law into her own hands, as Inuit are accustomed to do, is consequently alien to her, making her feel like "some other, unfamiliar, self" and as soon as she comes to her senses, she is not able to follow Inuit custom. She is, after all, *qalunaat* as well as Inuit. And it is her *qalunaat* blood that makes her determined to discover the truth:

> she thought about what she had said to Derek, about Joe's story needing an ending, and realized she'd been wrong. Joe was Inuit and Inuit lives were like sundogs or Arctic rainbows, they ran not in lines but in circles. […] It was she, with her *qalunaat* blood, who demanded resolution. It was she who could only find her way to a singular truth. (327)

In contrast to Joe, who was fully Inuit and therefore experienced life as running in circles, Edie's *qalunaat* genes seem to force her to experience life as linear. It is precisely the demand for a resolution that makes Edie search for the truth in the first place, and her *qalunaat* background is therefore just as important for uncovering the truth as her Inuit upbringing. At the same time, it is her Inuit search method that finds her the concrete evidence. While the need to find resolutions may be a *qalunaat* custom, and her literary and political knowledge may be unconvincing, it is her assimilation of such knowledge with Inuit methods and intuition that enables her to solve the mystery. Edie's hybrid identity is thus a key to the novel's resolution.

Constable Derek Palliser is another important character in the novel, with five chapters dedicated to his perspective and activities. Derek is a chain-smoking lemming enthusiast with a broken heart, who has "let himself be pushed around too often" (45) and, in contrast to Edie, he seems to be a spineless figure. Like Edie, Derek is of mixed heritage, and to him this is a source of ambivalence:

> For a small minority in Kuujuaq, he'd always been an object of derision on account of his mongrel blood: part *qalunaat*, part Inuit and, almost unforgivably, part Cree, the Inuit's natural enemy. He'd grown up with the idea that he was someone who probably didn't belong anywhere, but that didn't mean he liked being reminded of the fact. (49)

His heritage is clearly associated with something negative, and within himself he is both Inuit and the Inuit's enemy, "a cur, a mixed blood, someone fashioned at the borders out of the scraps no one else wanted" (53). His feeling of not belonging is clearly connected to his job, which makes him an outsider in his own community:

> Among a certain crowd he was still seen as some kind of collaborator. And there were plenty of folk up here who didn't see the need for a legal system at all, considered it just another southern import. They didn't want to know that more often than not Derek kept the southern legal system off their backs. (94)

Derek is caught between the legal system and his local community; representing the system but at the same time painfully aware that "the only rules that mattered a damn were the ones the land imposed on those struggling to carve a living from it" (60). Even though he clearly undergoes development throughout the novel, and in the end appears to be more energetic and active than at the beginning, his hybrid identity is not a quality that aids him in the investigation or allows him to see things from several perspectives. Based on his job as a police constable, he is obviously mostly associated with his *qalunaat* side, and possible Inuit influences remain hidden. He seems to exist primarily in a binary opposition with Edie, representing the law and rationality where she represents Inuit tradition and intuition. At the same time, his passivity and lack of initiative strengthen Edie's vigour and determination, qualities that oppose the clichés of femininity I discussed above.

Performing gender in an Arctic setting

Edie is a hybrid figure in more ways than one, and how she performs gender makes it possible to talk about *gender hybridity*, a gender confusion that Fabienne Darling-Wolf refers to as "a broad set of acts, behaviours, and attitudes that do not conform to the socio-culturally constructed meaning assigned to each gender in a particular society or culture at a particular point in time" (Darling-Wolf 2008, 67). Throughout the novel, Edie is a tough figure, a hardy woman who is not easily scared, since it "[takes] a lot to frighten an ex-polar bear hunter" (2). Something that does scare her, however, is "just how badly she wanted Joe to be able to go ahead with his nursing training" (2). Edie's love for her ex-stepson Joe is clearly the major motivation for her actions in the novel, and after his death her main goal is to clear his name and find his murderers.

In relation to Joe, she can be emotional and gentle, but most of the time she is tough and fearless, identifying more with her hunter background than with the norms of femininity. She lacks domestic skills (304), and even though the elders in her community do not approve of a woman guiding men and are always looking for an excuse to unseat her, "they knew she was the best damned guide in the High Arctic" (5). In addition, she takes pride in the certainty that "no one knew better than Edie how to hunt" (3). But "the fierceness and ego that had once made her such a good hunter" also made her "a difficult wife", in her ex-husband Sammy's opinion (62).

Non-domestic, tough female characters have become a commonplace in crime fiction, and in this respect Edie does not differ from other heroines of recent feminist crime fiction, such as Dana Stabenow's Kate Shugak, Sara Paretsky's V.I. Warshawski, Sue Grafton's Kinsey Millhone and Val McDermid's Kate Brannigan. What makes Edie different, however, is her hybrid background and the Arctic context; therefore, her gender performance cannot be understood exclusively on the basis of a western notion of gender, but should be understood in relation to ethnicity.

As Keri E. Iyall Smith states in *Hybrid Identities*, "a hybrid identity might form as a result of false dichotomy, where an identity that seemingly only has the capacity to occupy two forms is actually shown to encompass another form" (Smith 2008, 6). For example, a person who feels that the two forms of gender identity are too narrow may want to oppose this binary and become something more. On the basis of the premise that notions of masculinity and femininity are to some extent "social constructs predicated on a binary opposition" (Darling-Wolf 2008, 64), a way to oppose this premise would be to inhabit this "other form" that Iyall Smith mentions. In *White Heat*, this opposition becomes evident in the way in which Edie incorporates both feminine and masculine traits in her personality: When Joe dies, it "stirred in her some compulsion which she was unable to ignore. Call it a hunter's sense, intuition, mother love, whatever, she didn't care" (128). Here, "mother love" is the only strictly gendered expression, being generally a female feeling. But hunters in Inuit societies are mostly men (see e.g. Billson & Mancini 2007, 35), and "a hunter's sense" can therefore be interpreted as a more masculine feeling. In the quoted passage, these two feelings operate together, and Edie does not care whether the compulsion she feels is masculine or feminine. In this way, the dichotomies that characterize "those constitutive categories that seek to keep gender in its place" (Butler [1990] 1999, 47) are dissolved, and gender instead becomes fluid. Furthermore, a third position is potentially created, from which it is possible to perform gender in a less binary manner, since there are no strict divisions between masculine and feminine in the way in which Edie acts and performs.

This gender hybridity does not necessarily say anything about life as a female Inuit in the Arctic, or about gender norms in Arctic communities, but the ways in which gender is portrayed in the novel potentially challenge notions of gender in the Arctic and may open up opportunities for new discussions and discourses about the Inuit.

Enriching hybridity or inverted discourses?

In *White Heat*, the inhabitants of Autisaq are portrayed as taking part in a modern society that is struggling with both local and global problems, and the Inuit are presented as complex figures in a complex world. The portrayal of the outsiders is, however, less multifaceted and there seems to be a chasm between the insiders and the outsiders, with the indigene being depicted as superior to the non-Native. But, instead of establishing a dichotomized pattern in which all the outsiders are bad and all the insiders are good, the novel – to a certain extent – allows for nuances, especially when it comes to drugs: Drug and alcohol abuse is something to which insiders and outsiders alike surrender, and Robert Patma, who could have represented the incarnation of egoism and evil as Joe's murderer, becomes the border figure illustrating these nuances:

> As [Edie]'d sat beside him last night, listening to his breathing, and contemplating putting a stop to it, the idea had come to her that he was nothing, an addict, but then Sammy had come into the room and she'd thought of herself, of her ex and of Willa. *At some point in our lives, hadn't all of us been the same?* Whatever Robert Patma had done, he wasn't so different from the people she loved. (346)

Robert Patma is just like Edie's loved ones, he is like Edie herself, a human who allowed the drugs to hold him down during a weak period in his life. At the same time, Patma is the only one of the outsiders who is included within such an extenuation; the Russians and the Americans remain one-dimensional and bad throughout. The Russians in particular are presented in an unfavourable light, as the quintessence of evil, as cynical murderers or egocentric oil barons without scruples. The Manichean value judgements thus shift, but the discourse does not, at least not in the portrayal of the whites who live outside the Arctic.

But is it even possible for a crime novel to overcome discourses that have to do with dichotomized representations of the enemy? One of the genre's limitations is exactly the tradition of schematic patterns of "good guys" and "bad guys", insiders and outsiders. Within the plot structure, in line with such limitations, the novel's depiction of outsiders in the Arctic allows prevalent discourses, in which dichotomies between locals and foreigners are clear-cut and the latter are seldom portrayed as complex or dynamic, to remain unchallenged. Instead, the binary images of the enemy are reinforced.

Laura Smyth Groening claims that postmodern fiction, "for all its determination to centre the marginal, often perpetuates many of the same stereotypes as does 200-year-old historical fiction" (Groening 2005, 17). For example, the trope of the dead and dying Indian (or Inuit) culture is persistent, and it is most often aligned with the past, whereas Euro-Canadian culture, "depicted as the master of technology, becomes synonymous with the future" (Groening 2005, 20). Moreover, the former has a tendency to be associated with women, while the latter is often connected to men, so that women stand for the past, tradition and the land, while men stand for progress and the future. In *White Heat*, this is literally the case, since Edie's mother is indigenous, related to the past, while her father is white, associated with technology (American movies) and the future. This same pattern is repeated in the relationship between Edie and Derek. In this respect, the novel perpetuates traditional patterns instead of challenging them.

But the role of modernity is ambivalent. Based on her research into social problems in the Arctic, in her blog McGrath presents numbers and percentages for unemployment, suicides, violence and so on (McGrath 2012). This information is reflected in the description of social problems in Autisaq, described as "a fog of drink, boredom, unwanted pregnancies, low expectation and educational underachievement" (95). On the one hand, many of the locals have problems with alcohol, drugs or both and, as mentioned above, this is presented in the novel as one of the main motives for locals to get involved with the criminal outsiders. On the other hand, *White Heat* is based on the notion of an Arctic modernity with a highly modern main character who is not trapped in a romanticized, ancient tradition. Edie is an educated, smart and tough woman capable of using both her western knowledge and her Inuit cognition to solve the crime puzzle, although traditional forms of knowledge are clearly privileged. As Inuit *and* white, Edie is the personification of an enriching hybridity, an example of how the Arctic is part of the modern world. The novel also disrupts dichotomies through the way in which Edie performs gender. With her insistence on being both a hunter and a motherly person, both masculine and feminine, or rather something more, she transgresses binaries.

McGrath's novel has a sympathetic and well-meaning agenda. It portrays an Inuit community of strong indigenous women in a land traditionally associated with masculinity, and where the Inuit are not necessarily portrayed as victims. Nevertheless, a power relationship between the Inuit and the whites is still evident, and hybridity does not fully manage to create something new out of the opposites. Insiders and outsiders rarely learn anything from one another and no one except Edie manages to overcome the chasm between what is portrayed as two separated groups. The novel's portrayal of the whites largely sustains a reversed binary system, picturing the whites as evil, egocentric or both. But the genre's tendency to simplify is also extended to Inuit representatives of the wider society, such as Mayor Simeonie, and Constable Derek Palliser, who are described as respectively corrupt and ignorant, even though the latter gradually becomes more thoughtful throughout the novel. It is only through the hybrid character of Edie Kiglatuk that a different Arctic truly emerges, where gender is allowed to be fluid and western and Inuit values exist side by side.

References

Anderson, Alun. 2009. *After the Ice: Life, Death and Politics in the New Arctic.* London: Virgin Books.

Berg, Anne-Jorunn. 2014. "What Happened to the Cyborg? On Feminist Theories of Materiality." *The Norwegian Journal of Gender Research* 38 (2): 110–126.

Bhabha, Homi K. (1994) 2004. *The Location of Culture*. Hoboken, NJ: Taylor & Francis.

Billson, Jane M., and Kyra Mancini. 2007. *Inuit Women: Their Powerful Spirit in a Century of Change.* Lanham, MD.: Rowman & Littlefield.

Brody, Hugh. 1987. *Living Arctic: Hunters of the Canadian North.* London: Faber and Faber.

Butler, Judith. (1990) 1999. *Gender Trouble: Feminism and the Subversions of Identity.* New York and London: Routledge.

Christian, Ed. 2001. "Introducing the Post-Colonial Detective: Putting Marginality to Work." In *The Post-Colonial Detective*, edited by Ed Christian, 1–16. New York: Palgrave.

Collignon, Beatrice. 2006. *Knowing Places: The Inuinnait, Landscapes and the Environment*. Edmonton: CCI Press.

Darling-Wolf, Fabienne. 2008. "Disturbingly Hybrid or Distressingly Patriarchal? Gender Hybridity in a Global Environment." In *Hybrid Identities: Theoretical and Empirical Examinations*, edited by Keri E. Smith and Patricia Iyall, 63–79. Boston, MA: Brill.

Dodds, Klaus. 2016. *The Scramble for the Poles: The Geopolitics of the Arctic and Antarctic*. Cambridge: Polity Press.

Ekman, Kerstin. 1996. *Blackwater*. London: Vintage.

Friedman, Susan Stanford. 1998. *Mappings: Feminism and the Cultural Geographies of Encounters*. Princeton, NJ: Princeton University Press.

Government of Canada. 2010, Aug. 18. "Apology for the Inuit High Arctic relocation." Accessed Nov. 21, 2016. https://www.aadnc-aandc.gc.ca /eng/1100100016115/1100100016116

Groening, Laura Smyth. 2005. *Listening to Old Woman Speak: Natives and AlterNatives in Canadian Literature.* Montréal: McGill-Queen's University Press.

Hansson, Heidi. 2010. "Arctic Crime Discourse: Dana Stabenow's Kate Shugak Series." In *Arctic Discourses*, edited by Anka Ryall, Johan Schimanski, and Henning Howlid Wærp, 218–239. Newcastle: Cambridge Scholars.

Howard, Roger. 2009. *The Arctic Gold Rush: The New Race for Tomorrow's Natural Resources*. London: Continuum.

Høeg, Peter. 1993. *Miss Smilla's Feeling For Snow.* Translated by Tiina Nunnally. New York: Farrar, Straus and Giroux.

McGrath, Melanie. 2006. *The Long Exile*. London: Fourth Estate.

McGrath, M. J. 2011. *White Heat*. London: Mantle.

McGrath, Melanie. "Arctic homicide 10x national average." Accessed Apr. 4, 2016. http://www.melaniemcgrath.com/blog/arctic-homicide-10x-national-average.

Pettersson, Lars. 2012. *Kautokeino, en blodig kniv* [Kautokeino, a bloody knife]. Stockholm: Ordfront.

—. 2014. *Slaktmånad* [The month for slaughtering]. Stockholm: Ordfront.

—. 2016. *Mörkertid* [Polar night]. Stockholm: Ordfront.

Sale, Richard. 2010. *The Scramble for the Arctic: Ownership, Exploitation and Conflict in the Far North*. London: Francis Lincoln.

Smith, Keri E. Iyall. 2008. "Hybrid Identities: Theoretical Examinations." In *Hybrid Identities: Theoretical and Empirical Examinations*, edited by Keri E. I. Smith and Patricia Leavy, 3–11. Boston, MA: Brill.

Steinberg, Philip E. 2015. *Contesting the Arctic: Politics and Imaginaries in the Circumpolar North*. London: I. B. Tauris.

Stevenson, Lisa. 2006. "Introduction." In *Critical Inuit Studies,* edited by Pamela Stern and Lisa Stevenson, 2–22. Lincoln, NE and London: University of Nebraska Press.

Thisted, Kirsten. 2002. "The Power to Represent: Intertextuality and Discourse in *Smilla's Sense of Snow*." In *Narrating the Arctic*: *A Cultural History of Nordic Scientific Practices*, edited by Michael Bravo and Sverker Sörlin, 311–342. Canton, MA: Science History Publications.

Thompson, Jon. 1993. *Fiction, Crime, and the Empire: Clues to Modernity and Postmodernism.* Urbana and Chicago, IL: University of Illinois Press.

Truc, Olivier. 2014. *Forty days without Shadow: An Arctic Thriller*. London: Trapdoor.

Tucker, Louise. 2007. "The Last Place of Innocence: Melanie McGrath Talks to Louise Tucker." Appendix to *The Long Exile*, 1–22. London: Harper Perennial.

Notes

[1] For example Dodds 2016, Steinberg 2015, Sale 2010, Howard 2009, Anderson 2009.

[2] In the following, page references to the novel are given in parentheses.

"A PLACE IN THE SUN": HISTORICAL PERSPECTIVES ON THE DEBATE ON DEVELOPMENT AND MODERNITY IN GREENLAND

KIRSTEN THISTED

"[U]nderstanding interconnections in the present will require understanding them in the past *and* re-constructing our understandings *of* the past" (Bhambra 2007, 154). With these words, Professor of Sociology Gurminder K. Bhambra argues that the conceptual framework of sociological discourse on modernity should be deconstructed as concerns its Eurocentric claims on modernity as *a European* phenomenon, invented endogenously within the cultural-geographical sphere of Europe. Retelling the history of modernity in the perspective of colonialism and the resulting *connected histories* brings to the fore the otherwise silenced interconnections "of peoples and places that transcend the boundaries established within the dominant approaches" Bhambra 2007, 152). This new perspective will not only provide a better understanding of the present, but hopefully "provide a clearing from which we can begin to look at the world again and begin to create new forms for the future" (Bambra 2007, 11).

From a similar standpoint, this chapter will add a historical perspective to the present debate on development and modernity in Greenland. The chapter invites the readers back to the time when Greenland was a Danish colony, with the intention of offering insight into how Greenlandic and Danish players at the time dealt with the issue of the modernization of Greenland. The focus of analysis will be Mathias Storch's *Strejflys over Grønland* (Gleams of light over Greenland) from 1930. This book was written by a Greenlander in Danish, as a response to a Danish publication, Sophie Petersen's *Grønland i hverdag og fest* (Greenland in everyday life and celebrations) from 1928. Petersen's book reflected contemporary points of view and perceptions in the Danish discourse on Greenland in religious circles and among people who had previously worked in Greenland. Storch's book reflected perceptions in Greenlandic society, as these were also expressed in the two Greenlandic newspapers, *Atuagagdliutit* and *Avangnâmioq.*[1] Unfortunately, Storch's book has slipped into oblivion, even in Greenland. The book demonstrates the fallacy of the preconception of Greenlanders as bearers and defenders of tradition against modernity and can certainly be seen as an argument for the description of Greenlandic history and modernity as a history that is connected with Denmark/Europe rather than as an "Other" of Danish/European history. Including such silenced and forgotten voices into remembrance and history is definitely relevant for the ongoing process of envisioning new futures for Greenland.

In their own and others' perception, Greenlanders are associated with the Inuit, a term that is currently used politically as a common denominator for a diversity of ethnic subgroups of Inuit origin who define themselves as belonging to the category of indigenous peoples. As many have pointed out, "indigenous" is an ambiguous term, not least because it still carries so many connotations that previously clung to the term "primitive".[2] In this discourse, *modernity* is the defining characteristic of white Europeans and North Americans, while *tradition* is the defining characteristic of indigenous peoples. This has led to a mindset that treats tradition and modernity as mutually exclusive, incapable of co-existing. Human beings have become bearers of this perceived incompatibility, so that as a group they must be categorized as one *or* the other. Upon reflection, most people would recognize this is not the way it works in the real world,[3] but the persistent discourse generates its own reality in the form of conventions, institutions and practices.

The discourse is prevalent in the United Nations' Declaration on the Rights of Indigenous Peoples (UN 2007), and it persists even after the Brundtland report (UN World Commission on Environment and Development, 1987) introduced the term "sustainability" with the aim of combining environmental, economic and social development. Nevertheless, reports on "Arctic Sustainable Development" are full of statements about indigenous people's "traditional knowledge" and "the indigenous cultural heritage that allows man to coexist with nature" (Keskitalo 2004, 164). Even in the face of change, tradition is held up as the defining characteristic of indigenous peoples – but now through its *absence,* as the gradual disappearance of so-called "traditional ways" is seen as a *loss*, presumed to be the root cause of the widespread social problems that exist in most of these communities (Keskitalo 2004, 165).

In a sense, this narrative of loss has long since been analysed and explained as an external discourse, "where non-indigenous people can be left nostalgic and regretful for the destruction that their colonial forbears wrought on pre-colonial societies, without having to attend to the (post)colonial difficulties which they now face" (Thomas 1994; quoted in Porter 2016, 39). Nevertheless, the discourse continues to frame the positions available to the peoples in question, if they want to have a voice and, not least, a chance to have their voice heard. This also applies to Greenland, which over the past century has undergone profound social, industrial and political development. Today, Greenland is a self-governing part of the Danish Kingdom. Self-government was implemented in 2009, replacing Greenlandic Home Rule (1979–2009). Before that, Greenland was a Danish colony, from 1721 until 1953, in which year it was included in Denmark by a change in the Danish constitution. Thus, Greenland has gradually developed in the direction of an independent nation state (Gad 2016), and the country is now thoroughly urbanized and modernized (Sejersen 2015). Still, the discourse of indigeneity is prevalent in the Greenlandic debate. Professor of political science Carina Keskitalo quotes then Home Rule Prime Minister Lars Emil Johansen for the following statement at a conference about whaling:

> Some have the perception that this category [aboriginal subsistence whaling in the International Whaling Commission, where Greenlandic whaling is placed] is a living whaling museum. We do not! We do not export the whale meat, but use our whale resources internally in Greenland as we use our fish resources. [...] The income is used to cover the cost of modern and efficient killing-methods. (Johansen 1997; quoted in Keskitalo 2004, 79)

What Keskitalo seeks to highlight here is not so much Johansen's attempt at negotiating the role of indigenous peoples as museum custodians but "the degree to which the argument nevertheless needs to draw on traditional elements" (Keskitalo 2004, 79). The changes in the Arctic areas notwithstanding, the Inuit and other indigenous peoples continue to be relegated to the role of "active stewards of Arctic wildlife" that has been assigned to them by the discourse on indigenous peoples combined with the discourse on "the Arctic" (Keskitalo 2004, 99).

The term "indigenous people" is not mentioned in the Act on Greenland Self-Government, and there is a reason for this. The act goes far beyond the UN Declaration on the Rights of Indigenous Peoples, which attempts to regulate the relationship between indigenous peoples and states, seeking to protect the former from the latter. The utopia that is set up in the declaration is the right to self-government in internal and local affairs – not national independence, which is the utopia of the Act on Greenland Self-Government (Thisted 2013, 234–35).[4]

Greenlandic Self-Government has meant a new start for Greenland, including the demand for a revision of Greenlandic history based on

Greenlandic sources and Greenlandic perspectives (Heinrich 2014; Thisted 2017). Most historical accounts begin after the Second World War when the United Nations was formed, and it became clear that Greenland could not continue as a Danish colony (Beukel, Jensen, Rytter and Elo 2007). This is also the period highlighted by the Greenlandic Reconciliation Commission, whose mission reflects the need to clarify what actually happened in the aftermath of colonialism, including the Greenlanders' own involvement in the introduction of modernity, as stated on the commission's website.[5] However, Greenland's road to modernity and independence has its beginning long before this point.

It is a firmly established Danish narrative that Denmark has donated Greenland's future, symbolized by modernity and development. That was the point made by Danish prime minister Hans Hedtoft in an important speech in 1948, where the Danish political discourse about Greenland changed from one of colonialism and exclusion to one of equality and inclusion of the Greenlanders into the former colonial state (Hedtoft 1948, 40; Andersen and Thisted forthcoming). The story of the Danes as the donors of modernity has been repeated over and over again, not only in political speeches but also in movies and literature (Thisted 2003a; Gant 2003). Hedtoft was up against strong voices within the Danish administration of Greenland, which claimed that the Greenlanders were far from ready for modernity but still needed to be protected against the modern world by the colonial power (Heinrich 2012). However, Hedtoft's speech acceded to Greenlandic requirements for modernization, equality, and the integration of Greenland into the Kingdom of Denmark. Since long before the Second World War, development and modernization had been the buzzwords of the Greenlandic elite, who envisioned a new Greenland where sealing and hunting would be replaced with fishing, sheep farming and mining, and Greenlanders would take up leading positions within trade and administration. Mathias Storch's book is a striking example of this line of thinking.

Likewise, the debate between Storch and Petersen gives an insight into a time when the notion of Greenland as part of an "Arctic" region had not yet been constructed and thus had not come to frame the discourse. The construction of "the Arctic" has blurred distinctions between places and peoples. The case demonstrates how any talk about "the Arctic" as a whole will necessarily remain superficial. Understanding Arctic modernities requires an in-depth understanding of the geographical and climatic conditions and historical developments in each particular area. Therefore, before embarking on an analysis of the books by Petersen and

Storch, some considerations on Greenland and its place in "the Arctic" are required.

Greenland and "the Arctic"

When Keskitalo insists on putting "the Arctic" in quotation marks, it is because she believes the Arctic should be considered a social construct, in line with nations and corresponding imagined communities and geopolitical units. She demonstrates how Canada has played a leading role in articulating the Arctic as a region, especially by establishing Universities of the Arctic and the Arctic Council. Specifically Canadian conditions have thus had a crucial impact on the discourse about the Arctic. However, Greenland differs fundamentally from Canada and the hegemonic discourse on the Arctic in a number of ways. First, it has a warmer climate on higher latitudes. Second, it is not characterized by unclear relations between an "indigenous" and a "local" (non-indigenous) population. Danes have not settled in Greenland in large numbers,[6] and when a Dane marries a Greenlander, the offspring is considered Greenlandic. Generally, Danes consider themselves *guests* in Greenland. Third, Greenland is incorporated into the Danish realm, and Danish sovereignty is performed by certain rituals, for instance by the queen wearing the Greenlandic national costume when she visits. Phrases such as "Greenland has internationalized Danish science" or "Greenland has made Denmark so much bigger", occur frequently in prefaces and speeches. However, Danes have not incorporated Greenland into their own identity-building narrative. Although Greenland has often provided the wilderness where Danish boys go to become men (Thisted 2002; Thisted 2003b; Oxfeldt 2003), Greenland does not appear as a key signifier in the semantic field defining the meaning of "Danish". Greenland instead played into the Danish self-image of being helpful and non-exploitative, like the role that aid to developing countries plays for the Scandinavian sense of identity (Gad 2008; Jensen 2012).

Not even Greenland should be seen as a single, uniform space. Within Greenland, there are huge differences between North and South, West and East. As a political community, the entity of "Greenland" is of a very recent date. West Greenland especially differs markedly from the hegemonic discourse on "the Arctic". West Greenland was colonized early by Danes and Norwegians, who came to Greenland to engage in trade and missionary work, and the many colonies that were settled along the vast coastline (the towns of modern-day Greenland) included countless trading stations (clusters of houses with a shop) and settlements (clusters of houses without a shop). Thus, one could travel the length of the coast from Upernavik in the North to Nanortalik in the South without ever failing to find shelter – as celebrated in one of Greenland's two national anthems, Jonathan Petersen's *Nuna asiilasooq* / A mighty, rocky country (ca. 1916):

> It is so lovely
> When travelling
> The length of the coast,
> North and south,
> Everywhere, one finds people
> And settlements to visit [7]

This description, however, did not include East Greenland, which was not colonized until the 1880s, or Thule in the north-west corner of West Greenland.[8] This place had long since lost communication with the rest of Greenland but had connections with the Inuit in Canada. West-Greenlandic Christian missionaries arrived here in 1909, and the first trading station was established in 1910.

In a Greenlandic context, the widespread image of the white, inhospitable Arctic is thus reserved for the northernmost part near the North Pole, and the uninhabited North East Greenland. These parts were open to the heroic explorers who filled in the last "white spots" on the Greenlandic map in the years leading up to the First World War.[9] Historically, these regions and the related activities were mainly labelled using derivatives of the word *polar*: polar regions (*polaregne*), polar research (*polarforskning*), polar heroes (*polarhelte*) and polar expeditions (*polarekspeditioner*).[10] Greenland's west coast, on the other hand, was included in the *North Atlantic*, a Scandinavian term for the body of waters that enabled traffic by ship between Norway, Denmark, the Faroes, Iceland, Greenland, the Shetlands and Scotland. In the "polar" regions, ice blocked the passage, and sledges were a more useful means of transportation.

Thus, when Danish polar heroes such as Knud Rasmussen (1879–1933) and Peter Freuchen (1886–1957) wrote about the "primitive" Eskimos, they were referring to the isolated people of East Greenland, especially Thule and Arctic America. On his famous Fifth Thule Expedition (1921–24) Knud Rasmussen together with two Inuit from Thule sledged through Arctic America, Canada and Alaska with a detour into Siberia. Today, these two Inuit are considered representatives of the "Greenlandic Inuit". At the time of the expedition, however, the Thule Inuit were just as exotic to the West Greenlanders as were the people of Canada, Alaska and Siberia. When Sophie Petersen and Mathias Storch speak of "Greenland" and "Greenlanders", they are referring to the long-ago Christianized people of Danish West Greenland.

To promote education and knowledge about Greenland

Sophie Petersen (1885–1965) grew up in Copenhagen, where her father managed a coffee store. Although the family was of limited means, Petersen, who was an only child, attended a modestly priced private school for girls. In 1911, she graduated from the University of Copenhagen with an MA in natural history and geography, majoring in geology – a highly unusual combination for a woman at the time. Throughout her life she taught at upper secondary schools in Denmark but took recurring sabbaticals to go on lengthy journeys that formed the basis for books, articles and lectures. Her book *Danmarks gamle Tropekolonier* (Denmark's former tropical colonies) from 1946 stands as her principal work.[11] However, *Grønland i Hverdag og Fest* has a special place in her body of work as the first in what became a series of travel books, where she found her voice in a hybrid of travel account, general information and textbook. Sophie Petersen first visited Greenland in 1920 on a three-month study trip. After this trip, she was seen as an expert on all things Greenlandic, and the following year she was among the select group invited for the celebration of the bicentenary of the colonization of Greenland. Petersen draws on these experiences in her book, which helps make it an engaging read, despite its didactic aim.[12]

No sooner had Sophie Petersen returned from her visit to Greenland than she joined the debate on Greenland's future. Her engagement included a stinging rebuke to the before- mentioned Peter Freuchen,[13] who in lectures and articles had criticized the behaviour of Danes in Greenland. Sophie Petersen became known as a more moderate, uncontroversial voice in the Danish debate on Greenland. Petersen wrote *Grønland i Hverdag og Fest* at the request of the Danish publishing house C.A. Reitzels Forlag, which may have been inspired by a proposal from former Minister of the Interior C. N. Hauge at a meeting at Det Grønlandske Selskab (Greenlandic Society) in March 1927. At the meeting, Hauge had raised the need for a textbook on contemporary Greenland for both private and school use. Many expedition reports and scientific treatises had been published, and in 1921, the expansive and thorough *Grønland i Tohundredaaret for Hans Egedes Landing* (Greenland at the bicentenary of Hans Egede's landing, edited by G. C. Amdrup) came out, but the latter was far too voluminous and costly for popular education. *Jorden og Menneskelivet* (Earth and mankind, edited by M. Vahl and G. Hatt, 1922–27) contained a lengthy section on Greenland – but this was only available as part of the four-volume book. Peter Freuchen's book *Grønland, Land og Folk* (Greenland, country and people, 1927) was too politically oriented – and too critical – to serve as "*the* book" on Greenland. Although Reitzel aimed to keep *Grønland i Hverdag og Fest* affordable, the significance of the issue nevertheless called for a certain standard, and the book appears as a thoroughly beautiful and "solid" tome, printed on heavy paper and lavishly illustrated with black-and-white photos in quality reproduction. This made the book too expensive to buy as classroom sets, but the aim was at least to make it available in libraries. On 28 February 1930, the Ministry of Education sent out a circular letter asking the state schools to state whether they had acquired the book, as the Ministry was considering buying a number of books for distribution. This gave the book the highest seal of approval.

Mathias Storch (1883–1957) grew up in a family living on hunting and fishing in a small settlement in Egedesminde (Aasiaat) District in Disko Bay, North Greenland. Normally, he would have grown up to be a hunter, but because he showed an aptitude for academic learning he was enrolled at Ilinniarfissuaq, Godthaab Seminarium (Godthaab College of Education) which he attended from 1900 to 1906. At the time, training as a catechist (an assistant to the – Danish – priest) was the form of higher education available in Greenland. In rare cases, an additional stay in Denmark would be granted. This happened to Storch. After his graduation, he spent three years in Denmark and was ordained a priest in 1910. In 1920, as the first Greenlander, he was given a temporary appointment as provost in North Greenland. In 1927, the provosts' districts of North and South Greenland were merged into one, and Storch became the vice provost for the entire country, still with a residence in Ilulissat in North Greenland. In his youth, Storch was active in the so-called *peqatigiinniat* movement, a sort of evangelical revival movement that combined Christianity, popular education and national revival (Thuesen 1988). Storch's novel *Singnagtugaq* (The dream), the first novel to be published in Greenlandic, was influenced by this movement. Storch was a prolific writer, and especially in his youth he was also politically active. In 1921, he was appointed to the Danish Government's "Greenland commission", charged with revising the governance act for Greenland. In 1927–33 he served on the provincial council for North Greenland. When the Danish king visited Greenland in 1921 to celebrate the bicentenary for the mission's arrival, Storch was awarded the Cross of the Order of the Dannebrog.

Strejflys over Grønland was written in Danish in late 1928/early 1929 during a stay in Denmark. With the title, Storch makes two references. One is that he is either not able or does not intend to provide a complete description of the country, the people or history but merely seeks to shed gleams of light on selected aspects and events. The other is that the light that everyone is speaking of bringing to Greenland in the form of knowledge, development and progress is only, seen from a Greenlandic perspective, evident as sporadic gleams of light. Storch's book intends to let the Danes know that Greenlanders, like all other peoples, desire "a place in the sun" – and that this desire cannot be pushed aside for ever (Storch 1930, 124).

For a Greenlander to speak his mind like this was by no means common at the time, although Storch's novel had been published in Danish only a year after its release in Greenlandic, under the title *En grønlænders drøm* (A Greenlander's dream) in Knud Rasmussen's translation (1915). Rasmussen was sympathetic to the Greenlanders' desire for influence and education, but from today's point of view, his preface to the book seems condescending and patronizing. Rasmussen clearly did not feel that the text would be able to speak for itself but feels compelled to explain and make excuses for its shortcomings (Thisted 2005). Since then, as outlined above, Storch had risen to a much more influential position in society. Nevertheless, he was unsure whether his book would ever be published, as evident from his humble, self-doubting words: "If this is printed" (Storch 1930, 103). Storch's book is thus in no way framed by the story of commissioned work and government approval that characterizes Petersen's book. Nevertheless, his manuscript appears to have been received with interest. First, he contacted the Danish publisher Gyldendalske Boghandel og Nordisk Forlag, which in a letter dated 23 March 1929 welcomes his initiative – especially if Storch would investigate the possibilities of any financial support. The publisher only returned the manuscript at his own request, probably because Storch had found an alternative avenue for publication. The book was released the following year by Levin & Munksgaards Forlag with an introduction dated 23 October 1929, penned by Hother Ostermann, who is also credited on the title page as the person in charge of "bringing about" the publication.

Hother Ostermann (1876–1950) had served as a priest in Greenland in 1901–1921 and was Storch's predecessor in Ilulissat. The two men were in close contact, and over time they developed a close personal relationship.[14] Ostermann was an authority on Greenlandic issues and had already authored several books on Greenlandic history, including *Den grønlandske Missions og Kirkes Historie* (The history of the Greenlandic mission and church, 1921) and had also delivered weighty contributions to the previously mentioned magnum opus *Grønland i Tohundredaaret for*

Hans Egedes Landing. Ostermann was thus a key ally for Storch, and undoubtedly facilitated his contact with the publisher. He too adds an introduction, which is mainly centred on Storch's CV as a way of conveying to the reader that the author is an important man.

Storch too clearly thought it was important to convince the reader that his opinions were worth paying attention to. As customary in books on Greenland, he opens with a description of "The country and life in nature" – but his main purpose with the introduction is to position himself as an educated and cultivated man, at eye level with the intended reader. The text is brimming with imagery, comparisons and metaphors, presenting the author as a man who is at home in a world of poetry, music, painting and film. The reader is encouraged to embrace all these avenues for imagining Greenland, a country that is unlike anything the reader might encounter in Denmark. The cries of seagulls in Greenland are thus nothing like what the reader may have heard by the lakes in Copenhagen but rather "like a slow melodious tone that [...] may range from the highest to the deepest notes or conversely – as when a master joyfully engages his instrument to invoke a swell of notes that so grips one's imagination that one fully forgets oneself for a time" (Storch 1930, 21).[15] In this way, Storch positions himself as a sort of cultural translator: he is capable of seeing his country from an outside perspective, as he knows Danes do, but the main potential of his text is that he simultaneously commands an inside familiarity with the country and is able to convey this perspective to his reader.

We cannot know how Storch felt about the way Ostermann's introduction frames his text. He probably realized it was the price he had to pay to be published. In addition to presenting Storch as an important man, Ostermann seeks to soften the sting of his criticism, presumably in the hope of making the book's message easier to swallow. First of all, he attributes the reforms achieved until then far greater positive significance than Storch does. Ostermann thus inscribes Mathias Storch's life in a narrative embedded in the same structure that Storch uses to frame his own text: past, present, future: "This man's career, which is in itself fairly unusual, is a characteristic example of the tremendous progress in modern-day Greenland in contrast to the past, when everything stood still".[16] (Ostermann in Storch 1930, 9). In Ostermann's view, the criticism voiced in *En grønlanders drøm* pertains to the past, while the present represents the realization of the dream portrayed in the novel: "And now we may marvel to see that much of what was dreamed in "A Greenlander's dream" has either appeared or is about to appear in reality in Greenland".[17] (Ostermann in Storch 1930, 11). This gives the impression that the problems Storch describes in his new book are of a more recent date and are in the process of finding their solution. This makes it seem highly likely that events will unfold according to Ostermann's desires and hopes: that Storch "may one day be able to give us a third book to tell us that now they [the problems] have been solved!"[18] (Ostermann in Storch 1930, 12). Ostermann thus strikes the same optimistic tone as Petersen's book – an outlook that is not shared by Storch, who takes a more sombre view of his country's future.[19]

Greenlandic modernity

On one key issue, Storch and Petersen are in complete agreement: Greenland needs development in the form of new economic opportunities. Although the Danish policy had always been to keep the Greenlanders occupied as seal hunters (Thomsen 1998; Rud 2016), at this point, everyone had to realize that seal hunting had been so reduced that an alternative was needed. There was less clarity about the issue of the trade monopoly. Over time, Denmark had consistently defended the monopoly, arguing that it was a necessary means of protecting the Greenlanders. The Danes provided for the Greenlanders while protecting them from being taken advantage of and falling prey to the kind of exploitation that had occurred in other colonized countries (Rink 1862; Marqvardt 1999, 9ff.). Around this time, there was pressure from Greenlanders, Danes and others to open Greenland up. Everyone knew it was only a matter of time before reforms would have to be initiated. As Petersen writes, "However, we [Danes] may some day be forced to provide open access to the country, and then it would be better had we in time begun to give the Greenlanders a certain trade education" (Petersen 1928, 180–81).[20] Petersen thus argues that the lifting of the monopoly requires preparation, and that there is no time to waste in initiating the process. Here, too, she is in complete agreement with Storch, just as he agrees with Petersen that simply lifting the monopoly without preparing the Greenlanders to take charge would have negative ramifications for the population (Petersen 1928, 113–14).

In a Greenlandic context, modernity at this time is primarily a matter of new economic opportunities, better and healthier living conditions, not least in the form of new and healthier housing, and more up-to-date education to enable the Greenlanders to handle commerce and administration. In both Greenlandic and Danish sources, this transformation of the Greenlanders' economic possibilities and background is seen as a decisive break with the past. Greenland is thus embedded in a perception of time that the conceptual historian Reinhardt Koselleck characterizes as specific to "the modern": a temporality in which the present is experienced as an overcoming of the past, in the sense that we experience a qualitative newness or progress – a "new time" (*Neuzeit*) – in which the future is an open horizon of expectation (Koselleck 2004). This form of temporality is most clearly articulated in Ostermann's introduction, as in the quoted passages about Storch's career and books as testimony to the rapid development in Greenland, in contrast to earlier times when "everything stood still".

Bringing Greenland into "the modern" means that something will by necessity be lost. Both Petersen and Storch identify this "something" with a Danish nostalgia for the image of Greenland that they know and romanticize. This applies to Sophie Petersen herself, as she openly acknowledges, when she arrives in Nuuk and notes to her chagrin that the modern Greenlandic woman wears her hair up, "as she has seen Danish ladies wear it",[21] instead of wearing it in "the bun that we know so well from the picture books of our childhood"[22] (Petersen 1928, 20). Like so many other travellers in Greenland, Sophie Petersen was repulsed when she saw too many elements of her own world on display in Greenlandic homes – the bric-a-brac on the shelves of European furniture was "hideous"[23] (Petersen 1928, 26) in contrast to a soap stone lamp, which was "a profoundly genuinely Eskimo object"[24] (Petersen 1928, 24). On the other hand, she realizes that this is a necessary process that cannot and should not be stopped. Storch too is well aware of the Danish view of Greenland, and how this view is inscribed in a European critical discourse on "Western civilization" in general and modernity in particular:

> The transition from the nerve-wracking conditions in the densely populated Europe to the wide open spaces may often be the cause of the delight [that visitors feel upon encountering Greenland]. It can be wonderfully liberating to go from a place where each step costs money, and where unrelenting competition stirs up unrest and worry – to wake up one day, after a dozen days of "seasickness" or "sensations", removed to very different and completely new conditions characterized by sublime serenity, where one can be one's own master and in command of one's own time.[25] (Storch 1930, 19-20)

By keeping the description in abstract terms, Storch avoids elaborating on who exactly is experiencing the so-called "free life" in Greenland. He clearly was not speaking of himself; his days were packed with work, as was also the case for the Danes who came to Greenland, judging by the letters and reports they sent home. The notion of the free life comes from the image of the unfettered hunters, whose life, however, Storch knew far too well to paint it in rosy hues. Storch thus employs the cliché of the free hunters with a different aim, as we shall see.

As a geographic location, "old Greenland" (that is, colonial Greenland) conformed perfectly to the criteria of the particular combination of something at once alien and familiar that characterizes "the exotic" (Huggan 2001, 13). Greenland was alien by virtue of its unique nature, yet still held a somewhat familiar quality, because the Inuit had long since been Christened and brought under an administrative framework introduced by the Danes. In the colonies, moreover, many Greenlanders were of so-called "mixed blood" and perceived by the Danes as being closer to the Danish culture. The notion of this mix of races and cultures played a key role in the Danish understanding of the Greenlanders' transition to modernity. In her section on schooling, Sophie Petersen conveys the widespread perception of the time that the Greenlanders' racial qualities place limitations on their capacity to learn. In the following, she struggles to make up her mind whether it is the conditions of the school, with more or less qualified teachers, that determines the potential for learning, or whether it is down to essential qualities.[26] Eventually, her presentation seems to favour the latter:

> Anyone visiting many Greenlandic schools will obviously find significant differences in whether the pupils are taught by a young, college-educated catechist who has received quality training or by an older catechist, whose education dates back to before the educational college became what it is now; one will therefore naturally often notice an excessive emphasis on rote learning and no effort to promote understanding. [...]
>
> Were we now to stay a little to observe the teaching, we would soon discover that Greenlandic children are adept at rote learning and show great interest; however, they have as yet little practice in independent thinking, an ability that is best developed in children of mixed marriages. They are naturally very rebellious and deeply spoiled from home but easy to bring into line, and it *is* possible to achieve good discipline among them.
>
> Greenlandic children as a rule are very fond of going to school and eager to learn to read and write; for the latter especially, they show wonderful aptitude, however, they always struggle with arithmetic; the Eskimo race has little sense of numbers, and beyond what they can calculate using their fingers and toes, they do very poorly.[27] (Petersen 1928, 84–85)

As is evident from the previously quoted passage from the end of the book, however, Petersen eventually concludes that the Greenlanders absolutely must learn to come to terms with numbers. Time and again, however, it is emphasized that those of mixed blood, "that is, Eskimos with some Danish blood",[28] take the lead in society and have the capacity to spearhead the transition to the new conditions (Petersen 1928, 176).[29]

Storch was himself of "mixed blood", his last name coming from a "bluff Jutlander"[30] who had immigrated in the late 1700s and married a Greenlandic woman (Ostermann in the preface to *Strejflys over Grønland*) (Ostermann in Storch 1930, 9). Greenlanders, on the other hand, were much less focused on the issue of "mixed blood". Storch never mentions it; he perceives himself as Greenlandic. Being of mixed blood was the rule rather than the exception, and as previously mentioned, Greenlanders of mixed blood were not seen to make up a special racial category, but were generally regarded as Greenlanders. Storch only raises the question in order to argue why Denmark has a responsibility for Greenland:

> Over the years, Denmark has mixed so much blood with Greenland, that one may speak of Denmark's obligations to Greenland not only by the *right of history* but also by *the right of blood.* Denmark can thus not evade this obligation without branding itself dishonest. So far, fortunately, we have evidence that this is not what Denmark intends to do.[31] (Storch 1930, 112)

The context for this statement is the sale of the Danish West Indies to the United States in 1917.[32] Although Storch does not mention it explicitly, this was an issue that Greenlanders felt strongly about at the time. They knew what they had but might have severe doubts about what they would get, if Denmark decided to hand over Greenland as well. Naturally, to Storch the issue is not whether Greenlanders can learn arithmetic. He is bold enough to take this for granted.[33] Storch's main issue is independence in the sense of co-determination and influence on the country's administration:

> Anyone who is familiar with Greenland's history cannot deny that every time the Greenlanders have been consulted, there has always been progress for Greenland.
>
> However, all these measures [to consult with the Greenlanders] were made by *necessity,* not voluntarily [by the Danes]. And now, one must hope and believe that experience has sufficiently taught those in power that the Greenlanders should be consulted much more than has been the case so far. And now, with all the talk of opening Greenland, the Provincial Council must make haste in initiating a planned effort to make the Greenlanders independent. The situation leaves much to be desired in this regard.[34] (Storch 1930, 111–112)

Time and again, he returns to this point: Denmark has taken a small, free people under its guardianship. "And it *must* restore its independence"[35] (Storch 1930, 110). Thus, the Danes' own favourite image of the free hunting culture provides leverage for political independence in Storch's argumentation.

Moreover, at the stroke of a pen, the statement deletes the emotional economy based on Danish self-sacrifice and Greenlandic gratitude by which the colonial power legitimized its presence in Greenland. The Danes consistently talk about acting "to the benefit of Greenland"[36] (Storch 1930, 124). But according to Storch that is indeed their *obligation*. In fact, the Greenlanders were doing just fine until the Danes showed up. They were trading with the whalers and thus acquired the goods they needed. The Danes put an end to this with colonization and the trade monopoly (Storch 1930, 110). This was clearly not something the Greenlanders had asked for. And the Greenlanders always had to pay. If they were in debt, the money was collected, regardless how hopeless the debtor's situation was (Storch 1930, 94). Only recently had a new policy been introduced, which had Greenlanders paying *the same* for their goods as buyers in Denmark. In contrast to the Danish perception of the low prices in Greenland – a notion that was generated and promoted by books such as Sophie Petersen's – goods had previously been considerably more *expensive* than the same items were in Denmark (Storch 1930, 101). The new building enterprises, which led to a trade deficit in the 1920s, are paid for by the cryolite fees (Storch 1930, 124–125).[37] Storch summarizes: "Therefore I think that it is a careless handling of facts to speak of the big drain that Greenland represents, as merely a burden for Danish taxpayers"[38] (1930, 125).

When Storch feels compelled to portray the future as an endless series of problems to be addressed, these problems are all related to factors that lie beyond the Greenlanders' control: climate change, altered settlement patterns due to colonization and the over-exploitation of seals by other nations, mainly Norwegians. Storch thus seems to see modernity in a similar light to Talal Asad's concept of "conscripts of Western Civilization" (Asad 1992), which David Scott recasts as "conscripts of modernity" (Scott 2004): the colonized people were subjected to this condition, whether they wanted it or not, and there is no turning back time. The colonized are thus in no way strangers to modernity but very much part of it, and to portray the colonized as if they still lived in stagnant tradition merely adds insult to injury. Moreover, in Storch's presentation, the Greenlanders too have made important choices. First of all, this applies to their conversion to Christianity. And precisely in the field of spiritual development, Storch sees great progress, as this is an area where the Greenlanders are taking the lead. Greenlandic priests have thus been much more committed to the effort than Danish clergy, because they realized that it "was about the interests of their own community"[39] (Storch 1930, 98). That is what Storch sees as the solution: the Greenlanders themselves have to take the lead.

A Greenlandic *uchronotopia*

In his analysis of major turning points in history, so-called "zero hours", where a crisis necessitates a completely new perspective, the historian Hagen Schulz-Forberg introduces the term *uchronotopia*: a combination of Mikhail Bakhtin's *chronotopoi* – narratives that follow their own historical logic in representing time and space framed by the genre the narrative is part of – and *utopia*: the idea of a perfect time to come. It is Schulz-Forberg's point that this process goes well beyond changing our perceptions of the future – our notion of the future is closely related to our notion of the past, as Schulz-Forberg argues, with Heidegger. The future, or rather, our envisioning of it, sets the framework for reinterpretations and redefinitions of the past, which are needed in order to legitimize and give meaning to the new course (Schulz-Forberg 2013).

Storch and Petersen both describe Greenland as a society in transformation, but only Storch represents the crisis as being so profound that it is necessary to plot a new *uchronotopia.* The difference in the two presentations thus primarily hinges on the balance of power between Danes and Greenlanders. While Petersen does not appear to imagine any major changes to this balance, such changes are part and parcel of Storch's vision. Thus, the representation of the past becomes a key issue – and a source of disagreement.

Sophie Petersen inscribes Greenland's colonial history into the firmly established narrative about Denmark as a benign colonial power that has always had the welfare of Greenlanders at heart (Thisted 2014). Thus, it reads in the opening to her book:

> Denmark's colonization and exploration of Greenland are unparalleled in the history of colonization, it has inspired admiration and been tremendously important for our reputation in the cultural word, and time and again, outside observers have emphasized that Denmark is the only country that has succeeded in protecting an indigenous people from extinction.[40] (Petersen 1928, 5)

Towards the end of her book, there is a slight note of reservation, but the conclusion remains the same:

> In the course of time, mistakes *may* have been made; but actions have *always* had the best interests of the indigenous population at heart, and under the shielding monopoly, so far, the Eskimos have been protected from the dangers that have destroyed, among others, their kinsfolk in Labrador. Denmark has therefore, *above other nations*, earned the right to continue in the future to help the Greenlandic population, safeguard their interests and teach them to handle independence.
>
> *Generation after generation*
> *We took into our care*
> *The dream of a people's heart by the shores of the Arctic Ocean*
> *May Dannebrog* [the flag of Denmark] *stand guard over Greenland!*[41]
>
> (Petersen 1928, 181; italics in original)

In Petersen's presentation, Greenland's future is thus Denmark's *right:* a right rooted in history. Storch puts his view of the historical record in somewhat different terms:

> The fact is that with the progress of enlightenment, Greenlanders become increasingly convinced that it is better for Greenland *to seek closer ties to Denmark, which has never deliberately harmed Greenland during the two hundred years it held sovereignty over the country.*[42] (Storch 1930, 120–121; italics in original)

Notably, the phrase "which has never deliberately harmed Greenland" is significantly more reserved than Sophie Petersen's statement. It implies that Greenland actually *was* harmed, despite the best intentions of the Danes. Mistakes had stemmed from ignorance on the part of the Danes – and a lack of willingness to consult with the Greenlanders. Thus, the past is burdened by flaws and shortcomings, not superficial but severe, which need to be corrected in the future. And importantly, it is not a choice that is up to the Danes. While Petersen's text deals solely with Denmark's actions and continued right to act, Storch speaks of the Greenlanders' own desires and concerns: the *Greenlanders* are becoming increasingly convinced that it is in Greenland's best interest to seek closer ties with Denmark. This implies that the right to choose and make decisions rests with the Greenlanders themselves.

In his analysis of argumentation strategies in the twenty-first century in relation to Greenland and the Arctic, the anthropologist Frank Sejersen has introduced the concept of *double agency.* Agency, according to Sejersen, can be broken down into two "aspects". The first aspect concerns people's involvement in existing structures: stakeholder participation and integration (Sejersen 2015, 49–50). The second aspect, however, deals with people's possibility of creatively shaping new situations and transformation – and this involves rights and self-determination on a higher level:

> The second aspect [...] is people's possibilities of actively pursuing creative, flexible and innovative strategies that create change and transform society in directions that lie within a horizon of expectation and possibilities related to the group in question. This aspect emphasises *rights-holder possibilities* and *self-determination* and involves expanding the frameworks of choices and decisions. (Sejersen 2015, 50)

While Sophie Petersen seems prepared to assign the Greenlanders the first aspect of agency, Mathias Storch's ambitions clearly go beyond that to include the second aspect. Storch envisions a future in which the Greenlanders themselves take charge of their country: as the future unfolds in the dream invoked in the title of his earlier novel (Storch 1914; Thisted 2005). Greenland is the land of the Greenlanders, and thus they have the right to decide their own future. The narrow constraints imposed on Greenlandic self-determination therefore have to be expanded or, better yet, lifted entirely.

Differences and similarities

At no time does Storch envision that the development of Greenland will proceed without the Danes. Nor would that have been a realistic option at the time. Under the circumstances, all Storch can do is hold the Danes to their word of helping and elevating the Greenlanders. This has to be pursued through *cooperation,* not by rendering the Greenlanders powerless. Therefore, he rounds off the book with the following words:

> In closing, let me quote Minister Hauge's words, which should never be forgotten: *'But Greenland is our last colony, and it is not for sale nor should it be offered up, for any price.'* On the other hand, the system of the "closed-off country" should not mean: rendering the Greenlanders powerless. Instead, it should mean: *promoting independence through a systematic effort.*[43] (Storch 1930, 129–130; italics in original)

However, Storch struggles to discern any signs of this systematic promotion of Greenlandic independence. To Storch, the lack of cooperation between Danes and Greenlanders is the biggest obstacle to a positive development for Greenland. In Greenland, the two peoples live side by side but apart, with no exchange and communication (Storch 1930, 77). That has been the case throughout history, and traders have played a particularly problematic role, because they lacked understanding of the national language and the population's mindset (Storch 1930, 74) – and, moreover, because many of them were scoundrels, out to cheat and defraud the local population. Among other examples, Storch mentions the widely detested use of measuring tubs for weighing in the blubber. Instead of giving the hunters an honest deal, the merchants relied on a variety of tricks, including bottomless tubs placed over a deep pit (Storch 1930, 61). The Greenlanders were powerless to act against this abuse: "After all, the men at the trading station held the power of life and death over the population"[44] (Storch 1930, 63). Hence the Greenlandic desire to separate trade and administration.

In his view of history, Storch follows H. J. Rink (1819–1893), who was the inspector of South Greenland in 1855–1858, and who introduced the so-called board of guardians, which involved the Greenlanders in minor aspects of governance. Rink took a highly critical view of the Danish administration, which in his assessment had undermined the authority of the traditional community leaders: the strong men and shamans of the settlements, without putting anything else in their place. This, he felt, had led to stagnation and decline for the Greenlandic people (Rink 1866). Storch is aligned with this narrative but adds that over time, the boards of guardians stagnated, until after the turn of the century, when a new generation of civil servants "revived the Greenlandic administration under new forms"[45] (Storch 1930, 77). Unfortunately, these "new forms" fall short of their promise: instead of promoting cooperation between Danes and Greenlanders, the initiatives aiming to reform political, administrative and judicial organs instead largely had the *opposite* effect, pulling Greenlanders and Danes even further apart to live in separate worlds (Storch 1930, 90–91). Storch's book cautions the Danes that this state of affairs cannot continue.

The indignation evident in Storch's text is thus mainly aimed at the hypocrisy displayed by the Danes, not least when they claim to treat Danes and Greenlanders as equals. As an example, Storch mentions a passage in Petersen's book, where she claims that Greenlandic priests are now paid exactly the same as their Danish colleagues. That simply is not true, he points out. Their salary is only half that of Danish priests. This inequity applies throughout society. The situation is particularly difficult for Greenlanders employed as outpost managers. They have a huge responsibility but their salary is so low that they are almost bound to wind up deep in debt. As a specific example of inequity, Storch brings up Rodebay (Oqaatsut near Ilulissat in Disko Bay):

> The position of factor was held by a former sealer, who received an annual salary of 3-400 kroner. Later, the outpost was managed by a permanently employed young man, who received 400 kroner a year. When he fell into debt, he was replaced by a Danish builder, who received a salary of 3,000 kroner. That was quite a difference! […] *And then they claim that Greenlanders cannot handle a position of responsibility.*[46] (Storch 1930, 94–95; my italics)

Storch describes the practice of paying Danes eight times the salary of a Greenlander, rather than paying a Greenlander a salary he can survive on, as "ill-advised thrift!"[47] (Storch 1930, 95). The real accusation here, however, is the hypocrisy of first bringing a man into a completely unmanageable situation and then using his failure as proof of Danish superiority over Greenlandic ineptitude.

Similarly, it is not hard to point out inefficiencies in Greenlandic institutions, as they are rendered de facto inoperable. First of all, Storch points out that the unlimited power held by the director of the Greenland Department is inappropriate. There is no point in seeking to develop Greenlandic institutions when the director of the Department "reigns supreme and acts in a way that bypasses these institutions"[48] (Storch 1930, 117). Instead, the director should be a conduit for the Greenlandic institutions, "a spokesman to convey Greenlandic requests to the Danish government"[49] (Storch 1930, 118). Storch also considers it crucially important to oblige the Greenlandic request to merge the two provincial councils into one. First of all, the provincial councils only have an advisory role and thus hold no actual power. Second, cases are delayed endlessly by being shuttled back and forth between the two provincial councils, from the north to the south and back again, as well as back and forth between Greenland and Denmark (Storch 1930, 118–119). Just as Danes and Greenlanders should cooperate, so too should the councils in North Greenland and South Greenland, because "such an exchange of thoughts between the representatives of the two parts of the country would render the legislative process for Greenland more fruitful"[50] (Storch 1930, 119). The decision-making balance has to be shifted in Greenland's favour (Storch 1930, 122).

The sharpest sting in Storch's text may lie in his use of the word *adscription* (*stavnsbånd*) to characterize the Greenlanders' situation and his consequent demand for the *abolition of adscription* (*stavnsbaandsløsning*) (Storch 1930, 115; italics in original). In Danish history, the abolition of adscription in 1788, which transformed the status of peasants from feudal serfs to free citizens, was a key event on the path from feudalism to modern democracy. Regrettably, this civil liberty was never instituted in

Greenland, Storch claims. Greenlanders do not have freedom of movement, and although the law makes Greenlandic civil servants Danish citizens, thus placing them under the Danish justice system, they are not allowed to seek employment outside Greenland. Not least in this area, Storch calls for equal rights.

In later Greenlandic history, especially from the late 1960s, the emphasis was on a culturally determined *difference* between Greenlanders and Danes, but for Storch and his contemporaries, the focus is on emphasizing the Greenlanders' fundamental *similarity* with the Danes. Greenlanders and Danes may be different peoples, each with their respective history and language, but since colonization has placed them in the same society, they should, as far as possible, live under the same terms and, most importantly, get to know each other, for example by learning each other's languages. It is no good for the Danes to feel superior to the people they are pretending to aid and support. The Greenlanders' sense of inferiority does not stem from any inherent inferiority but from the fact that they hit a wall when they hope for a European education to be able to "achieve the same spiritual level as other peoples in the world"[51] (Storch 1930, 114–115).

Storch therefore thinks that the most obvious and only viable path for Greenland is to become part of Denmark and for all Greenlanders to become Danish citizens. He is firmly convinced that "in the not too distant future, Greenland will be part of the Danish realm, like the Faroe Islands, so that Greenland will choose representatives for the Danish Parliament"[52] (Storch 1930, 120). In 1929, when these words were written, this vision seemed nowhere as obvious as it may today, since the United Nations' call for a dissolution of colonies made Storch's wish come true, in 1953. A measure made by *necessity,* not voluntarily by the Danes, to quote Storch's words about how change usually happened in Greenland.

A place in the sun – like the great realm of the Danes

Despite the celebration of nature and the wide-open, empty spaces in the introduction to his book, Storch describes any and all initiatives to modernize Greenland in the most positive terms. The early signs of industrialization are unreservedly welcomed. Despite the cryolite success story, Storch has little faith in mining as the way forward for Greenland (Storch 1930, 123–124). At this time, fishing is seen as the most promising option, and as in Storch's novel, the vision outlined in this later book clearly involves turning Greenland into a modern society in Denmark's image – but with the Greenlanders as the people controlling the development and reaping the benefits. This is the content of the metaphor "a place in the sun"[53] (Storch 1930, 124). Thus, the metaphor refers to a life that embraces both modernity and democracy/self-determination.

A central element in the text is an anecdote, where Storch relates a conversation he had the previous summer with a man from the northernmost reaches of Greenland, an area that had, as yet, seen little development. The man had told him of his visit down south, including a visit to Holsteinsborg (Sisimiut) where he had seen the fish cannery plant: "Eventually, he distilled his impressions into the following words: 'ímaqa sôrdlo qavdlunât nunarssuat' =: so maybe this is like the great realm of the Danes!' Yes, there really *have* been changes for the better!"[54] (Storch 1930, 99). The final words are the author's comment and summarize the narrator's point: that modernization is a beneficial development. The suffix -ssuat in qavdlunât nunarssuat not only implies that the country of the Danes is big but also that it is mighty. Perhaps even superior to Greenland. But the Greenlanders are keen to gain a share in this status and power!

At this point, at least in Storch's presentation, Greenlandic identity has not become stuck in a form of negative reflection of all things Danish. Such a discourse does not emerge until after 1953, particularly from the late 1960s: the notion that Greenlandic identity must by necessity be defined as non-Danish. Regrettably, this latter position turned Greenlanders into co-creators of the imagined opposition between tradition and Greenlandic identity versus modernity and Danish identity.

Although Storch hardly imagined that seal hunting would end completely, he does not share Petersen's evident nostalgia for the hunting culture. Petersen basically has difficulty envisioning a Greenlander who is not also, at least partially, a hunter and a fisherman. Although the objective of her book is to "offer a popular presentation on today's Greenland with its difficulties and problems, which are such as must necessarily arise in a country whose population is entering a new cultural phase, and where ancient occupations are no longer adequate but due to be supplanted by new ones,"[55] she still summarizes her chapter on occupations and the economy as follows:

> It is a big question what shall provide Greenland's main livelihood in the future. Fishing may provide sustenance; but it will not provide the by-products in the form of skins and blubber that are necessary for clothing and fuel in such a cold country. Therefore, it would be better, if at all possible, to *combine* seal hunting and fishery, so that the ideal future Greenlander would be a man who went seal hunting when there were seals about, and fishing when there were fish.[56] (Peterson 1928, 40)

Sophie Petersen still imagines a closed-off country that is mainly self-sustaining. In its choice of illustrations too, *Grønland i Hverdag og Fest* maintains this focus on the hunting and fishing culture. The cover of the boarded edition features a coloured pen-and-ink drawing of a kayaker with a shooting screen in the foreground and an iceberg in the background: an iconic image of Greenland. Inside the book, there are images of kayaks, kayaking equipment, mud huts, seal skinning, blubber scraping, women's boats, traditional skin tents etc., etc. Only three small photos, much smaller and of much poorer quality than most of the other photographs in the book, illustrate fishing, one photo depicts the graphite mine Amitsoq, and a single photo was taken by the sheep breeding station in Julianehaab. Thus, although the reader is told, on the one hand, that Greenland is developing and changing, he or she is still likely to be left with the safe assurance that Greenland still exists as a picturesque fairy-tale land, up there, in the high north.

Whether Sophie Petersen personally chose the illustrations is impossible to know. It is safe to assume that Storch had no say over that aspect of the production, and the cover of his book, which has very few illustrations, features a motif from "old Greenland", with skin tents and sledges. It would also have been a matter of which photos were actually available. The hunting culture was very well documented, while there were likely to have been far fewer photos of the emerging, modern Greenland. Jette Bang (1914–1934) was the first photographer to document modern Greenland – and even among her pictures and films, (partially staged and reconstructed) images of the hunting culture were the ones most widely reproduced and remembered (Jørgensen 2014).

The vision of Greenland that Storch advocates thus had a long and arduous path ahead of it to reach Danish awareness. And after 1953, developments went exactly as Storch had feared. That was the reason why he did not call for an immediate elimination of the monopoly but argued instead that the monopoly should be dismantled *from within,* so that the country's own people would be ready to take over governance of the country (Storch 1930, 125). After 1953, on the surface, Greenland soon came to resemble Denmark, with modern housing blocks, hospitals, roads, port facilities etc. Tuberculosis was eradicated, and life-time expectancy went up. However, the Greenlanders were *not* seated at the head of the table in the decision-making processes. Hence the anger that fuelled the calls for home rule.

Today, Storch's text offers itself as a mainstay in the formulation of a new narrative focusing on the Greenlanders' own struggle for independence: A narrative in which the struggle for independence is linked to development and modernity, since it is modernity that makes independence conceivable and achievable.

On a superficial level, Petersen's and Storch's books appear as complete opposites. From a contemporary point of view, however, they are surprisingly similar as concerns the fundamentally positive view of the arrival of modernity in Greenland. Although both authors view modernity as something originating in Europe, they do not see it as *belonging* to Europe; modernity may be appropriated by Greenland without making Greenland any less "Greenlandic". Both books thus portray the Greenlanders as active *co-creators* of modernity, quite unlike the image of the Inuit as passive victims of modernity that gained hegemony throughout the 1970s and 80s, also in Greenland.

References

Amdrup, G.C., ed. 1921. *Grønland i Tohundredaaret for Hans Egedes Landing* [Greenland two hundred years after the arrival of Hans Egede]. Meddelelser om Grønland vol. 60-61. Copenhagen: C.A. Reitzel.

Andersen, Astrid Nonbo and Kirsten Thisted, forthcoming. "'Whatever a colony is' – 1945-1953 as a Zero Hour in Danish–Greenlandic Relations." In *Zero Hours. Politics of Time in Global Perspectives, 1940s-1970s*, edited by Hagen Schulz-Forberg. London, New York and Shanghai: Palgrave Macmillan.

Asad, Talal. 1992. "Conscripts of Western Civilization." In *Dialectical Anthropology: Essays in Honor of Stanley Diamond,* vol. I, *Civilization in Crisis,* edited by Christine Gailey, 333–51. Gainesville: University Press of Florida.

Beukel, Erik, Frede Jensen, P. Rytter and Jens Elo. 2007. *Afvikling af Grønlands kolonistatus 1945-54* [The dismantling of the colonial status of Greenland 1945-54]. Copenhagen: Dansk Institut for Internationale Studier.

Bhambra, Gurminder K. 2007. *Rethinking Modernity. Postcolonialism and the Sociological Imagination.* Basingstoke: Palgrave Macmillan.

Drivenes, Einar-Arne and Harald Dag Jølle. 2006. *Into the Ice: The History of Norway and the Polar Regions.* Oslo: Gyldendal Norsk Forlag.

Freuchen, Peter. 1927. *Grønland, Land og Folk* [Greenland, land and people]. København: Pios Boghandel.

Gad, Ulrik Pram. 2008. "Når mor/barn relationen bliver teenager. Kompatible rigsfællesskabsbilleder som (dis)integrationsteori" [When the mother/child relationship becomes a teenager. Compatible images of the realm as (dis)integration theory]. *Politica* 40 (2): 111–133.

—. 2016. *National Identity Politics and Postcolonial Sovereignty Games: Greenland, Denmark, and the European Union.* Monographs on Greenland vol. 353. Man and Society vol. 43. University of Copenhagen: Museum Tusculanum Press.

Gam, Mikael. 1968. "Den grønlandske Folkeskoles Udvikling" [The development of the Danish primary schools]. *Årbog for dansk skolehistorie* 1968: 32–48.

Gant, Erik. 2003. "Grønlandske overgange. Fra *Qivitoq* til *Tukuma*" [Greenlandic transitions. From *Qivitoq* to Tukuma]. Kosmorama Vol. 232: *Film fra Nord*: 125–142.

Hedtoft, Hans. 1948. "Grønlands fremtid. Statsminister Hans Hedtofts tale for Studenterforeningen og Det Grønlandske Selskab lørdag den 27. november 1948" [The future of Greenland. Prime Minister Hans Hedtoft's speech to the Students' Association and the Greenland Society on Saturday, Nov. 27, 1948]. *Det Grønlandske Selskabs Årsskrift* 1949: 22–40.

Heinrich, Jens. 2012. *Eske Brun of det moderne Grønlands tilblivelse* [Eske Brun and the creation of modern Greenland]. Nuuk: Forlaget Atuagkat ApS.

—. 2014 . "Forsoning i den grønlandske optik" [Reconciliation from a Greenlandic point of view]. *Jyllandsposten,* May 22, 2014.

Huggan, Graham. 2001. *The Postcolonial Exotic: Marketing the Margins.* London and New York: Routledge.

Jensen, Lars. 2012. *Danmark – Rigsfællesskab, tropekolonier og den postkoloniale arv* [Denmark – present realm, tropical colonies and the postcolonial heritage]. Copenhagen: Hans Reitzels Forlag.

Johansen, Lars Emil. 1997. "Address by Lars Emil Johansen, Prime Minister, Greenland Home Rule." In *Whaling in the North Atlantic,* edited by G. Petursdottir. University of Iceland: Fisheries Research Institute.

Jørgensen, Anne Mette. 2014. "A Gentle Gaze on the Colony: Jette Bang's Documentary Filming in Greenland 1938–9." In *Films on Ice: Cinemas of the Arctic,* edited by Scott MacKenzie and Anna Westerståhl Stenport, 235–244. Edinburgh: Edinburgh University Press.

Keskitalo, E.C.H. 2004. *Negotiating the Arctic: The Construction of an International Region*. New York and London: Routledge.

Koselleck, Reinhardt. 2004. *Futures Past: On the Semantics of Historical Time.* New York: Columbia University Press.

Kuper, Adam. 2003. "The Return of the Native." *Current Anthropology* 44 (3): 389–395.

Latour, Bruno. 1993. *We have never been Modern.* (*Nous n'avons jamais été modernes*, 1991.) Translated by Catherine Porter. Cambridge, MA: Harvard University Press.

Marquardt, Ole. 1999. "Grønlænderne og vestens civilization – træk af Rink-tidens grønlandspolitiske discussion" [Greenlanders and Western civilization – features of the political discussions of the Rink era]. *Grønlandsk kultur- og samfundsforskning* 1998/99: 7–26. Nuuk: Ilisimatusarfik.

Ostermann, H. 1921. *Den grønlandske Missions og Kirkes Historie* [The history of the Greenlandic mission and church]. Copenhagen: Lohse.

Oxfeldt, Elisabeth. 2003. "Profanering og identitet i Klaus Rifbjergs *Nansen og Johansen*" [Profanation and identity in Klaus Rifbjerg's *Nansen and Johansen*]. *Bøygen: Organ for nordisk språk og litteratur* 3 (3): 57–66.

Petersen, Sophie. 1928. *Grønland i Hverdag og Fest* [Greenland in everyday life and celebrations]. Copenhagen: C.A. Reitzels Forlag.

—. 1946. *Danmarks gamle Tropekolonier* [Denmark's former tropical colonies]. Copenhagen: H. Hagerups Boghandel.

Porter, Libby. (2010) 2016. *Unlearning the Colonial Cultures of Planning.* London and New York: Routledge.

Rink, H.J. 1862. "Om Aarsagen til Grønlændernes og lignende, af Jagt levende, Nationers materielle Tilbagegang ved Berøringen med Europæerne" [About the reasons for the decline of the Greenlanders and other hunting peoples when encountering the Europeans]. *Dansk Maanedsskrift ekstrahefte.*

—. 1866. "Indledning" [Introduction]. In *Eskimoiske Eventyr og Sagn* [Eskimo fairy tales and legends], edited by H.J. Rink, 9–44. Copenhagen: C.A. Reitzels Boghandel.

Rud, Søren. 2016. *Colonialism in Greenland: Tradition, Governance and Legacy*. London: Palgrave/Macmillan.

Schulz-Forberg, Hagen. 2013. "Introduction: Time and Again Towards the Future: Claims on Time as a New Approach for Global History." In *Zero Hours: Conceptual Insecurities and New Beginnings in the Interwar Period,* edited by Hagen Schulz-Forberg, 15–52. Brussels: P.I.E.-Peter Lang.

Scott, James. 2004. *Conscripts of Modernity: The Tragedy of Colonial Enlightenment*. Durham and London: Duke University Press.

Sejersen, Frank. 2015. *Rethinking Greenland and the Arctic in the Era of Climate Change*. London and New York: Routledge.

Storch, Mathias. (1914) 1974. *Sinnattugaq* [The dream]. Nuuk: Atuakkiorfik.
—. 1915. *En Grønlænders Drøm* [A Greenlander's dream]. Translated by Knus Rasmussen. Copenhagen: Grønlandsk Litteraturselskab.
—. 1930. *Strejflys over Grønland* [Gleams of light over Greenland]. Copenhagen: Levin & Munksgaards Forlag.
Thisted, Kirsten. 2002. "The power to represent. Intertextuality and discourse in *Miss Smilla's Sense of Snow*." In *Narrating the Actic. A Cultural History of the Nordic Scientific Practices*, edited by Michael Bravo and Sverker Sörlin, 311–342. Canton, MA: Watson Publishing International, Science History Publications, USA.
—. 2003a. "Danske Grønlandsfiktioner. Om billedet af Grønland i dansk litteratur." [Danish fictions of Greenland. The image of Greenland in Danish literature]. *Kosmorama* 232: 32–67.
—. 2003b. "'Bjørnefedtet … Johansen.'" Klaus Rifbjergs Nansen og Johansen i postkolonialt perspektiv" [The bear grease … Johansen. Klaus Rifbjerg's Nansen and Johansen i a postcolonial perspective]. *Spring, Tidsskrift for moderne dansk litteratur* 20: 53–73.
—. 2005. "Postkolonialisme i nordisk perspektiv: relationen Danmark-Grønland" [Postcolonialism in Nordic Perspective: The Danmark-Greenland relation]. *Kultur på kryds og tværs*, edited by Henning Bech and Anne Scott Sørensen, 16–42. Aarhus: Klim.
—. 2009. "Where once Dannebrog waved for more than 200 years." *Review of Deveopment and Change.* 15 (1&2): 147–172.
—. 2012. "Grønland i Hverdag og Fest – Kolonialisme, nationalisme og folkelig oplysning i mellemkrigstidens Danmark" [Greenland on everyday and festive occasions. Colonialism, nationalism and public information in Denmark in the interwar years]. In *Malunar Mót*, edited by Eidun Andreassen, Malan Johannesen, Anfinnur Johansen and Turid Sigurdardóttir, 460–478. Tórshavn: Faroe University Press.
—. 2013. "Discourses of Indigeneity: Branding Greenland In The Age Of Self-Government And Climate Change." In *Science, Geopolitics and Culture in the Polar Regions – Norden beyond Borders*, edited by Sverker Sörlin, 227–258. Farnham, Burlington: Ashgate.
—. 2014. "Imperial Ghosts in the North Atlantic: Old and New Narratives about the Colonial Relations between Greenland and Denmark." In *Colonialism across Europe: transcultural History and National Memory*, edited by Dirk Göttsche and Axel Dunker, 107–134. Bielefeld: Aisthesis Verlag.
—. 2017. "The Greenlandic Reconciliation Commission: Ethnonationalism or a Quest for a Post-Post-Colonial Identity." In *Arctic Environmental Modernities: From the Age of Polar Exploration to the Era of the*

Anthropocene, edited by Lill-Ann Körber, Scott MacKenzie and Anna Westerståhl Stenport, 231–246. Palgrave Studies in World Environmental History. London: Palgrave.

Thomas, N. 1994. *Colonialism's Culture: Anthropology, Travel, and Government*. Cambridge: Cambridge University Press.

Thomsen, Hanne. 1998. "Ægte grønlændere og nye grønlændere: om forskellige opfattelser af grønlandskhed" [Genuine Greenlanders and new Greenlanders: About different perceptions of being Greenlandic]. *Den jyske historiker* 81: 21–55.

Thorleifsen, Daniel. 1999. "Det lukkede land." [The closed country]. In *Kalaallit Nunaat. Gyldendals bog om Grønland* [Kalaallit Nunaat, Gyldendal's book about Greenland], edited by Bent Günter and Aqissiaq Møller, 217–222. Copenhagen: Gyldendal.

Thuesen, Søren. 1988. *Fremad, Opad. Kampen for en modern grønlandsk identitet* [Forward, upward. The struggle for a modern Greenlandic identity]. Copenhagen: Rhodos.

UN 1987. Report of the World Commission on Environment and Development: Our Common Future. New York: World Commission on Environment and Development.

UN 2007. United Nations' Declaration on the Rights of Indigenous Peoples. New York: United Nations General Assembly.

Vahl, M. og G. Hatt. 1922–27. *Jorden og Menneskelivet. Geografisk Haandbog* [Earth and human life. Geographical Handbook] vols. *1–4.* Copenhagen: I.H. Schultz Forlagsboghandel.

Notes

[1] *Atuagagdliutit* is Greenland's oldest newspaper, published since 1861 and printed in Godthaab (Nuuk). *Avangnâmioq* (The Nord Greenlander) was published from Godhavn (Qeqartarsuaq). Both are considered a type of newspaper, although their release were far from daily.

[2] See, e.g., Kuper's provocative contribution "The Return of the Native", Kuper 2003.

[3] Cf. Bruno Latour 1993 [1991]. *We Have Never Been Modern.* "We" referring to the supposedly long–since modernized peoples of the West. If "we" have never been modern, the dichotomy between the modernized West and the traditional Rest makes no sense.

[4] After the implementation of Greenland Self-Government in 2009, Lars Emil Johansen has advocated that Greenlanders no longer need to define themselves as an indigenous people, see Thisted 2013:235. However, the category of indigenous peoples still conveys certain legal rights, not least with regard to whaling, and the

various forums for indigenous peoples also make up an important part of the Greenlanders' international contacts and networks.

[5] http://saammaatta.gl/da/Baggrund,last accessed 31. Dec. 2016. About the Greenlandic Reconciliation Commission see also Thisted 2017.

[6] Not in modern times, that is. The Vikings settled in Greenland around 1000, but disappeared around 1400.

[7] Nuannersoqaqaaq/angallavigigaanni /: tamaat sineriak :/:avannamut kujammullu/inunnik naapitsiffik / nunaqarfinnillu/uninngavissalik. Here translated from the Danish translation authored by Mads Lidegaard, included in the Danish "Højskolesangbogen" (song book for the Danish adult education colleges) as no. 390.

[8] Thule was the name given to the place by Knud Rasmussen (see below), who established the below-mentioned trading post on a private basis. Thule was not included in the Danish state until 1937.

[9] In addition, of course, there is the Greenlandic ice sheet covering roughly 80% of the surface of Greenland. Here, no one has ever wanted or been able to live, and it has therefore continued to offer the kind of physical and personal challenges that once attracted explorers and now appeals to adventure tourists.

[10] Denmark shares this with Norway, where the adjective "polar" is mainly used about the Arctic Ocean north of Norway, the North Pole itself and the archipelago of Svalbard in the high Arctic, not far from the coast of North-East Greenland. See, e.g. Drivenes og Jølle 2006.

[11] On this, see Thisted 2009.

[12] For a more detailed discussion of the book's structure and style, see Thisted 2012.

[13] Freuchen was Knud Rasmussen's colleague and expedition partner. In 1910 he co-founded a trading station in Thule with Knud Rasmussen, which he managed until 1919.

[14] As evident from letters from Ostermann to Storch in Mathias Storch's private archive (The Greenlandic National Archives 10.00.10).

[15] "som en langsom meloditone, der (...) kan spænde fra de højeste til de dybeste toner eller omvendt – som naar en mester leger paa sit instrument og fremkalder en tonevælde, der sætter éns fantasi i saadan svingning, at man forglemmer sig selv en stund." All quotations from Storch's and Petersen's books are my own.

[16] "Denne mands karrière, der i sig selv er ret usædvanlig, belyser paa en karakteristisk maade det rivende fremskridt i den nye tids Grønland i modsætning til fortiden, hvor alt stod stille"

[17] "Og nu har det formet sig saa forunderligt, at det, som der drømmes om i "En Grønlænders drøm", for en meget stor del er traadt ud eller er ved at træde ud i virkeligheden i Grønland."

[18] "engang kunne skænke os en tredje bog, som fortæller om, at nu er de [problemerne] løst!"

[19] This decision to apply a light and optimistic view of the cause one seeks to promote was also clearly part of the authors' strategy at the time. Peter Freuchen, who reviewed the manuscript prior to printing, advises Storch to choose a lighter

tone, since "Never yet have pessimists prevailed in the long term" ("Det er endnu aldrig set, at Pessimister i det lange Løb har held med sig") (Letter dated 28 Feb. 1929, Mathias Storch's private archive, The Greenlandic National Archives 10.00.10). Whether Storch took his advice and tempered his tone is not known, since the original book manuscript does not appear to have been preserved.

[20] "Men noget andet er at vi [danskerne] godt en skønne Dag kan blive tvunget til at lukke Landet op, og saa var det dog vist alligevel bedre, at vi i god Tid var begyndt at give Grønlænderne en vis Handelsopdragelse."

[21] "som hun har set de danske Damer bære det"

[22] "den Top, som vi kender saa godt fra vor Barndoms Billedbøger"

[23] "frygteligt"

[24] "noget saa ægte eskimoisk"

[25] "Overgangen fra de nervepirrende forhold i det tætbefolkede Europa til den frie natur kan ofte være skyld i denne begejstring [som de tilrejsende føler ved mødet med Grønland]. Det kan nemlig være vidunderligt befriende at komme fra et sted, hvor hvert skridt koster penge, og hvor den stundesløse konkurrence skaber uro og bekymringer – for saa, efter en halv snes dages "søsyge" eller "fornemmelser", en skønne dag at vaagne, hensat til helt andre og nye forhold, hvor der hersker ophøjet ro, hvor man kan være sin egen herre og hvor man selv raader over sin tid."

[26] The Greenlandic school was closely linked to the mission from the outset. Teaching was planned by the priests and carried out in collaboration with Greenlandic catechists and untrained "readers". In 1905, Greenland got its first School Act, which placed the school system under auspices of the Danish ministry for church and school affairs, the so-called "Kultusministerium". In 1920–21, a commission was in charge of all issues pertaining to Greenland, and a new act replaced the act of 1905. Compulsory schooling was introduced (previously, there had been compulsory education), and generally, there were attempts to improve schooling in Greenland. The new act introduced Danish as a foreign language, and boarding schools were established for the students who were seen to show the greatest academic aptitude. These initiatives laid the basis for the rapid development (and Danification) of Greenlandic schools after 1953 (Gam 1968).

[27] "Hvis man aflægger Besøg ved mange grønlandske Skoler, vil man selvfølgelig finde, at der er stor Forskel paa, om Eleverne undervises af en ung, seminarieuddannet Kateket, som selv har faaet en god Uddannelse, eller af en ældre Kateket, hvis Uddannelse ligger langt tilbage i Tiden, før Seminariet blev det, det nu er; man vil derfor naturligvis tit mærke, at der lægges alt for meget Vægt paa Udenadslæren og intet gøres for Forstaaelsen. [...} Hvis vi nu bliver lidt og hører paa Undervisningen, vil vi hurtigt opdage, at Grønlænderbørnene har let ved at lære udenad og er meget interesserede; men at de endnu er for lidt opøvede i selvstændig Tænkning, en Evne, som er bedst udviklet hos Børn af blandede Ægteskaber. De er af Naturen meget rebelske og stærkt forkælede hjemmefra, men lette at ave, og der kan skabes god Disciplin iblandt dem. Grønlænderbørn holder i Almindelighed meget af at gaa i Skole og vil gerne lære at læse og skrive; især det sidste er de overordentlig dygtige til, derimod kniber det altid stærkt med Regning;

den eskimoiske Race har meget lidt Talbegreb, og udover, hvad man kan bruge Fingrene og Tæerne til at beregne, gaar daarligt."
[28] "d.v.s. Eskimoer med dansk Blod i sig"
[29] By contrast, Petersen is clearly proud and happy on behalf of the Scandinavians because of their ability to "keep the race clean" ("holde racen ren") (Petersen 1928, 76). Eskimos are clearly considered inferior, so that a mix with the Scandinavians would have dragged the latter down, while the Greenlanders in turn are elevated by their present mix with European blood.
[30] "djærv Jyde"
[31] "Danmark har gennem aarenes løb blandet saa meget blod med Grønland, at man ikke alene kan tale om Danmarks forpligtelser overfor Grønland med historisk ret, men ogsaa med blodets ret. Danmark kan altsaa ikke løbe fra denne forpligtelse uden at stemple sig selv for uærlighed. Hidtil har vi heldigvis beviser paa, at Danmark ikke vil gøre dette."
[32] The Danish West Indies, consisted of three islands in the Caribbean: Saint Thomas (annexed 1972), Saint John (annexed 1675) and Saint Croix (purchased from the French West India Company in 1733). The intention was to exploit the profitable triangular trade, where firearms and other manufactured goods were exported to Africa in exchange for slaves, who were then transported to the Caribbean to work the sugar plantations. Finally, sugar and rum was shipped to Denmark. After a rebellion, slavery was officially abolished in 1848. In 1852, the sale of the increasingly unprofitable colony was first debated in the Danish parliament. The islands were eventually sold for 25 million dollars to the United States in 1917, and the islands were renamed "the United States Virgin Islands".
[33] Storch attributes differences in European and Greenlandic culture to physical circumstances. The failure of Greenlanders to develop a written culture was thus not due to a lack of ability but due to the lack of a need for writing, since there was no trading to speak of in pre-colonial Greenlandic society (Storch 1930, 36).
[34] "De, som kender Grønlands historie, kan ikke nægte, at hvergang man har taget Grønlænderne med paa raad, har der altid været noget fremskridt for Grønland. Men alle disse forholdsregler [når man har taget grønlænderne med på råd] blev truffet af nødvendighed, ikke af lyst [fra danskernes side]. Og nu maa man haabe og tro, at erfaringen tilstrækkeligt maa have lært de styrende, at Grønlænderne maa tages med paa raad i langt større udstrækning, end man hidtil har gjort. Og nu under den meget tale om Grønlands aabning bør styrelsen skynde sig at tage et planmæssigt arbejde op for at selvstændiggøre Grønlænderne. Der staar nemlig meget tilbage at ønske i saa henseende."
[35] "Og det maa gøre det selvstændigt igen"
[36] "til gavn for Grønland"
[37] In 1865, the right to mine cryolite was handed over to the private company Kryolit Mine og Handels-Selskabet A/S. The firm was required to pay the Danish state a fee of 20%. Cryolite mining continued until 1990 (Thorleifsen 1999, 222).
[38] "Jeg tror derfor, at man omgaas letsindigt med kendsgerningerne, naar man taler om det store underskud, som Grønland er, kun en byrde for den danske skatteyder."

[39] "gjaldt deres eget samfunds interesser"

[40] "Danmarks Kolonisation og udforskning af Grønland er uden Sidestykke i Nationernes Kolonisationshistorie, den har fremkaldt beundring og betydet overordentlig meget for vort Omdømme i den kulturelle Verden, og Gang paa Gang er det udefra blevet fremhævet, at Danmark er det eneste Land, som har forstaaet at beskytte et Naturfolk mod Udryddelse."

[41] "Der *kan* her i Tidens Løb være begaaet Fejlgreb; men der er altid handlet med den indfødte Befolknings Vel for Øje, og under det skærmende Monopol er det hidtil lykkedes at beskytte Eskimoerne mod de farer, som bl.a. har ødelagt deres Stammefrænder i Labrador. Danmark har derfor *frem for andre Nationer* erhvervet sig Retten til ogsaa i Fremtiden at hjælpe den grønlandske Befolkning, varetage dens Interesser og oplære den til Selvstændighed. *Slægt efter Slægt/Vi tog i Varetægt/Et Folkehjertes Drøm ved Polarhavets Strøm/Maa Dannebrog staa vagt over Grønland!*" The stanzas at the end are from a poem by the eskimologist William Thalbitzer (1873–1958).

[42] "Forholdet er nemlig det, at Grønlænderne, efterhaanden som oplysningen skrider frem, bliver mer og mer overbeviste om, at det er bedst for Grønland *at knytte sig nærmere til Danmark, som med vilje aldrig har gjort Grønland nogen fortræd i de to hundrede aar, det har haft højhedsret over landet."*

[43] "Tilsidst skal jeg anføre minister Hauges ord, som aldrig bør glemmes: *"Men Grønland er vor sidste koloni, og denne er ikke til salg og bør ikke blive falbudt for nogen pris."* Men på den anden side bør systemet "det lukkede land" ikke betyde: umyndiggørelse af Grønlænderne. Det må derimod betyde: *selvstændiggørelse gennem planmæssigt arbejde."*

[44] "Handelens folk havde jo 'hals og haandsret' over befolkningen."

[45] "atter kom liv i den grønlandske administration under nye former"

[46] "Som stedets bestyrer fungerede en forhenværende sælfanger, honorarlønnet med 3 à 400 kr. om aaret. Senere bestyredes stedet af en fastansat ung mand med en løn af 400 kr. aarlig. Og da han kom i gæld, erstattedes han med en dansk haandværker, der lønnedes med 3000 kr. Der var dog en forskel! [...] Og saa siger man, at Grønlænderne ikke kan klare sig i en selvstændig stilling."

[47] "daarlig sparsommelighed!"

[48] "regerer enevældigt og arbejder udenom disse institutioner"

[49] "en talsmand for de grønlandske ønsker hos den danske regering"

[50] "en saadan tankeudveksling mellem de to landsdeles repræsentanter vilde frugtbargøre lovgivningsarbejdet for Grønland"

[51] "naa samme aandelige standpunkt som andre mennesker i verden"

[52] "ikke altfor langt ude i fremtiden, at Grønland gøres til en del af det danske rige ligesom Færøerne, saaledes at Grønland vælger repræsentanter til den danske rigsdag"

[53] "en plads i solen"

[54] "Tilsidst samlede han sit indtryk i de ord: 'ímaqa sôrdlo qavdlunât nunarssuat' =: det er maaske ligesom Danskernes store land!' Ja, der *er* virkelig ogsaa sket forandringer til det gode!"

[55] "give en populær Skildring af Øjeblikkets Grønland med dets Vanskeligheder og Problemer, der er saadanne, som nødvendigvis maa opstaa i et Land, hvis Folk er ved at gaa ind i en ny Kulturfase, og hvor ældgamle Erhvervsgrene ikke længer slaar til, men skal afløses af nye,"

[56] "Det er et stort Spørgsmål, hvad der i Fremtiden skal være Hovederhvervet i Grønland. Fiskeriet kan nok skaffe Føden; men det giver ikke de Biprodukter i Form af Skind og Spæk, der er nødvendig til Klæder og Brændsel i saa koldt et Land. Derfor ville det være det bedste, hvis det var muligt, at *forene* Sælfangst og Fiskeri, saa at Idealet af en Fremtidsgrønlænder var en Mand, der gik paa Sælfangst, naar der var Sæler, og paa Fiskeri, naar der var Fisk."

INDEX